Computational Methods in Photochemistry

MOLECULAR AND SUPRAMOLECULAR PHOTOCHEMISTRY

Series Editors

V. Ramamurthy
Professor
Department of Chemistry
University of Miami
Coral Gables, Florida

Kirk S. Schanze
Professor
Department of Chemistry
University of Florida
Gainesville, Florida

Computational Methods in Photochemistry

edited by
Andrei G. Kutateladze
University of Denver
Denver, Colorado, U.S.A.

CRC Press
Taylor & Francis Group
Boca Raton London New York

CRC Press is an imprint of the
Taylor & Francis Group, an **informa** business

A TAYLOR & FRANCIS BOOK

CRC Press
Taylor & Francis Group
6000 Broken Sound Parkway NW, Suite 300
Boca Raton, FL 33487-2742

First issued in paperback 2019

ISBN-13: 978-0-8247-5345-0 (hbk)
ISBN-13: 978-0-367-39291-8 (pbk)
Library of Congress Card Number 2004058222

Library of Congress Cataloging-in-Publication Data

Computational methods in photochemistry / [edited by Andrei G. Kutateladze].
p. cm. -- (Molecular and supramolecular photochemistry ; 13)
Includes bibliographical references and index.
ISBN 0-8247-5345-3 (alk. paper)
1. Photochemistry--Textbooks. 2. Photochemistry--Mathematical models. I. Kutateladze, Andrei
G. II. Series.

QD708.2.C638 2005
541'.35--dc22   2004058222

Visit the Taylor & Francis Web site at
http://www.taylorandfrancis.com

and the CRC Press Web site at
http://www.crcpress.com

Contents

Contributors

Michael J. Bearpark
Department of Chemistry, King's College, London, U.K.

Michal Ben-Nun
Department of Chemistry and The Beckman Institute, University of Illinois at Urbana-Champaign, Urbana, Illinois

Lluís Blancafort
Institut de Química Computacional, Departament de Química, Universitat de Girona, Campus de Montilivi, Girona, Spain

Weston Thatcher Borden
Department of Chemistry, University of Washington, Seattle, Washington

Nina P. Gritsan
Institute of Chemical Kinetics and Combustion and Novosibirsk State University, Novosibirsk, Russia

Zdenek Havlas
Institute of Organic Chemistry and Biochemistry, Academy of Sciences of the Czech Republic and Center for Complex Molecular Systems and Biomolecules, Prague, Czech Republic

Martin Klessinger
Organisch-Chemisches Institut der Westfälischen Wilhelms-Universität, Münster, Germany

Mojmír Kývala
Institute of Organic Chemistry and Biochemistry, Academy of Sciences of the Czech Republic and Center for Complex Molecular Systems and Biomolecules, Prague, Czech Republic

Benjamin Levine
Department of Chemistry and The Beckman Institute, University of Illinois at Urbana-Champaign, Urbana, Illinois

Leslie Manohar
Department of Chemistry and The Beckman Institute, University of Illinois at Urbana-Champaign, Urbana, Illinois

Todd J. Martinez
Department of Chemistry and The Beckman Institute, University of Illinois at Urbana-Champaign, Urbana, Illinois

Josef Michl
Department of Chemistry and Biochemistry, University of Colorado, Boulder, Colorado

François Ogliaro
Department of Chemistry, King's College, London, U.K.

Massimo Olivucci
Dipartimento di Chimica dell'Università di Siena, Siena, Italy

Seth Olsen
Department of Chemistry and The Beckman Institute, University of Illinois at Urbana-Champaign, Urbana, Illinois

Jane Owens
Department of Chemistry and The Beckman Institute, University of Illinois at Urbana-Champaign, Urbana, Illinois

Matthew S. Platz
Department of Chemistry, Ohio State University, Columbus, Ohio

Jason Quenneville
Department of Chemistry and The Beckman Institute, University of Illinois at Urbana-Champaign, Urbana, Illinois

Michael A. Robb
Department of Chemistry, King's College, London, U.K.

Adalgisa Sinicropi
Dipartimento di Chimica dell'Università di Siena, Siena, Italy

Alexis Thompson
Department of Chemistry and The Beckman Institute, University of Illinois at Urbana-Champaign, Urbana, Illinois

Alessandro Toniolo
Department of Chemistry and The Beckman Institute, University of Illinois at Urbana-Champaign, Urbana, Illinois

Frank Weinhold
Theoretical Chemistry Institute and Department of Chemistry, University of Wisconsin, Madison, Wisconsin

Howard E. Zimmerman
Department of Chemistry, University of Wisconsin, Madison, Wisconsin

1

Some Theoretical Applications of Organic Photochemistry: Excited State and Open Shell Examples

HOWARD E. ZIMMERMAN

Chemistry Department
University of Wisconsin,
Madison, Wisconsin

CONTENTS

1.1   INTRODUCTION

This chapter covers the development of theoretical organic photochemistry in the writer's research group, from its primitive beginnings to its present state. A series of examples are selected from the author's research as exemplifying basic principles and computational methodology. The development of increasingly powerful theoretical and mechanistic treatments is shown with examples.

1.2   THE BEGINNINGS OF UNDERSTANDING PHOTOCHEMICAL MECHANISMS BASED ON EXCITED STATE STRUCTURE: CONCEPTS INTRODUCED IN 1961

In 1961 photochemical reaction mechanisms commonly consisted of "bond-switching" diagrams or electron pushing using "dot-dot" or "plus-minus" representations. These depicted which bonds were broken and which were formed but did not relate molecular transformationsto the excited-state structures reacting. Sometimes reactions were described in such a fashion (e.g., "bond-crossing") despite considerable informa-

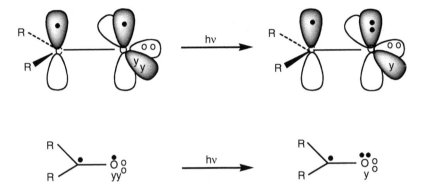

Figure 1.1 The n-π* excitation process of carbonyl compounds. Note: The open circles represent sp-hybrid electrons, the "y" electrons are those in a p_y orbital, and solid circles are π-system electrons. Electron pairs are shown as usual as bonds.

tion regarding excited-state electronic structures being available. Thus, it is curious that at the time so little attempt was made to correlate electronic structure to reactivity.

1.3 *n-π* REACTIVITY; TWO VARIATIONS: P_Y BEHAVIOR AND π-SYSTEM PROCESSES

A good starting point is the n-π* excited state of carbonyl compounds. In 1961, singlet and triplet electronic structures of simple ketones and aldehydes were known but had not been utilized in understanding photochemical reactivity. With this challenge in mind, I suggested using both the three-dimensional and the two-dimensional structures depicted in Figure 1.1. The three-dimensional picture has stereochemical value, whereas the two-dimensional representation permits rapid drawing. It was noted that in using the latter, one would still think of the three-dimensional counterpart.[1]

    For example, the Norrish type I cleavage involves loss of an alkyl group bonded to a ketone carbonyl moiety. It had been recognized that the alkyl group is lost as a free-radical species leaving an acyl radical, but this had been unrelated to the ketone excited-state structure. The suggested mechanistic transformation is depicted in Scheme 1.1. Here the p_y

orbital has a delocalized "one-electron hole" conjugated with the co-planar sigma bond. The string of three basis orbitals — p_y and the two of the sigma bond being broken — comprise a system isoconjugate with allyl. The carbonyl carbon to alkyl bond order is thus less than unity. This weakened bond then is broken.

More generally, using these structures, it proved possible to write mechanisms for a large number of the photochemical reactions known at the time. The structures are written in terms of hybrid and atomic orbitals as a basis, but the verbal terminology was that of the states and their main configurations.

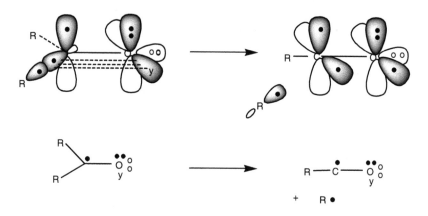

Scheme 1.1 The Norrish type I reaction in three dimensions and in two-dimension shorthand notation.

One reaction of particular interest was the Santonin to Lumisantonin rearrangement as depicted in Equation 1.1. This had been described by Barton and Gilman[2] in a series of elegant publications, but with no mechanism except for a description as "bond crossing." Clearly, the reaction involved a dramatic rearrangement with specific stereochemical changes. Superficially it appears as if the keto carbonyl group at carbon 3 (C-3) has moved to C-2, that the methyl group originally at C-4 has moved to C-1, and that C-10 bearing a methyl group has become inverted in configuration.

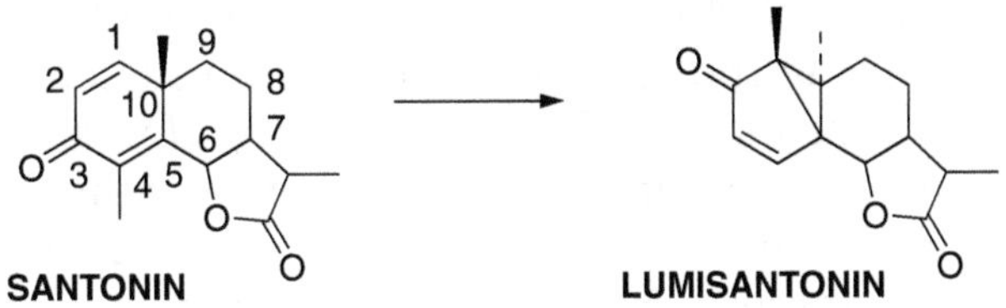

SANTONIN   LUMISANTONIN   (1.1)

Clearly, a bond-switching mechanism of the type in Equation 1.2 is unrelated to the electronic structure of the excited state and does not provide a basis for understanding the reaction or predicting further excited-state processes.

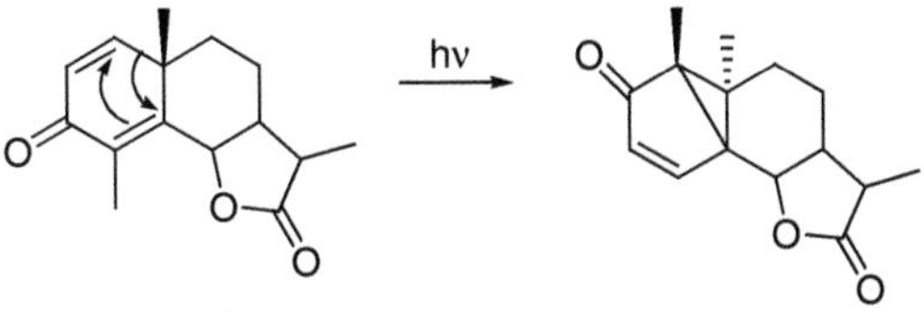

(1.2)

Our research began with a simpler 2,5-cyclohexadiene as shown in Scheme 1.2.[3] Using the n-π^* configuration as outlined in Scheme 1.1, but utilizing the shorthand notation here for simplicity, we can understand the rearrangement of 4,4-diphenyl-2,5-cyclohexadiene to afford the bicyclic photoproduct actually obtained. The final step involves a "type A zwitterion" whose rearrangement is known to proceed as shown.

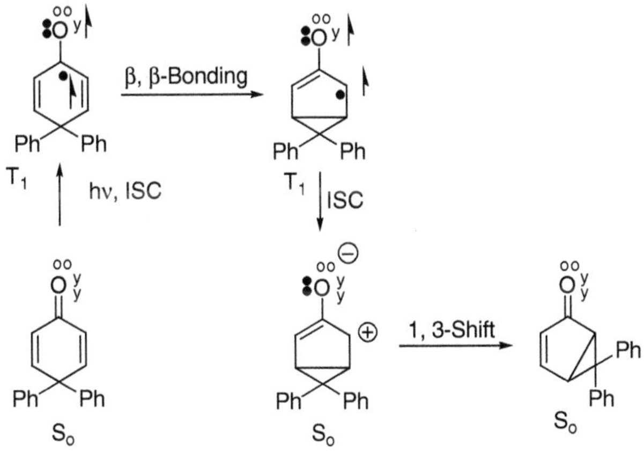

Scheme 1.2 The Type A rearrangement.

Application of this mechanism to the Santonin photo-chemistry leads to the correct Lumisantonin skeleton. We are then left with the matter of stereochemistry. However, writing the Santonin mechanism in three dimensions but with the same electronic transformations, we arrive specifically at the experimental stereochemistry of Lumisantonin as depicted in Scheme 1.3.[1]

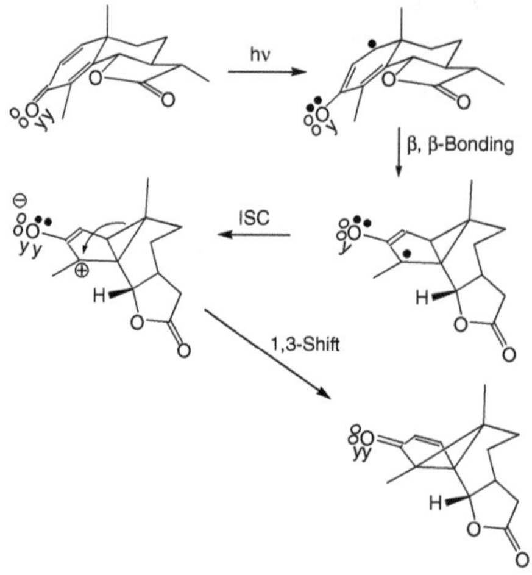

Scheme 1.3 The Santonin rearrangement in three dimensions.

More generally, making use of reactant electronic struc-tures as derived from Scheme 1.1, it proved possible to write detailed mechanisms for a variety of n-π^* and other photo-chemical reactions.[1] However, in the case of the Type A dienone rearrangement, we still need to understand the driv-ing force for β,β-bonding of the dienone triplet.

The purpose of the preceding description of a few trans-formations of the many studied is to lay the groundwork for the development of the subsequent theoretical studies.

1.4 INITIAL BUT MORE QUANTITATIVE TREATMENTS

To begin this discussion, it is worthwhile to recognize that in the early days of organic photochemistry, the best methods were one-electron methods (i.e., Hückel treatments). A good word needs to be put in for that methodology despite its inadequacy for most quantitative questions today. Thus, Hückel and graph theory are identical, and one needs to recognize that these are purely topological. When applied to molecular systems, the only energies obtained are those derived from the topology. Nevertheless, the topological energetic component quite often, if not most often, is dominant, with electron–electron interactions contributing second-order effects.

Hence, one often can analyze a reacting system and reach a tentative conclusion by considering the system's topology.

1.5   EARLY COMPUTATIONAL RESULTS: TOPOLOGICAL-HÜCKEL TREATMENTS

In the case of the Type A rearrangement of 2,5-cyclohexadienones, simple one-electron computations[4] of the Mulliken–Wheland–Mann[5] type were performed that involve iterative modification of resonance integrals. These led to results that were qualitatively similar to triplet configuration interaction computations carried out in parallel. The results are shown in Schemes 1.4a and 1.4b.

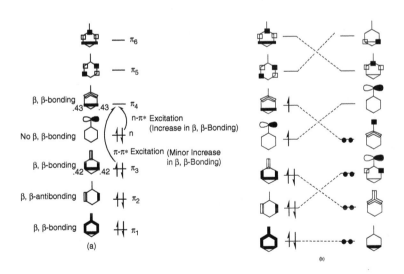

Scheme 1.4   (a) (left) Excitation processes for 2,5-dienones. (b) (right) Bonding correlations with conversion of the n-π^* state to the Type A zwitterion (far right).

Scheme 1.4a shows that n-π^* excitation involves electron promotion from an MO (i.e., "n") that is negligibly bonding between the β-carbon p-orbitals to π_4, which is β,β-bonding as a consequence of the wave function having like signs at these carbons. (This idea of a bond order between noncontiguous orbitals is just equivalent to first-order perturbation theory.) In contrast, π–π^* excitation from π_3 to π_4 involves a negligible enhancement in β,β-bond order.

A correlation diagram for formation of the Type A zwitterion, in which the β-carbons have overlapped to form a sigma bond, is shown in Scheme 1.4b. The one-electron computations were initially carried out assuming sigma-pi orthogonality but were checked using the extended Hückel method.[6] One interesting item deriving from the correlation diagram is that reversion of the ground state (i.e., S_0) zwitterion product is symmetry forbidden as correlating with a doubly excited configuration of the dienone reactant. A second point is that the n-π^* configuration has two singly occupied molecular orbitals crossing en route to zwitterionic product. It was noted[4,7] that this affords a route to ground-state product. In this instance, if the reactant is taken as the n-π^* triplet and the product as the S_0 zwitterion, then intersystem crossing (ISC) must be involved rather than the conical intersection implied in that publication.[7] Also, one really should not write a simple one-electron correlation diagram connecting species of different multiplicities. Nevertheless, the reasoning is qualitatively correct.

Before the present-day powerful computers and quantum mechanical packages were available, we made use of a DEC PDP-11/55 for SCF-CI computations.[8] The hard drive was used for virtual memory. The study began with derived configurational wave functions taken as linear combinations of Slater determinantal functions. The proper linear combinations were used for the ground state and for singly excited configurations. Triplet configurations were also considered. Finally, the proper matrix elements for interaction between singly excited configurations were set up. A basis set of p-orbitals was used with x-ray geometry for the molecules of interest ó 1-acetonaphthone, 2-acetonapthone, acetophenone, cyclohexenones, benzophenone, benzoquinone, biacetyl, 2,5-cyclohexadienones, and formaldehyde. The agreement with the literature on singlet and triplet (where known) excited states was quite good.

Then the computations on the 2,5-cyclohexadienones were used to obtain β,β-bond orders of the five electronic states listed in Scheme 1.5. As noted above, such bond orders between nonconjugated centers are equivalent to a perturbation computation and provide information on tendencies of the two centers to bond. There are positive bond orders for three of these states. However, it is known that the Type A rearrangement occurs selectively from a triplet, and only the n-π* triplet has a positive bond order, thus providing theoretical support for the reaction mechanism.

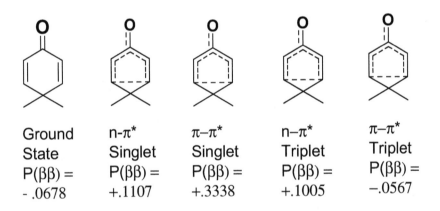

Ground State	n-π* Singlet	π-π* Singlet	n-π* Triplet	π-π* Triplet
$P(\beta\beta) = -.0678$	$P(\beta\beta) = +.1107$	$P(\beta\beta) = +.3338$	$P(\beta\beta) = +.1005$	$P(\beta\beta) = -.0567$

Scheme 1.5 β,β bond orders of the electronic states of dienones.

1.6 THE BARRELENE TO SEMIBULLVALENE REARRANGEMENT

Another example of an early and simplistic computation is the barrelene to semibullvalene transformation. In Scheme 1.6 the overall transformation is shown to involve two triplet diradical intermediates, diradical 1 and diradical 2. Experimental support exists for the rapid symmetrization of diradical 2 as depicted.

In 1967, ideal methods for obtaining three-dimensional coordinates were unavailable. These were obtained by use of graph paper and a millimeter ruler. Then the extended Hückel method of Hoffmann and Lipscomb[6] was applied to each of the species shown in Figure 1.2. Although the method clearly did not take multiplicity into account, it did afford approximate one-dimensional reaction hypersurfaces for S_0 and the singly excited species.[9] The plot is given in Figure 1.2. Much later, in the last decade, complete active space self-consistent field (CASSCF) results were compared.

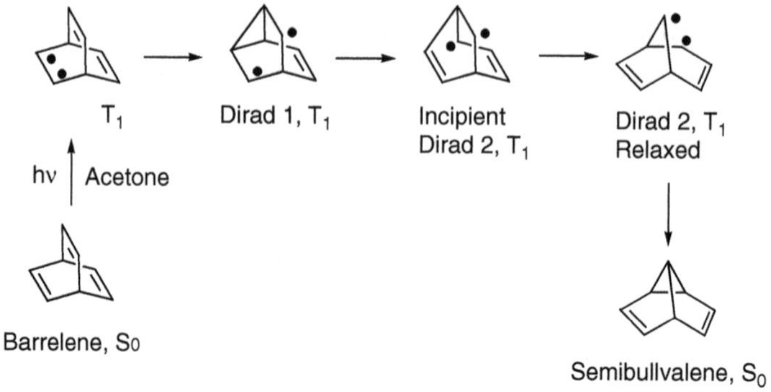

Scheme 1.6   The barrelene to semibullvalene transformation.

One interesting aspect is the lack of a major excited-state energy barrier. Such a barrier was obtained for an alternative mechanism in which two vinyl bridges bonded initially, followed by scission of two sigma bonds. This alternative mechanism was excluded experimentally. However, the extended Hückel picture differs only quantitatively from the SCF-CI computation we carried out many years later (vide infra).

1.7   MORE MODERN COMPUTATION OF THE BARRELENE TO SEMIBULLVALENE CONVERSION

In our early studies noted above, we carried out computations on the barrelene to semibullvalene transformation. With the availability of Gaussian[10] and Gamess[11] methods, more sophisticated computations were possible. Computations with RHF/6-31G* for the singlets and ROHF/6-31G* for the triplet were

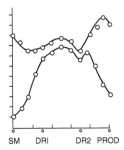

SM   DRI   DR2 PROD

Figure 1.2 Extended Hückel treatment of barrelene to semi-bullvalene reaction.

followed by CASSCF/6-31G*, afforded the results depicted in Figure 1.3.

There are several items of particular interest. First, this fairly sophisticated CASSCF computation confirms the prediction by the early, very primitive one-electron computations that there are two triplet energy minima corresponding to diradical 1 and diradical 2. The second is the near degeneracy of the triplet states with S_0 at the geometries of diradical 1 and diradical 2; the CASSCF computations place the triplets close to the ground state at these points.

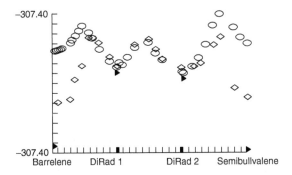

-307.40

-307.40
Barrelene   DiRad 1   DiRad 2   Semibullvalene

Figure 1.3 Triplet hypersurface for conversion of barrelene to semibullvalene (ellipsoids). The diamonds correspond to ground-state points. Solid triangles are selected S_1 points. The ordinate is in Hartrees (−307.54 to −307.40).

A different but fascinating aspect derives from the attempted computation of the S_0 state corresponding to diradical 1. On attempted geometry optimization, S_0 of barrelene itself was regenerated. Thus, the computer had run a reaction converting diradical 1 to barrelene. This is another example of the well-known[12] phenomenon in which S_0 1,4-diradicals cleave bond 2,3 in a Grob-like fragmentation. This suggests that any intersystem crossing from T_1 to S_0 early along the reaction route should lead back to ground-state reactant barrelene.

In general, for a successful photochemical reaction, intersystem crossing of triplets to S_0 and internal conversion of S_1 in reactions need to occur after considerable geometric transformation toward product. Thus conical intersections play an important role in organic photochemistry. However, for a successful reaction the aim is the occurrence of these late rather than early. The importance of conical intersections was recognized even earlier by this author[4,7] and Michl.[13] However, for a successful reaction precise location of conical intersections has become practical primarily due to the work of Robb et al.[14]

1.8 DETAILED REACTIVITY OF INTERMEDIATE DIRADICALS IN THE BARRELENE TO SEMIBULLVALENE REARRANGEMENT

Photochemistry is controlled in two ways.[1,4,7] First, there is the avoidance of excited-state energy barriers, which plays a major role. Second, the placement of conical intersections and avoided crossings determine reaction success vs. radiationless decay. A conical intersection positioned close to the Franck–Condon geometry and before to an appreciable barrier favors decay to the reactant ground state and a low or zero quantum yield. A conical intersection positioned after the first energy barrier on the excited state hypersurface facilitates reaction, epecially if the conical intersection appears close to a product geometry.

The Di-π-Methane rearrangement provides an illustrative case and is of intrinsic interest relative to the two main reaction intermediates — diradical 1, and diradical 2. This discussion begins with the independent generation of diradical 1.[15] Scheme 1.7 presents the thermal and photochemical behavior of three azo compounds related to barrelene, benzobarrelene, and 2,3-naphthobarrelene.

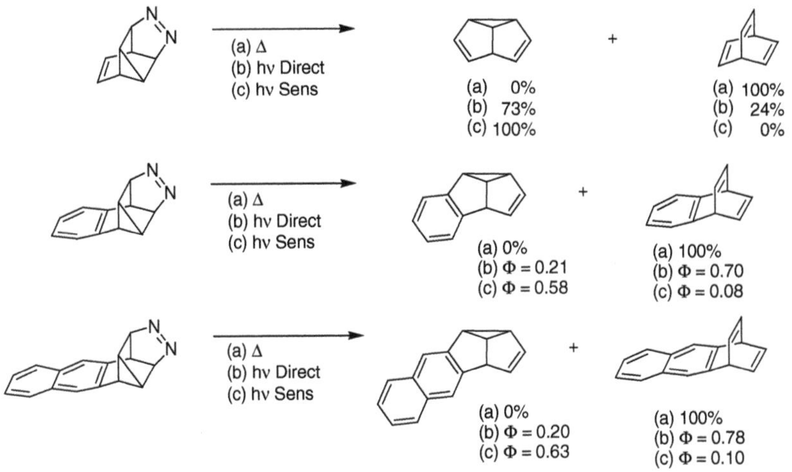

(a) Δ
(b) hv Direct
(c) hv Sens

(a)　0%
(b)　73%
(c) 100%

(a) 100%
(b)　24%
(c)　0%

(a) Δ
(b) hv Direct
(c) hv Sens

(a) 0%
(b) Φ = 0.21
(c) Φ = 0.58

(a) 100%
(b) Φ = 0.70
(c) Φ = 0.08

(a) Δ
(b) hv Direct
(c) hv Sens

(a) 0%
(b) Φ = 0.20
(c) Φ = 0.63

(a) 100%
(b) Φ = 0.78
(c) Φ = 0.10

Scheme 1.7　Generation of diradical 1.

From this initial study[15] it was clear that the loss of nitrogen by thermolysis leads to the ground state. Whether this is a concerted process (i.e., a reverse [2 + 2 + 2]) or generates S_0 of diradical 1, which then fragments to barrelene, is debatable. In any case, the transition structure for the concerted process is just diradical 1 loosely bonded to a molecular nitrogen molecule. What is clear is that sensitization leads to diradical 1 and preferentially to the semibullvalenes rather than radiationless decay to ground state.

In a continuing study[16] evidence was adduced that diradicals 1 and 2 are energy minima and are nonequilibrating under the usual reaction conditions of the di-π-methane rearrangement. It also proved possible to identify a higher energy triplet diradical 1, that is, T_2. Two independent routes to T_1 and T_2 were found experimentally. These were applied to dimethylbenzobarrelene and to dimethylnaphthobarrelene; the latter example is shown in Scheme 1.8. Independent of the source of diradical 1, the same ratio of the two isomeric naphthosemibullvalenes was obtained. This clarifies that there is a common intermediate (i.e., an energy minimum).

Another striking observation was that all the T_1 diradicals 1 are subject to a Jahn–Teller distortion in which one of the π-bonds was stretched. A final observation came from spin-orbit coupling computations that showed appreciable SOC only in diradical 2. The triplet computations on diradicals 1 and 2 were done at the CASSCF(6,6)/6-31G* level, whereas the spin-orbit coupling computations were carried out with a modified[16] Gamess[11] with a multiconfigurational (i.e., MCSCF) singlet interacting with the three components of the triplets.

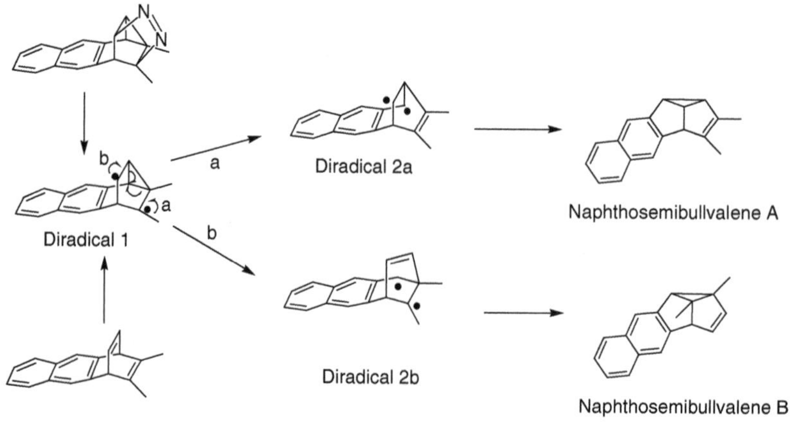

Diradical 1

Diradical 2a

Naphthosemibullvalene A

Diradical 2b

Naphthosemibullvalene B

Scheme 1.8

Table 1.1 gives the spin-orbit coupling results. There is a fourfold greater SOC for diradical 2 than for diradical 1, thus accounting for the enhanced intersystem crossing from diradical 2 with this species then closing to form semibullvalene. The programming in Gamess[11] was modified[16] so that in a "do loop" the individual contributions from pairs of hybrids were collected separately.

TABLE 1.1   Spin-Orbit Coupling in Barrelene Diradicals I and II (cm⁻¹)

Diradical I. RMS Spin-Orbit Coupling: 0.758

	L_x	L_y	L_z
[(S = 0.0, MS = 0.0)//(S = 1.0, MS = 1.0)]	0.5360	−0.0001	0.0000
[(S = 0.0, MS = 0.0)//(S = 1.0, MS = 0.0)]	0.0000	0.0000	−0.0001
[(S = 0.0, MS = 0.0)//(S = 1.0, MS = −1.0)]	−0.5360	0.0001	0.0000

Diradical II. RMS Spin-Orbit Coupling: 3.264

	L_x	L_y	L_z
[(S = 0.0, MS = 0.0)//(S = 1.0, MS = 1.0)]	−0.8079	−0.0887	0.0000
[(S = 0.0, MS = 0.0)//(S = 1.0, MS = 0.0) >	0.0000	0.0000	3.0544
[(S = 0.0, MS = 0.0)//(S = 1.0, MS = −1.0) >	0.8079	0.0887	0.0000

1.9 SINGLE-PHOTON COUNTING METHOLODOGY[17]

Another use of computation was in single-photon counting. This method of obtaining fluorescent lifetimes involves use of a lamp with, for example, nanosecond-width lamp flashes directed toward a fluorescent compound. At each time, t, a number of molecules are excited and begin to decay with emission. The emission intensity at time t is given by Equation 1.3. For simplicity of presentation here we assume a single negative exponential excited-state decay function as in Equation 1.4.

$$E_t = \sum_{n=0}^{t} I_n D_{(t-n)} \qquad (1.3)$$

$$D_t = ae^{-kt} \qquad (1.4)$$

Thus

$$E_t = \sum_{n=0}^{t} I_n ae^{e-k(t-n)} \, . \qquad (1.5)$$

Derivatizing E_t in Equation 1.5 with respect to k and a, we obtain K_t and A_t as in Equations 1.6 and 1.7. That is, substitution of Equation 1.4 into 1.3 and derivatizing E_t with respect to k, one obtains the dependence, K_t, of emission intensity at time t relative to the decay rate constant k. A_t results similarly.

$$K_t = \partial E_t / \partial k = -a \sum_{n=0}^{t} I_n(t-n)e^{-k(t-n)} \qquad (1.6)$$

$$A_t = \partial E_t / \partial a = \sum_{n=0}^{t} I_n e^{-k(t-n)} \qquad (1.7)$$

Then, in vector-matrix form, the errors in E at the various decay times are given in the vector **ΔE**, and the values of K_t and at these times are given in the 2 × n matrix **V** in Equation 1.8.

$$\begin{bmatrix} \Delta E_1 \\ \Delta E_2 \\ \Delta E_3 \\ \vdots \\ \Delta E_N \end{bmatrix} = \begin{bmatrix} A_1 & K_1 \\ A_2 & K_2 \\ A_3 & K_3 \\ \vdots & \vdots \\ A_N & K_N \end{bmatrix} \begin{bmatrix} \Delta a \\ \Delta k \end{bmatrix} \qquad \text{OR} \qquad \overline{\overline{\Delta E}} = \overline{\overline{V}}\,\overline{\overline{\Delta}} \qquad (1.8)$$

All the items in the matrix equation are known except for the vector **Δ**, which affords the deviations of the preexponential *a* and the rate constant *k*. With more decay points than unknowns, the problem is overdetermined. Least-squares treatment computationally affords the unknown **Δ** in the first iteration. These corrections are then used for adjusting the rate constant and preexponential for the next iteration. Further iterations lead to convergence and afford final values of the two desired constants.

The same method works with other decay functions. With a double exponential, the dimensions of **V** and **Δ** are increased accordingly.

1.10 THE META EFFECT

This research was inspired by a remarkable observation by Havinga et al.[18] that in photolysis of the two phenolic phosphates shown in Equation 1.9, the meta isomer hydrolyzed more rapidly than the para counterpart. This not only is the reverse of what was observed in the dark but is counter to all intuition of ground-state organic chemistry. In the ground state, one expects withdrawal of electron density from the ortho and para sites, and thus it seemed strange that the meta isomer was more reactive than the para isomer. Havinga et al.[18] noted at the time that this, indeed, was contrary to normal (i.e., ground-state) resonance theory and had no rational explanation.

(1.9)

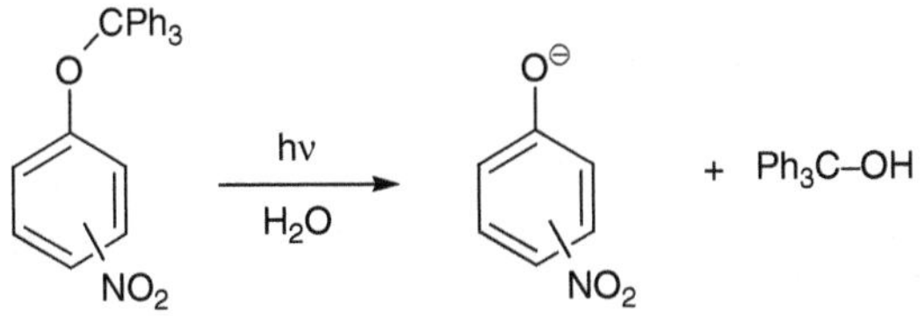

(1.10)

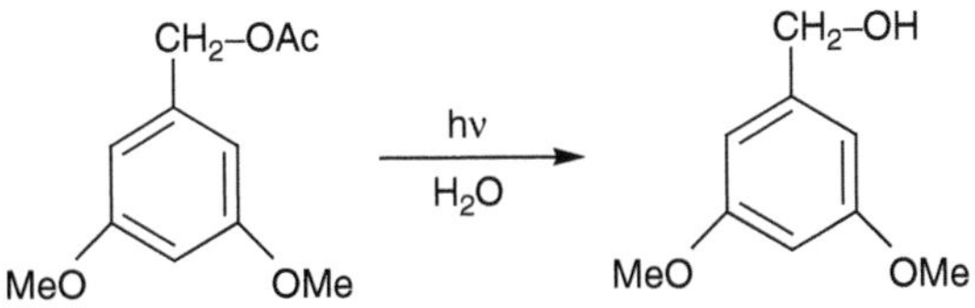

(1.11)

These unusual results led me to consider the generality of the phenomenon and the mechanistic rationale. Two further examples[19a,19b] are shown in Equations 1.10 and 1.11. That in Equation 1.10 is parallel to the phosphate reaction of Havinga et al., except that a trityl cation is expelled rather than a phosphate moiety. And, it was again found that the meta isomer was more reactive than the para isomer.

The mirror-image example[19b] is found in Equation 1.11, where there is expulsion of an acetate anion. Here, meta-electron donors accelerate the reaction, and the dimethoxybenzyl derivative shown was one of the best in exhibiting the meta effect.

Again, the excited-state behavior is the reverse of the usual ortho–para electron transmission taught in undergraduate courses. The meta effect may be pictured qualitatively in resonance terms, as in Figure 1.4. Electron transmission is then pictured as to and from the meta site. A more quantitative picture was simply obtained from Hückel computations.

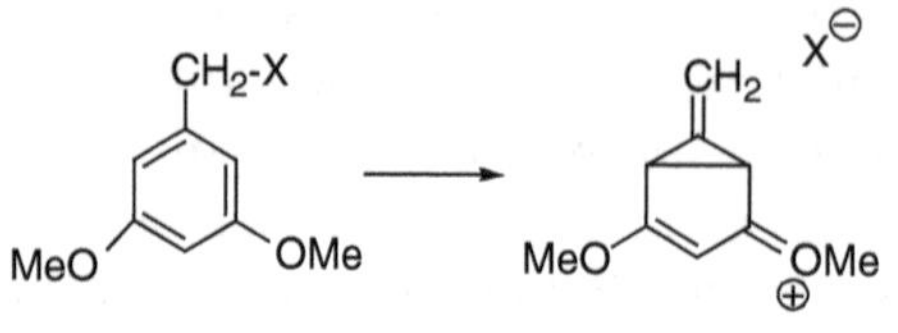

Figure 1.4 Meta expulsion.

The simplest model consists of a benzylic species in which the benzylic carbon simulates an electron donor if it provides two electrons and an acceptor if it has none. Hence for the electron donor example, the system has eight electrons; for the electron-withdrawing–group example, the system has six electrons. The densities computed in 1963 for the benzylic species' molecular orbitals are shown in Scheme 1.9.

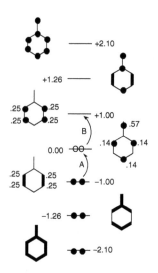

Scheme 1.9 A one-electron picture of excitation in six-electron and eight-electron systems. Source of change in electron density meta and para to the substituent.

With eight electrons, electronic excitation (note arrow B) moves the density from an MO that has no weighting and no density at the meta position to one that has a density of 0.25.

The para position loses density. For the six-electron system, excitation (note arrow A) moves density from a molecular orbital with meta density to an MO with no meta density. However, the para position gains density.

That was in 1963. Much more recently,[20] Gaussian CASSCF computations with an active space of eight were carried out. The S_1 energies for three systems were obtained: the p-methylbenzyl, the m-methoxybenzyl, and the 3,5-dimethoxybenzyl molecules. In each case the computation was for the ion pair in which the counter ion was acetate. The radicals were paired with an acetoxy radical. The energies of these species are given in Table 1.2.

The ion pairs are of lower energy than the radical pairs for each of the para, meta, and dimethoxy species. Also, the m-methoxy cationic systems are of lower energy than the p-methoxylated system, with the 3,5-dimethoxybenzylic cationic system being the lowest in energy. Of interest also is the observation that para substitution lowers the energy of the radical species. Thus, the computations are in agreement with stabilization of S_1 benzylic cationic species by m-methoxy groups and also with the conclusion that heterolysis is

TABLE 1.2 Summary of S_1 Excited-State Energies for Benzylic Species

p-Methoxybenzyl			
Ion pair	−610.1234	Radical pair	−610.11704
m-Methoxybenzyl			
Ion pair	−610.14687	Radical pair	−610.11433
3,5-Dimethoxybenzyl			
Ion pair	−724.05620	Ion pair	−724.00483

Note: The energies are given in Hartrees (i.e., 727.5 kcal/Hartree).

preferred over homolysis for the meta isomers. Most noteworthy is that these considerably more sophisticated computations are consistent with the 1963 one-electron results.

Another approach to the problem involved the search for conical intersections between the S_1 and S_0 surfaces. The results for the 3,5-dimethoxylbenzylic systems are given in Table 1.3. Similar computations and results were carried out for the mono methoxylated systems.

It was found that conical intersections, with excited and ground states coming together, were a unique characteristic of the meta-methoxy cationic systems. With para substitution or in the case of the radical species (i.e., D0 and D1), conical intersections were not observed and the ground- and excited-state surfaces were considerably separated. In addition, solvation in the form of SCRF as implemented in Gaussian 98 was included in separate computations.

1.11 AN EXAMPLE WHERE SUPERFICIALLY IDENTICAL DIRADICALS GIVE DIFFERENT REACTIVITY

A particularly fascinating set of photochemical reactions is depicted in Scheme 1.10.[21] Reactant A produces photoproduct F and the mechanism proceeds via a diradical D. However, the photochemistry of F also proceeds via a diradical categorized as D but leads to A and C. Hence there is the paradox

TABLE 1.3 Energy Gaps and Conical Intersections for Dimethoxylbenzylic Systems

ΔE between S1 and S0 S1 geom	ΔE between D1 and D0 D1 geom
0.0025	0.0822
ΔE between S1 and S0 conical geom	ΔE between D1 and D0 conical geom
0.00005	0.00268
ΔE between S1 and S0 S0 geom	ΔE between D1 and D0 D0 geom
1.3966	0.0878

of the same diradical, written as D, leading to different products.

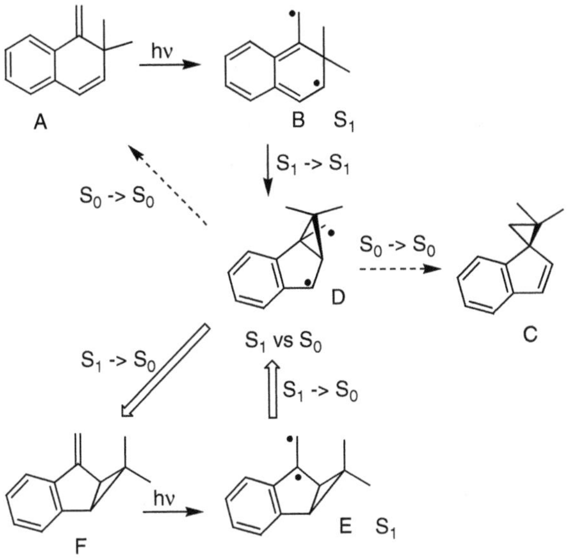

Scheme 1.10 The superficially same diradical species giving rise to two different reactions.

In this early study[21] SCF and SCF-CI computations were carried out by the technique described above on a DEC PDP-11/55 with the hard drive providing virtual memory. For these computations, a truncated set of basis orbitals was used. The truncation criterion utilized those orbitals that were part of the π-system and those that were involved in bond formation or loss of a bond.

However, the paradox proved to be only superficial. It was recognized that two singlet states of diradical D were involved: S_1 derived from reactant A and S_0 derived from reactant F. The MO correlation diagram is in Figure 1.5a. Here, only five of the MOs are shown. It is seen that in proceeding from reactant A there are no MO crossings and a singly excited configuration leads to diradical D in the S_1 configuration. In contrast, in starting with reactant F there is a HOMO-LUMO crossing permitting decay to an S_0 configuration of diradical D.

The corresponding state diagram (Figure 1.5b) resulting from configuration interaction (CI) computations reveals the

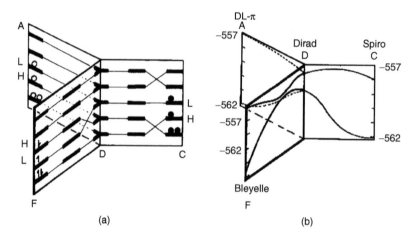

(a)   (b)

Figure 1.5 (a) (left) Correlation of the most germane MOs with H representing Highest Occupied Molecular Orbital (HOMO) and L representing Lowest Unoccupied Molecular Orbital (LUMO). (b) (right) Corresponding state correlation diagram (energies in e.v.).

avoided crossing already detected in MO terms above. It was noted that the avoided crossing was canted, and this is illustrated in Figure 1.6 as an enlarged view taken from Figure 1.5a. With such canting, one might anticipate a preference for decay to species F, and this was experimentally observed. Also, we note that in these cases of approximate computations, the

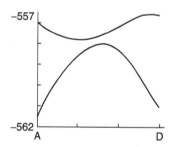

Figure 1.6 Canting of the avoided crossing or bifunnel.

difference between an avoided crossing and a real conical intersection where the surfaces meet is a matter of computational accuracy and thus not really differentiated in many cases.

1.12 DELTA-DENSITY MATRICES AND REACTIVITY

Most quantum mechanical computational packages include a one-electron density matrix as part of the output. The off-diagonal elements of these matrices may be construed as bond orders (or bond densities) and the diagonal elements as electron densities. One may define a delta-density matrix as the difference between the density matrices of two species. For example, the first density matrix might derive from a reactant and the second could be the density matrix of the corresponding excited state, however with the original geometry maintained.

The idea of using bond-order matrices derived from much simpler computations and using a ground state and the corresponding excited state goes back some time.[22] It is seen in Scheme 1.11 that promotion of one electron from MO 2 to MO 3 results in a marked decrease in bond order 1,2 but a large increase in bond order 2,3. Thus, one may predict a twisting of the molecule at bond 1,2 on excitation but a tendency toward planarization at bond 2,3. This reasoning was applied to a variety of photochemical processes and reactions in our publications.

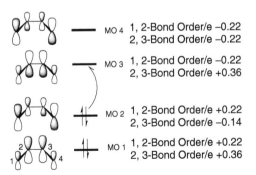

	MO 4	1, 2-Bond Order/e −0.22
		2, 3-Bond Order/e −0.22
	MO 3	1, 2-Bond Order/e −0.22
		2, 3-Bond Order/e +0.36
	MO 2	1, 2-Bond Order/e +0.22
		2, 3-Bond Order/e −0.14
	MO 1	1, 2-Bond Order/e +0.22
		2, 3-Bond Order/e +0.36

Scheme 1.11 Simple example of HOMO-LUMO butadiene excitation and bond-order change.

With much more powerful quantum mechanical computations available (i.e., Gaussian 98), the method was applied to
a variety of photochemical reactions[23] (note Scheme 1.12). The
expression in Equation 1.12 for the delta-density matrix elements includes overlap integrals to take care of basis set def-
initions. Weinhold NHOs[24] (i.e., hybrids) were used[23] in order
to permit easy analysis in terms of basis orbital pair bonds
comprising orbital pairs. Note: A refers to a reactant and B
refers to the corresponding excited state in this study.

$$\Delta D_{rt} = D^B{}_{rt}S^B{}_{rt} - D^A{}_{rt}S^A{}_{rt} \qquad (1.12)$$

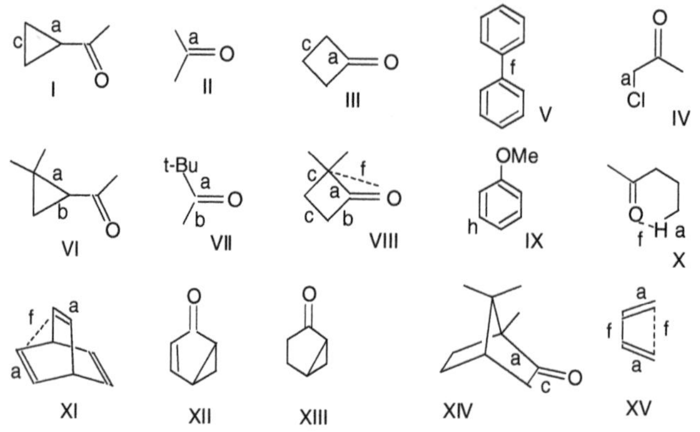

Scheme 1.12 Delta-density bond values.

However, the delta-density treatment proved more general
than dealing with just photochemical excitation.[25] Thus, if species A is a general reactant and B is a related species other
than the corresponding excited state, the same delta-density
treatment is valid. The new method has been applied to a
variety of organic transformations. It permits comparison of
bonding in species differing in multiplicity, and number of electrons as molecule A proceeds to form molecule B. The transformation may involve addition of an electron to form a radical
anion or loss of an electron to form a radical cation, or the
second molecule, B, may result from stretching of one bond.

We note that density matrices, and thus the delta-density
matrices, are spin free and thus afford measures of bonding

TABLE 1.4 Delta-Density Values for the Cyclopropyl Radical
Anion Formation[a]

Entry	Bonds	ΔD	Entry	One-Center Hybrid Orbitals	ΔD
1	C2–C4	−26	5	C(O)	+4989
2	C2–C3	−220	6	O(C)	+3071
3	C3–C4	−26	7	C2(C3)	−642
4	C=O	− 1452	8	C3(C2)	+1106
			9	C2(C4)	−652
			10	C4(C2)	+1000
			11	C4(C3)	+39
			12	C3(C4)	−7

[a] CASSCF $(8,8)/3$-$21G^*$ for the neutral molecule and CASSCF $(9,8)/3$-$21G^*$ for the radical anion.

independent of the states studies. From CASSCF computations, the density matrices used are one-electron functions. For open shell and triplet species, the spin-averaged density matrices correlate best with bond orders and are used.

The method is broad ranging in value. Photochemically, it predicts relaxation properties as well as the primary photochemical process. For radical anion and cation processes it is equally powerful. In the case of the metal reduction of cyclopropyl ketones in Equation 1.13, the opening of the three-membered ring is properly predicted, as is the regioselectivity; that is, bond 2,3 corresponds to the most negative delta-density matrix element (note Table 1.4). The method also correctly predicts the internal bond scission of the radical cation in Equation 1.14 (also see Table 1.5). Even in the case of ground-state processes such as the dehydronorbornyl solvolysis (note Equation 1.15), the prediction is in accord with organic intuition, namely, loss of electron density from the p-bond and formation of a 2,6-sigma bond as the carbon-to-chloride bond is stretched.

$$(1.13)$$

TABLE 1.5 Delta-Density Values for the Radical Cation 8

Bond or Orbital	CAS(5,6)[a]	CAS(7,7)[b]
C1–C5	−56	−1320
C1–C6	−594	−1008
C5–C6	−86	−152
C5(C1)	−676	−2205
C1(C5)	+538	−1379
C5(C6)	−308	−67
C6(C5)	+30	0
C1(C6)	+98	−1091
C6(C1)	−1824	−1906
O7(p_y lone pair)	−6444	−234
O7(sp lone pair)	−156	−1134

[a]CASSCF $(6,6)/3\text{-}21G^*//RHF/6\text{-}31G^*$ energy: −344.96684 a.u. (S0), CASSCF(5,6)/3-21G*: −344.69385 a.u (radical-cation).
[b]CASSCF $(7,7)/3\text{-}21G^*$: -344.69481 a.u. (radical-cation).

$$(1.14)$$

$$(1.15)$$

The method revealing which bonds are weakened and which bonds are strengthened gives a perturbation approach to the prediction of organic reactions in general, photochemical or otherwise. It does need to be recognized that, although the delta-density values obtained computationally are quantitative, their predictive powers at this point are qualitative. Our research is proceeding to ascertain the extent to which these matrices and derived quantities have quantitative relevance.

1.13 CONCLUSION

This article has attempted to depict some aspects of the development of organic theory over the decades as seen in our research group and to outline the parallel development of computational methodology from the days of the IBM 1604, the DEC PDP-8, and the PDP-11 to the Unix days of (e.g.) SGI, and also DEC VAX, and Alpha machines with Gaussian and Gamess capabilities and now to the ability of Linux PCs to carry out such *ab initio* computations.

ACKNOWLEDGMENT

Support of the author's research over the decades by the National Science Foundation is gratefully acknowledged. In these days of ubiquitous emphasis of support of research that will be of practical value in a year or two, such confidence in the value of basic research is particularly appreciated.

REFERENCES

1. (a) Zimmerman, H. E., *Seventeenth National Organic Symposium of the Am. Chem. Soc., Abstracts*, Bloomington, Indiana, 1961, pp. 31-41, (b) Zimmerman, H. E., Schuster, D. I., *J. Am. Chem. Soc.*, 84, 4527-4540, 1962, (c) Zimmerman, H. E., *Tetrahedron*, Suppl. 2, 19, 393-401, 1963, (d) Zimmerman, H. E. *in* "*Advances in Photochemistry*," Editors: Noyes, Jr. A., Hammond, G. S. and Pitts, J. N., Jr., Interscience, Vol. 1, 183-208, 1963.

2. Barton, D. H. R., Gilhan, P.T., *J. Chem. Soc.*, 82, 4596-4599, 1960.

3. (a) Zimmerman, H. E., Schuster, D. I., *J. Am. Chem. Soc.*, 83, 4486-4487, 1961, (b) Zimmerman, H. E., Schuster, D. I., *J. Am. Chem. Soc.*, 84, 4527-4540, 1962.

4. Zimmerman, H. E., Swenton, J. S., *J. Am. Chem. Soc.*, 89, 906-912, 1967.

5. Muller, N., Mulliken, R. S., *J. Am. Chem. Soc.* 80, 3489-3497, 1958.

6.   (a) Hoffmann, R., Lipscomb, W. N., *J. Phys. Chem.* 36, 2179-2189, 1962 (b) Hoffmann, R., Lipscomb, W. N., *J. Phys. Chem.* 36, 3489-3493, 1962 (c) Hoffmann, R., Tetrahedron, 22, 539-545, 1966.

7.   (a) Zimmerman, H. E., *J. Am. Chem. Soc.*, 1966, 88, 1564-1565., (b) Zimmerman, H. E., *J. Am. Chem. Soc.*, 88, 1566-1567, 1966.

8.   Zimmerman, H. E., Binkley, R. W., McCullough, J. J., Zimmerman, G. A., J. Am. Chem. Soc., 89, 6589-6595, 1967.

9.   (a) Zimmerman, Binkley, R. W., Givens, R. S., Sherwin, M. A. J., *Am. Chem. Soc.*, 1967, 89, 3932-3933, (b) Zimmerman, Grunewald, G. L., H. E., Paufler, R. M., Sherwin, M. A. *J. Am. Chem. Soc.*, 91, 2330-2338, 1969.

10. Gaussian: Gaussian 98, Revision A.9, Frisch, M. J., Trucks, G. W., Schlegel, H. B., Scuseria, G. E., Robb, M. A., Cheeseman, J. R., Zakrzewski, V. G., Montgomery, Jr., J. A., Stratmann, R. E., Burant, J. C., Millam, J. M., Daniels, A. D., Kudin, K. N., Strain, M. C., Farkas, O., Tomasi, J, Barone, V., Cossi, M., Cammi, R, Mennucci, B., Pomelli, C, Adamo, C., Clifford, C, Ochterski, J., Petersson, G. A., Ayala, P. Y., Cui, Q., K. Morokuma, K., Malick, D. K., Rabuck, A. D., Raghavachari, K., Foresman, J. B., Cioslowski, J., Ortiz, J. V., Stefanov, B. B., Liu, G., Liashenko, A., Piskorz, P., Komaromi, I., Gomperts, R., Martin, R. L., Fox, D. J., Keith, T., Al-Laham, M. A., Peng, C. Y., Nanayakkara, A., Gonzalez, C, Challacombe, M., Gill, P. M. W., Johnson, B., Chen, W., Wong, M. W., Andres, J. L., Gonzalez, C., Head-Gordon, M., Replogle, E. S., and Pople, J., A. Gaussian, Inc., Pittsburgh PA, 1998.

11. (a) Gamess, QPCE Program No. QG01, Quantum Chemistry Program Exchange, Indiana University, Schmidt, M. W., Baldridge, K. K., Boatz, J. A., Jensen, J. H., Koseki, S., Gordon, M. S.,Nguyen, K. A., Windus, T. L., Elbert, S. T., *QPCE Bull.* 1990, 10, (b) Schmidt, M. W. et. al., J. Comput. Chem., 14, 1347-1363, 1993.

12. Zimmerman, H. E., Armesto, D., Amezua, M. G., Gannett, T. P., Johnson, R. P., *J. Am. Chem. Soc.*, 101, 6367-6383, 1979.

13. Michl, J. Mol., *Photochem.*, 243-255, 257-286, 287-314, 1972.

14. (a) Robb, M. A., Bernardi, F., Olivucci, M., Pure App. Chem. 1995 67, 783-789, (b) Olivucci, M., Ragazos, In., Bernardi, F., Robb, M.A., *J. Am. Chem. Soc.* 115, 3710-3721, 1993.

15. Zimmerman, H. E., Boettcher, R. J., Buehler, N. E., Keck, G. E., Steinmetz, M. G., *J. Am. Chem. Soc.*, 98, 7680-7689, 1976.

16. Zimmerman, H. E., Kutateladze, A. G., Maekawa, Y., Mangette J. E., *J. Am. Chem. Soc.*, 116, 9795-9796, 1994.

17. (a) Zimmerman, H. E., Werthemann, D. P., Kamm, K. S., *J. Am. Chem. Soc.*, 95, 5094-5095,1973, (b)). Zimmerman, H. E., Cutler, T. P., *J.C.S. Chem. Commun.*, 598-599, 1975.

18. Havinga, E., de Jongh, R. O., Dorst, W., *Rec trav. Chim.* 75, 378, 1956.

19. (a) Zimmerman, H. E., Sandel, V. R., *J. Am. Chem. Soc.*, 85, 915-922, 1963, (b) Zimmerman, H. E., Somasekhara, S., *J. Am. Chem. Soc.*, 85, 922-927, 1963.

20. (a) Zimmerman, H. E., *J. Am. Chem. Soc.*, 117, 8988-8991, 1995 (b) Zimmerman, H. E., *J. Phys. Chem.*, 102, 5616-5621, 1998.

21. Zimmerman, H. E., Factor, R. E., *J. Am. Chem. Soc.*, 102, 3538-3548, 1980.

22. (a) Zimmerman, H. E., Gruenbaum, W. T., Klun, R. T., Steinmetz, M. G., Welter, T. R., *J.C.S. Chem. Commun.*, 228-230, 1978, (b) Zimmerman, H. E., Welter, T. R., *J. Amer. Chem. Soc.*, 100, 4131-4145, 1978, (c) Zimmerman, H. E., Klun, R. T., *Tetrahedron*, 43, 1775-1803, 1978, (d) Zimmerman, H. E., Steinmetz, M. G., *J.C.S. Chem. Commun.*, 231-232, 1978.

23. Zimmerman, H. E., Alabugin, I. V., *J. Am. Chem. Soc.*, 122, 952-953, 2000.

24. (a) Foster, J. P., Weinhold, F., *J. Am. Chem. Soc.* 102, 7211-7218, 1980, (b) Reed, A., Curtiss, L. A., Weinhold, F., *Chem. Rev.* 88, 899-926, 1988, (c) Weinhold, F., Schleyer R.V.R., Ed Wiley, New York, 3, 1792, 1998.

25. Zimmerman, H. E., Alabugin, I. V., *J. Am. Chem. Soc.* 121, 2265-2270, 2001.

2

Computational Investigation of Photochemical Reaction Mechanisms

LLUÍS BLANCAFORT[*], FRANÇOIS OGLIARO[****],
MASSIMO OLIVUCCI[***], MICHAEL A. ROBB[**],
MICHAEL J. BEARPARK[**], ADALGISA SINICROPI[***]

[*]Institut de Química Computacional,
Departament de Química,
Universitat de Girona,
Campus de Montilivi,
Girona, Spain

[**]Department of Chemistry,
Imperial College London,
South Kensington Campus,
London, U.K.

[***]Dipartimento di Chimica
dell'Università di Siena,
Siena, Italy

[****]Equipe de Chimie
et Biochimie Theorique,
Universite Henri Poincare,
Nancy, France

CONTENTS

2.1 INTRODUCTION

This article could have been subtitled "The Right Question for the Right Reason."[1] Part of our purpose is to review the computational methods widely available to the photochemist today, which the authors have used extensively over the past decade.[2] But we also have a second purpose: to give some guidelines (based on experience) concerning the kind of questions about photochemistry that these methods can usefully answer. The authors are often approached with the request, "Can you do some calculations on...?" The answer may be yes, but what is the question that really needs answering? Is it a question about the excitation spectrum (what state is that?) or reaction mechanism (why doesn't that product form?). Having the right question is at least as important as knowing how to use the available methods, as we hope to show using examples drawn from recent work.

Another subtitle might be "Beyond the Jablonski Diagram." Much of this article focuses on excited-state reaction paths away from the vertical excitation region and the corresponding importance of conical intersections between potential energy surfaces. To fully discuss this subject now would probably take a book, but there is much that is not yet ready to be set in stone. We aim to cover some recent and ongoing developments: for example, points far away from the minimum of a conical intersection may be mainly responsible for decay of excited states in some systems. This article builds on previous reviews[2] but differs in attempting to include more practical advice, at the expense of exhaustive reference lists available elsewhere. There are other approaches to studying conical intersections that differ from ours and have recently been reviewed (including[23,26b,50,51]). Because our approach is practical and mechanistic, we use a classical (or semiclassical) picture of reactivity throughout this review, rather than the more rigorous quantum one.

We have several target audiences in mind:

- People starting calculations as an adjunct to their experimental research in, for example, fast laser spectroscopy
- Theoreticians looking for a challenge different from accurate calculations of vertically excited states
- Anyone interested in how reaction paths on ground- and excited-state potential energy surfaces can be related to a photochemical mechanism

2.2 CALCULATING POTENTIAL ENERGY SURFACES

In this section, we briefly review the use of the complete active space self-consistent field (CASSCF) method for calculating excited states.[3] This method offers an acceptable compromise between accuracy and computational expense, but our main reason for choosing it is that it offers analytical gradients and second derivatives, which are essential for geometry optimization. As we discuss more fully below, CASSCF is often sufficient if one is interested in structure and mechanism (as we are here), but a more accurate treatment of dynamic electron correlation is often necessary for accurate energetics.

2.2.1   Before Starting the Calculation

2.2.1.1 Types of Excited States and Representation with CASSCF

CASSCF is full configuration interaction (CI)[53] with orbital optimization in a window of partially occupied active orbitals, the active space. To choose an appropriate active space for a problem requires a model of the target excited states: If the active window is not chosen correctly, then it is impossible to represent the excited state sought.

Because CASSCF is a full CI, it can be considered in either an atomic orbital (AO) basis (valence bond [VB] theory)[4] or symmetry-adapted molecular orbital (MO) theory.[5] The two pictures are equivalent, but the VB method is more powerful because (as we discuss more fully below) it can explain why geometries change in excited states and why two potential energy surfaces intersect. In this respect, the VB picture is more appropriate for the reactivity problems we discuss here, whereas MO theory is still key to spectroscopy.

We begin with a simple VB example. For polyenes, the "valence" π excited states are of two types: covalent (dot-dot), for example, S_1 butadiene, or ionic (hole-pair), for example, S_2 butadiene (Scheme 2.1).

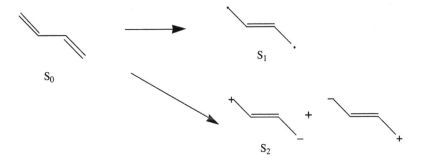

Scheme 2.1   VB representation of the S_1 and S_2 excited states of butadiene.

By valence excited states, we mean excited states where the orbitals are dominated by the valence (here *2s, 2p*) orbitals. But this idea is too simplistic in practice. For the covalent S_1 state of butadiene, an active space of four electrons in four π orbitals (which can be localized as *2p*) is sufficient. However, the ionic state[52] is dominated by configurations with two active electrons on the same center, and hence one electron will go into a *2p* orbital, whereas the second will go into a more diffuse *3p* orbital (Scheme 2.2) because of electron-electron repulsion. Thus, to describe the ionic S_2 state of butadiene correctly, we need four electrons in eight orbitals (a set of 4x *2p* and a set of 4x *3p* orbitals).

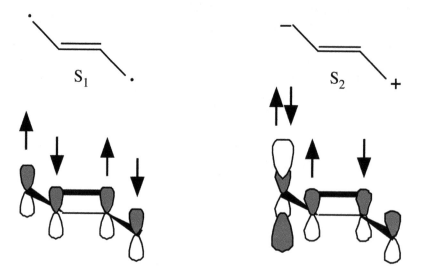

Scheme 2.2   The S_2 state of butadiene requires diffuse *p* orbitals.

However, there is a third type of excited state: Rydberg. For polyenes, the Rydberg states are like the ionic state just discussed,[52] but the electron goes into a much more diffuse atom-like orbital. Thus, diffuse basis functions are needed in the one-electron basis set and diffuse orbitals in the active space. Rydberg states are important spectroscopically, but decay will normally occur rapidly to valence or ionic states in a photochemical transformation. This can be illustrated with the photodissociation of NH_3.[6] The excited states of NH_3 are all Rydberg type. The lowest energy excitation is essentially a lone pair excitation into an np diffuse orbital. As the molecule loses an H atom, derydbergization occurs, as shown in Scheme 2.3.

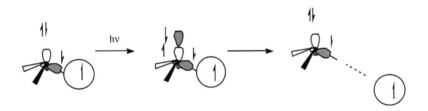

Scheme 2.3 Diffuse orbitals are needed to describe the photodissociation of NH_3.

In summary, to choose an active space for CASSCF, one needs a conceptual model of the target excited states. Part of this may come from having previously carried out some lower level of excited-state computation (CI singles [CIS] or time-dependent density functional methods [TDDFT]), but most of it must come from experience and chemical understanding.

2.2.1.2 Starting Orbitals

Hamilton and Pulay[7] showed that natural orbitals from an unrestricted Hartree Fock (UHF) computation may be very good orbitals for a ground-state CASSCF computation, with occupations very different from 2.0 or 0.0 defining the active space. This idea may not be as useful for an excited state, however, particularly if a different window of active orbitals is required from the one initially generated for the ground state. Fortunately, for an excited state, the CIS method using an restricted Hartree Fock (RHF) reference may give a good wave function, even if the energetics are unreliable. The natural orbitals of this CIS wave function could therefore be used effectively to start a CASSCF investigation of excited states. This is particularly useful for ionic excited states. Obtaining a balanced representation of several excited states might require manually averaging the densities from several CIS roots.

The alternative strategy is to start by inspecting the MOs obtained from a ground-state self-consistent field (SCF) computation. The major difficulty here is that the virtual orbitals of an extended basis SCF computation are completely unsuitable for the representation of an excited state. Thus, one should start from an SCF computation with a one-electron basis that does not have diffuse components (STO-3G or 3-21G), from which the orbitals can be visually inspected and an active space chosen. When the active space is correct, the reduced basis-set computation may be used as a starting point for an extended basis-set computation.

2.2.2 Convergence Problems and State Averaging

In small molecules with high symmetry, one can specify an excited state explicitly by symmetry and shell structure and occupancy. Thus, the first excited state might be obtained as the lowest root of a particular symmetry. Without symmetry, the nth excited state will simply be the nth root of the CI problem, but this can lead to convergence problems and conceptual difficulties. For example, the orbitals for the first excited state may be so poor for the ground state that the ground state actually becomes the second root of the CI secular equation. Then there is the problem of redundant orbitals. For example, if an active orbital has exactly double occupancy, then it is redundant because rotation with the core orbitals does not affect the energy, and the doubly-occupied orbital should move from the active to the core orbitals to avoid convergence problems. Consider the following situation. The ground state is described by a configuration $(a)^2(b)^2(c)^0$ and excited states by configurations $(a)^2(b)^1(c)^1$ and $(a)^1(b)^2(c)^1$. Clearly, when optimizing the state $(a)^1(b)^2(c)^1$ the b orbital is redundant. In what follows we explain why state averaging can solve some of these problems.

Convergence problems of the CASSCF wave function can be expected in situations of near state degeneracy (either real degeneracy or degeneracy arising from bad starting orbitals). This problem is especially severe when the degenerate states have different chemical character (for example, π, π^* and n, π^* states or covalent and ionic states) and when symmetry cannot be used to treat the states independently of each other. In these cases, optimization of the wave function for a given root can cause an increase in the energy of the lower root. This results in "flipping" between the two roots and oscillations in the energy instead of convergence.

The solution is state averaging between the two states.[8] State averaging is a poorly understood but simple idea. In a CASSCF computation, the CI vector defines the density matrix. The density matrix, in turn, determines the orbitals. In a state-averaged computation, several density matrices from several CI vectors representing different states are averaged. If the states are equally weighted, the order of the CI vectors is not important. In other words, in a state-averaged computation (with equal weights for each state), the states are treated as degenerate and the variable corresponding to rotation between these states is effectively removed. If the states are not degenerate, this is an approximation; if they are exactly degenerate, it is no approximation at all. Most commonly, two near-degenerate states are equally weighted.

Thus, the result of the state-averaged calculation is a set of state-averaged orbitals and the eigenvalues and CI vectors of the states of interest using those orbitals. The use of state-averaged orbitals implies an extra computational cost in any geometry optimization (conical intersection or not) because the coupled-perturbed (CP) multiconfigurational self-consistent field (MCSCF) equations must be solved to correct the gradients, which is not the case for the default CASSCF gradients. Use of incorrect gradients in the optimizations can lead to spurious results, although in practice the CP-MCSCF correction is often small and necessary only for very tight convergence.

State averaging is necessary (and almost unavoidable) in cases where the wave function for the upper of two degenerate states cannot be converged because of root flipping. However, the wave function for the lower state can usually be converged without any apparent problems, thus effectively ignoring the upper state. Care is required in these cases because it can lead to incorrect results. We mention here one particular case: spurious symmetry breaking of the wave function (also called doublet instability), which causes an artificial energy lowering.[9] This well-known problem has been intensively studied, for example, for formyloxyl radical and usually requires the use of extended CASSCF active spaces or inclusion of perturbation energy.[10] We have found this type of problem in organic radical cations, such as the radical cation of bis(methylene)adamantane (BMA). The use of state-averaged orbitals solves the instability problem here, keeping the minimal active space (three electrons in four orbitals in this case). BMA is a doublet, and the charge at D_2 symmetry is delocalized between the two π bonds of the molecule (see Figure 2.1). The D_2 structure is the formal transition structure (TS) for intermolecular electron transfer between two equivalent structures of C_{2v} symmetry, where the charge is localized in one of the π bonds.

At D_2 symmetry, there are two active-space orbitals of b_2 symmetry, and two active-space orbitals with b_3 symmetry, respectively. The ground state of the calculation is of $^2B_2$ symmetry and can be converged by restricting the symmetry of the determinants of the CASSCF. A separate calculation using state-averaged orbitals for the ground state and the first excited state raises the ground-state energy by 0.9 kcal/mol (3-21G* basis set), and the gap between the states is 0.5 kcal/mol. However, when symmetry-broken orbitals are used, the ground-state energy (no state averaging) is lowered with respect to the symmetry-restricted calculation by 19.4 kcal/mol because of spurious symmetry breaking. Thus, the charge localizes on one of the π bonds. At geometries of C_{2v} symmetry, where the symmetry between the two π bonds is lost, the energy lowering caused by spurious symmetry breaking can be avoided only with state-averaged orbitals between the ground state and the excited state. Figure 2.1 shows the reaction coordinate (linear interpolation) between the formal TS (D_2 symmetry) and the optimized minima for the charge-localized species (C_{2v} symmetry). The lower curve (diamonds) was obtained without state averaging, using symmetry-broken orbitals, whereas the

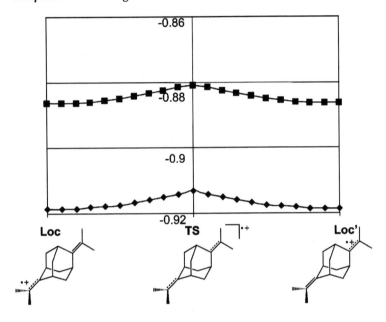

Figure 2.1 Reaction coordinate (linear interpolation) for electron transfer in BMA, CASSCF(3,4)/3-21G*. Upper curve (squares), state-averaged orbitals; lower curve (diamonds), without state-averaging. (From Blancafort, L., Jolibois, F., Olivucci, M., and Robb, M.A., *J. Am. Chem. Soc.*, 123, 722–732, 2001. With permission.)

upper curve (squares) gives the ground-state energies from state-averaged calculations with the excited state. Both curves appear to be smooth, but close inspection reveals that the lower curve, calculated without state averaging, has a small cusp at the TS. More importantly, the character of the wave function changes discontinuously for the lower curve because of the spurious charge localization at the TS. Similar behavior has been found in calculations on other radical cations,[11] and therefore we conclude that state-averaged orbitals are necessary in these types of problems to avoid spurious symmetry breaking and to obtain smooth reactivity profiles.

2.2.3 Checking the Finished Calculation

One of the classic problems of quantum chemistry is how to characterize the calculated wave functions and interpret them in terms of intuitive concepts. We have presented the CASSCF implementation[12] of a VB-based method[54] for the analysis of bonding in organic molecules. The method uses the spin-exchange density matrix P with a localized orbital basis, where the determinants of the CASSCF wave function become VB-like determinants with different spin-coupling patterns. The index P_{ij} evaluates the contributions of the determinants to the CASSCF wave function, and the result is used to generate resonance formulas. As an example, the method is applied to the characterization of excited states of indole[12] (Figure 2.2). The first excited state is covalent and is characterized by a decrease of the P_{ij} in the benzene moiety, similar to the B_{2u} excited state of benzene. In contrast, the ionic excited state has an inversion in the bonds of the pyrrole moiety induced by charge transfer to the benzene ring. The changes in the P_{ij} are reflected in the bond lengths of the geometries optimized for the different states.

2.2.3.1 Treatment of Dynamic Electron Correlation: Restricted Active Space Self-Consistent Field (RASSCF) and Multiconfigurational Second-Order Perturbation Theory (CASPT2)

The CASSCF method focuses on just a few active electrons in active orbitals chosen to give multiple configurations describing

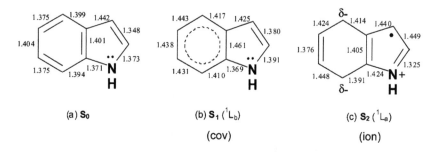

(a) S_0

(b) S_1 (^1L_b)

(cov)

(c) S_2 (^1L_a)

(ion)

Figure 2.2 Resonance structures for indole excited states and bond lengths/Å, from CASSCF(10,9)/6-31G* calculations.[12]

the required electronic state. Unfortunately, correlating only the active electrons often gives energies that are quantitatively incorrect. For reliable energy calculations of ionic excited states, one must go beyond the CASSCF level and include dynamic correlation between active and inactive electrons by using, for example, multiconfigurational second-order perturbation theory (CASPT2)[14] or multireference configuration interaction (MRCI).[15] (Because the dynamic electron correlation depends on the electron density, we would expect different correlation energies for, for example, the S_1 and S_2 states of butadiene shown in Scheme 2.2, which will affect the calculated energy difference between the two states.) Unfortunately, this treatment is usually costly in terms of computational effort. Another major problem with these methods at present is the lack of analytical energy gradients for geometry optimizations. This is a severe limitation because the gradient has to be computed numerically, which becomes more expensive and less accurate as the number of atoms increases. This makes geometry optimizations possible for only very small systems and makes extensive studies of potential energy surfaces expensive, if not unworkable.

The correlation energy usually affects only the energetics of the system, and numerical optimizations of excited-state minima and crossings with CASPT2[55] have shown that the changes in the optimized structures are small compared with the CASSCF results. However, in extreme cases (such as ionic excited states), correlation energy can also change the shape (topography) of the potential energy surface or the order of the excited states. For these cases, we have recently proposed[13] a computational strategy based on CASSCF that enables ionic and covalent $\pi\pi^*$ excited states to be described in a balanced way, providing excitation energies comparable with more accurate and expensive *ab initio* methods. This strategy depends upon (a) the restricted active space self-consistent field (RASSCF) method, in which dynamic correlation between the σ core and the valence π electrons is included by adding single σ excitations to all π configurations; and (b) the use of a new conventional basis set specifically designed for the description of valence ionic states. Full optimization of excited-state geometries and conical intersections between covalent and ionic excited states can now be routinely carried out.

2.2.3.2 TDDFT

With recent developments in time-dependent (response-based) density functional methods (TDDFT),[16] excitation energies can now be computed for many states at ground-state geometries optimized cheaply with DFT. For our purposes, there are still two problems with this approach. First, the information provided by a TDDFT calculation is almost orthogonal to that from a CASSCF calculation. TDDFT gives excitation energies and oscillator strengths, which can be compared with experiment, and the excitation energies will usually be better than from CASSCF because DFT treats dynamic electron correlation. However, the nature of the excited state may be difficult to determine with TDDFT if it is not dominated by singly excited configurations (cf. S_1 butadiene discussed above). By contrast, CASSCF gives a clearer picture of the nature of the excited state, both in terms of a sequence of single and double (and higher) excitations and also by using, for example, the VB picture described above. The second problem with TDDFT at present is the lack of experience with gradient codes for excited-state geometry optimization, which are central to the applications described below.

2.3 GENERAL INTRODUCTION TO SECTIONS 2.4 TO 2.6

Once we have correct starting wave functions for our excited states (and a good estimate of the vertical excitation energies), we can approach the study of the photochemical and photophysical mechanisms in three stages. The first stage is the optimization of the critical points on the various potential energy surfaces: minima, transition structures, and surface crossings (Section 2.4). The minima and transition structures can be characterized by frequency calculations, and the surface crossings can be characterized by examining the electronic states involved and the branching space that lifts the degeneracy. The second stage is to connect these points by running reaction path calculations (mainly Section 2.5). The picture thus obtained of the potential energy surface is very detailed but at the same time limited to the minimum energy path that no real trajectories (with nonzero kinetic energy) will follow. To get a more realistic simulation requires a third stage: dynamics calculations (Section 2.6). Here the molecule has thermal vibrational energy that allows it, in principle, to explore all regions of the potential energy surface, and one may discover that trajectories are taking "shortcuts" away from the optimized reaction paths or following paths that cannot be obtained from a reaction path calculation. Experience shows that stages 1 to 3 may have to be repeated several times, because stages 2 and 3 often indicate that stage 1 was incomplete.

Along these three stages there is clearly a progression in the sophistication and cost of the computational methods. Looking at the relationship with experiment, the more sophisticated the experiment we want to interpret, the more sophisticated the computational methods we need. In photochemical problems, this can be illustrated in relation to one of the principles of classical photochemistry, Kasha's rule.[17] This rule says that luminescence (singlet fluorescence or triplet phosphorescence) is observed exclusively from the minimum of the lowest (singlet or triplet) excited state. It assumes that whatever electronic states are excited initially, the interconversion of the states and the geometrical relaxation of the molecule to reach the lowest excited-state minimum is so fast that fluorescence can be detected from only that minimum. In fact, Kasha's rule is generally valid for experimental detection of fluorescence if the sensitivity of the detector is of the order of nanoseconds (10^{-9} s). In that case, any detected fluorescence will come from the lowest excited-state minimum, and the experiment will be well described by optimizing this minimum. However, modern laser equipment with pico- to femtosecond resolution (10^{-12} to 10^{-15} s) makes it possible to detect species with ultrashort lifetimes, such as transients or vibrationally locked intermediates. Computationally, some of these "traveling" species cannot be optimized as critical points. They need reaction path or even dynamics calculations to be identified correctly. An example is the photochemical electrocyclic ring-opening of cyclohexadiene,[18] which occurs in less than 200 femtoseconds (fs). Experimentally, three sequential phases with traveling times of 10, 43, and 77 fs were detected for this reaction. Reaction-path calculations made it possible to identify the three phases as π bond expansion, σ bond expansion, and decay to the ground state through a conical intersection, respectively.

In the context of photoreactivity, Kasha's rule states that the reaction starts from the minimum of the lowest excited state. However, one of the effects of the excitation is to produce vibrationally excited molecules. A more refined statement of Kasha's rule would therefore be that this vibrational energy is redistributed statistically (i.e., randomly) through the normal modes of the excited-state minimum before the photoreaction proceeds. A similar assumption is made in classical transition-state theory. However, examples of nonstatistical dynamics are known for organic ground-state reactions and have been well studied computationally.[19] The challenge for computational photochemistry is to translate these ideas to the excited state. In fact, because excited-state processes start with vibrationally hot molecules, such nonstatistical effects should be more frequent than in the ground state. For example, this will be the case when the decay coordinate from the Franck–Condon (FC) region and the coordinate for a given photochemical reaction from the excited-state minimum are the same or when the two coordinates are well coupled. In this case, the photoreactive vibrational modes will be excited preferentially, which will favor the corresponding reaction channel. An example is provided by excited-state calculations for the nucleic acid base cytosine.[20] The ultrafast decay observed for this molecule is caused by a mechanism whereby the decay from the FC geometry to the excited-state minimum on S_1 is followed by decay through a conical intersection. The computations suggest that the two processes follow the same reaction coordinate, namely, bond inversion of the π system (shown in Figure 2.3 as stretching of a carbonyl bond) and pyramidalization of a ring carbon. This probably explains how the barrier of approximately 4 kcal/mol between the minimum and the crossing can be crossed at an ultrafast rate. Again, dynamics calculations are required to prove the effects sug-

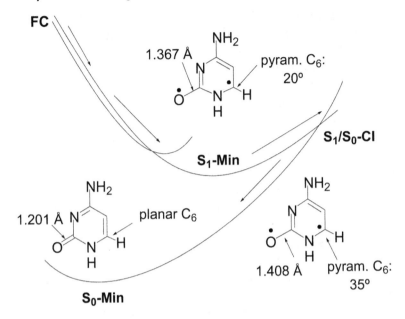

Figure 2.3 Reaction profile for the excited-state decay of cytosine through a conical intersection. (Adapted from Ismail, N., Blancafort, L., Olivucci, M., Kohler, B. and Robb, M.A., *J. Am. Chem. Soc.*, 124, 6818–6819, 2002. With permission.)

gested by the potential energy surface optimizations and reaction path calculations.

2.4 MINIMA, TRANSITION STRUCTURES, AND SURFACE CROSSINGS

2.4.1 General Features of Potential Energy Surfaces for Excited States: The Role of Conical Intersections for Reactivity

The simplest approach to the potential energy surface for any reaction is the optimization of the critical points: minima, transition structures, and surface crossings. With these elements and the reaction path calculations described in Section 2.5, it is often possible to represent the potential energy surface schematically and to rationalize problems of chemical reactivity satisfactorily. More complex problems of ultrafast photochemistry require the dynamics calculations described in Section 2.6.

The potential energy surfaces for excited-state reactions are extensions of the reaction profiles usually found in the ground state (two different minima connected by a transition structure). In photochemistry, several reaction profiles are connected by a state crossing. In Figure 2.4 we outline two reaction profiles to introduce some important concepts in the analysis of photochemical reactivity. We also give an overview of the conclusions that can be drawn from these calculations, together with the more difficult problems that must be addressed with dynamics.

In this first approach we follow Kasha's rule, that is, we consider that photochemical reactivity starts from the minimum of the lowest excited-state surface and focus on how the products are formed from there. Later we will discuss how the excited-state minimum can be reached from the FC region. There the molecule may go through a crossing between two excited states or a state switch along an avoided crossing. The concepts we introduce now will also be useful to discuss these issues.

In the topologies of the two surface crossings, the main distinction can be drawn between the sloped crossing of Figure

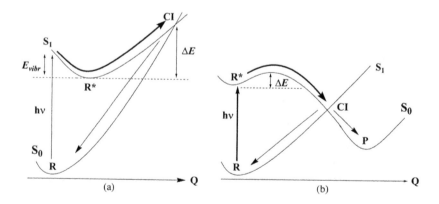

Figure 2.4 Excited-state reaction profiles for (a) sloped and (b) peaked intersections.

2.4a and the peaked crossing of Figure 2.4b. The classification comes from Atchity et al.[21] and follows the relative directions of the gradients and the position of the minima for the crossing states. Thus, the sloped crossing has two approximately parallel gradients and two minima on the same side of the crossing, on different electronic states (the two singlet states S_0 and S_1 in this case). In contrast to this, the gradients in the peaked crossing are approximately orthogonal to each other, and the two connected minima lie on the same electronic surface (S_0 in the example).

A rule of thumb for studies of photoreactivity is based on the distinction between sloped and peaked crossings. Thus, a reaction path through a sloped crossing will generally lead back to the reactant R (Figure 2.4a) and be photochemically unreactive (photophysical). However, this rule has exceptions when the crossing lies far away from the reactant geometry, in which case the ground-state path from a sloped crossing can lead to a product. However, at a peaked crossing the reaction path is branched between R and a product P. The photoreactivity will depend on the branching ratio at the crossing, which can be estimated only with dynamics. Cases where the ground-state reaction path that leads to the reactant R is preferred are called aborted (photophysical) reactions, because the path to the product P is interrupted at the crossing and the trajectory directed back to R.[22]

In simple cases, the reactions considered in Figure 2.4 are controlled by the barriers to access the crossings (ΔE). For a sloped crossing (Figure 2.4a), this is the energy difference between the reactant excited-state minimum R^* and the crossing. The peaked-crossing topology of Figure 2.4b follows the normal picture for thermally activated ground-state reactions, and the reaction is controlled by the barrier between R^* and the transition structure to the crossing. However, one additional factor that has to be considered in the interpretation of excited-state topologies is that the reactions start in the FC region (i.e., the geometry of the ground-state minimum), and this produces vibrationally hot molecules, that is, molecules with additional vibrational energy (E_{vibr} in Figure 2.4a). This is one reason why some excited-state reactions are ultrafast and occur in the femtosecond timescale, as shown in Figure 2.5.

In the gas phase (e.g., jet-cooled spectroscopy), the vibrational energy produced by the excitation cannot be dissipated into the solvent. It is available for the molecule to surmount the barriers, and this can make reactions that are formally thermally activated become ultrafast. Thus, in Figure 2.5a the excess vibrational energy is substantially higher than the barrier between R^* and the crossing, and this process will be ultrafast in spite of the barrier. This is the case for, for example, the cytosine example shown in Section 2.3. Another case of an ultrafast reaction occurs when there is no barrier between the FC region and the crossing, as shown for the peaked crossing of Figure 2.5b.

2.4.2 Characterization of Conical Intersections in Terms of Interaction between Electronic States

In the previous section we gave an overview of some general reaction profiles for photochemistry and highlighted the role of surface crossings. We will now give a description of surface crossings as conical intersections[2,23] in simple theoretical

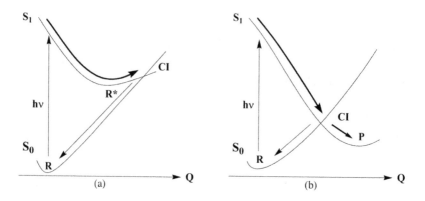

Figure 2.5 Reaction profiles for ultrafast excited-state reactions: sloped (a) and peaked (b) crossings.

terms, focusing on two aspects: (a) the interaction between the crossing states (ground state and excited state or two excited states) and (b) the nuclear coordinates associated with the intersection. These two elements generate some characteristic surface topologies, and in the following section we will show how these conceptual tools can help rationalize photochemical mechanisms.

The reaction profiles sketched in Figure 2.4 and Figure 2.5 consider one reaction coordinate. However, photochemical reactions can often not be rationalized with a single reaction coordinate because in the crossing regions there are two important coordinates that form the branching space of the reaction. Thus, at a crossing point between two states of the same multiplicity (we focus on singlet states here) there are two nuclear coordinates that lift the energy degeneracy, whereas displacement along the remaining n-2 vibrational coordinates leaves the states degenerate (strictly speaking, only to first order). The two degeneracy-lifting coordinates are the gradient difference x_1:

$$x_1 = \frac{\partial(E_1 - E_2)}{\partial q} \tag{2.1}$$

and the interstate coupling vector x_2:

$$x_2 = \left\langle \Psi_1 \left(\frac{\partial H}{\partial q} \right) \Psi_2 \right\rangle \tag{2.2}$$

The potential energy surface plotted along x_1 and x_2 has a conical shape, and therefore crossings of the same multiplicity are called conical intersections. The remaining n-2 coordinates form a hyperline, the intersecting space, which consists of an infinite number of crossing points. (For an illustration, see Figure 2.19 in Section 2.5.)

From the point of view of photochemical reactivity, the conical shape of the intersection is crucial to understanding the decay from the excited state and the product formation, because the reaction paths that lead to the ground-state products lie on the surface of the lower cone. Moreover, the one-dimensional profiles of Figure 2.4 and Figure 2.5 suggest that there will be only one or two decay paths from the crossing to the ground state, but the conical form allows for more paths to exist. In butadiene, for example, there are three paths on S_0.[25] Computationally, optimization of a conical intersection gives only the minimum of the intersecting space, but decay to the lower state in a real situation (as simulated by reaction dynamics) is possible in a much larger region of the degeneracy and at higher energies. This is an additional factor that makes some excited-state processes ultrafast, as it increases the probability of decay to the ground state. Moreover, the point of decay for a real molecule or trajectory (as opposed to the minimum of the crossing space) has an influence on the product formed after decay, because the local ground-state topology may be different than the one at the crossing minimum.

To understand the structure of conical intersections in more detail, it is useful to consider the electronic states (of the same multiplicity) involved in the crossing. It is possible to study this problem mathematically, using concepts from group theory that have been developed for the study of Jahn–Teller systems.[26] However, our aim is to make the concept useful to understanding photochemical mechanisms by presenting it in a more intuitive manner. We describe the crossing of two states that involve three electrons and three centers in a simplified version of the classic three-electron problem (H_3).[27]

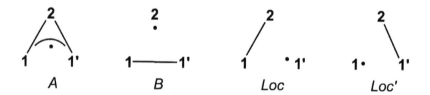

Scheme 2.4 H_3 resonance structures.

We start with a model peaked intersection. In our model structure, there are two equivalent centers labeled 1 and 1' and a third, different, center labeled 2. We consider two states

A and *B* characterized by the VB-like structures shown in Scheme 2.4. State *A* is an allyl-type structure, whereas state *B* has a bond between 1 and 1'. States *A* and *B* will be degenerate at geometries where the sides of the triangle formed by the three centers are approximately equivalent. This geometry will be our model conical intersection. We use two additional resonance structures, *Loc* and *Loc'*, which have bonds between centers 1 and 2, and 1' and 2, respectively. These resonance structures or states can be constructed as linear combinations of the two states *A* and *B*.

Now we consider the first degeneracy-lifting coordinate, the gradient difference x_1. It lifts the degeneracy by a displacement that favors one of the two states, lowering its energy and increasing the energy of the other state. Coordinate x_1 is shown in Figure 2.6a, together with the energy profile along the coordinate. It results from stretching of the 1–1' bond and compression of the 1–2 and 1'–2 bonds. One

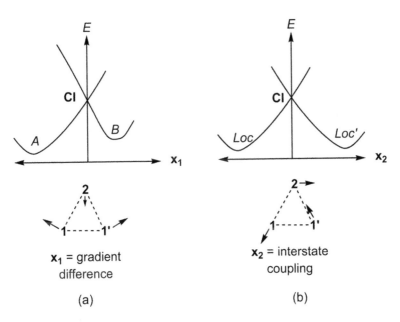

Figure 2.6 Energy profiles at a peaked conical intersection along (a) gradient difference and (b) interstate coupling coordinates.

direction of the coordinate (stretching of the 1–1' bond) stabilizes the allyl-type state (*A*), whereas the opposite direction (stretching of 1–2 and 1'–2 bonds and compression of the 1–1' bond) stabilizes state *B*. The second coordinate is the interstate coupling vector $\mathbf{x}_2$ (Figure 2.6b), which lifts the degeneracy by coupling (i.e., mixing) the two states. In the case of *A* and *B*, this leads to a localization of the 1–2 bond in one direction and the 1'–2 bond in the other direction. Note that this electronic effect (coupling of the states) is caused by a nuclear displacement along $\mathbf{x}_2$, which in this case is the antisymmetric stretching of the 1–2 and 1'–2 bonds. The overall conical shape is obtained by plotting the energy of the two states along the two coordinates. (In the surface of Figure 2.7, the two profiles of Figure 2.6 are cuts of the surface along the $\mathbf{x}_1$ and $\mathbf{x}_2$ coordinates.)

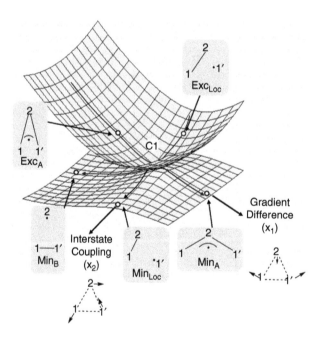

Figure 2.7 Potential energy surface along the two degeneracy-lifting coordinates $\mathbf{x}_1$ and $\mathbf{x}_2$ for a model peaked conical intersection.

In summary, following the two degeneracy-lifting coordinates x_1 and x_2 we obtain four critical points corresponding to the "minima" of the two crossing states, Min_A and Min_B, and the two structures with localized 1–2 or 1'–2 bonds, Min_{Loc} and $Min_{Loc'}$. These four structures lie on the lower surface of the cone and form a moat[27] around the tip of the cone. In an ideal case there will be four decay paths that lead from the tip of the cone to the minima of the crossing states. The paths are indicated by the arrows that come out of the tip of the cone in Figure 2.7 (structure $Min_{Loc'}$ and the corresponding decay path lie on the back of the cone and are not shown). As we show in Figure 2.7, there is a correlation between the structures on the lower and the upper part of the cone. The structures on the excited state (upper surface) have an unfavorable electronic structure with respect to their geometry, as shown for points Exc_A and Exc_{Loc}. Moreover, the lines on the surface are the profiles of Figure 2.6. They connect the excited-state structures to their ground-state electronic isomers Min_A and Min_{Loc} respectively. Thus, if one follows the decay of Exc_A through the tip of the cone, the straight path will lead to Min_A, whereas a different electronic isomer (for example Min_{Loc}) will be formed if the trajectory deviates from its original direction while crossing the surface.

The surfaces sketched above for the three-electron problem represent an ideal case. In a real situation, the minima of the states are embedded in the full space of coordinates and may be local minima or saddle points. Depending on the chemistry of the problem, there may also be barrierless paths from the ideal minima to other structures, and in these cases it is not possible to localize the structures in Figure 2.7 computationally. This problem is discussed in more detail in Section 2.5.

The analysis of an ideal sloped crossing (Figure 2.8) is very similar to the peaked case. The difference is that only one of the minima (e.g., Min_B) lies on the lower surface, whereas the other (e.g., Min_A) lies on the upper surface of the cone. The degeneracy-lifting vectors do not change, and there is still a lifting of the degeneracy along x_1. The difference with the peaked intersection case is that the energy for the two

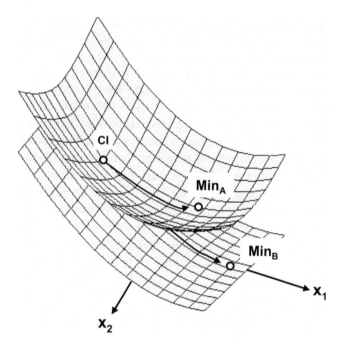

Figure 2.8 Potential energy surface along the two degeneracy-lifting coordinates x_1 and x_2 for a model sloped conical intersection.

states decreases along this coordinate by different amounts. In this case there will be fewer ground-state paths, and the lower-state minima that would correspond to Min_{Loc} and $Min_{Loc'}$ may disappear because of barrierless paths to Min_B.

2.4.3 Generation of Excited-State Reactive Species: Paths from the FC Geometry to the Excited-State Minimum

We now turn to the connection between the FC region and the lowest excited-state minimum that forms the starting point for reactivity. In the simplest case, the minimum of the spectroscopic state (the state that absorbs the excitation energy) and the lowest excited-state minimum (the minimum on the S_1 surface) are the same. When the spectroscopic state is a higher state, the initial decay will involve crossings between this excited state (and maybe others) to reach S_1. However, the decay can also involve a state switch where the wave function changes its chemical nature on the same surface. In this context we remind the reader that the states can be labeled simply in order of their energetics (the adiabatic surfaces S_0, S_1, S_2, etc.) or their chemical nature (e.g., covalent or ionic, π,π^* or n,π^*, etc.). The latter nomenclature — labeling a particular electronic isomer or diabatic state, which may be, for example, S_0 at one geometry and S_1 at another — is more useful to rationalize photoreactivity. However, some precision is necessary to avoid confusion, because the correspondence between a given chemical structure and an adiabatic surface is not always maintained, as the state switch shows.

One important point about photochemical reactions is that state switches between excited states occur in the region of topographies such as the peaked intersection region sketched above (Figure 2.4b). To understand the role of conical intersections in state switches, it is useful to compare the two-dimensional picture of Figure 2.6 with the simple one-dimensional model that is normally used to describe this phenomenon, with the help of an avoided crossing. (One dimensional means here that only one nuclear coordinate, the reaction coordinate, is considered.)

In the one-dimensional model of Figure 2.9, the dashed lines represent the diabatic states. These states are not the ones obtained from our *ab initio* potential energy surface computations, but they are idealized states whose electronic character does not change along the reaction coordinate. When the diabatic states cross, they couple, and this produces the real, adiabatic states (solid lines in Figure 2.9) that are no longer degenerate. In our calculations we use only the adiabatic states, and the calculated crossings are therefore real and not avoided. The important point is that the crossing (i.e., the conical intersection) and the avoided crossing (strictly speaking, the transition structure for the avoided crossing path) are different points on the potential energy surface and coexist. We show this for one example, the state switch between the two excited states of tryptophan.[28] With this example we also show how the reactive species is generated

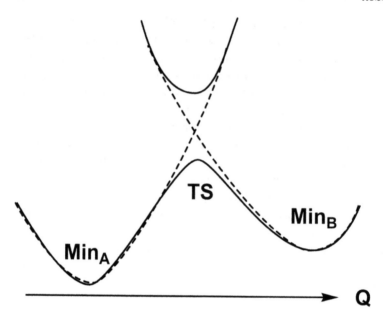

Figure 2.9 One-dimensional reaction profile for an avoided cross-ing.

from the spectroscopic species along the conical intersection topography (Figure 2.10).

The two low-lying excited states of tryptophan have different characters. One is covalent (^1L_b in the nomenclature of Platt[29]) and the other one has an ionic and charge-transfer character (^1L_a). At the FC geometry, the ^1L_b state is the first excited state (S_1) and ^1L_a is the second one (S_2). In fact, near the FC geometry there is a minimum for the ^1L_b state on the surface of S_1 (Min_{cov}). However, a second minimum corresponding to the charge-transfer state (^1L_a) was optimized on the surface of S_1 (Min_{CT}). The chromophore is polarized in the ^1L_a state, and Min_{CT} is stabilized by an intramolecular hydrogen bridge with the amino group. The two structures correspond to the minima Min_A and Min_B of the peaked intersection model given above (Figure 2.7; cf. Figure 2.10).

The intersection between the two states (CI–S_1/S_2) was optimized together with the TS between the two minima, which lies along the interstate coupling vector for CI–S_1/S_2. (The fourth critical point of the three-electron model discussed

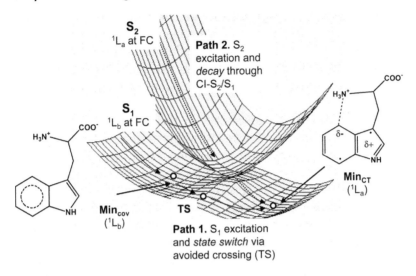

Figure 2.10 Potential energy surface for the covalent-to-ionic state switch in tryptophan along a peaked conical intersection.

above was not found.) Thus, the state-switching path from Min_{cov} to Min_{CT} is an avoided crossing path, but at the same time there is a real crossing point (intersection in the full space of coordinates) in the vicinity of the transition structure. This situation cannot be described correctly by a one-dimensional model.

The implications of the state switch topology sketched above for photochemical reactivity can be summarized using the tryptophan example. One of the photochemical reactions of tryptophan is an intramolecular hydrogen transfer from the ammonium group to the chromophore that ultimately quenches the fluorescence. Calculations show that this reaction originates at Min_{CT}. Spectroscopically, both low-lying states of tryptophan have nonnegligible oscillator strengths and can be excited experimentally, and therefore the hydrogen-transfer precursor Min_{CT} can be reached by two paths according to the computations. The first involves absorption to the covalent ^1L_b state (S_1 at the FC geometry) and a state switch to the charge-transfer state (^1L_a) along the avoided crossing path (solid arrows in Figure 2.10). The second path (dashed arrows) involves excitation to ^1L_a (S_2 at the FC geometry) and decay through CI–S_1/S_2. The calculated energetics for this topology are not definitive because of the limitations of the computational level (neglect of dynamic correlation with CASSCF; Section 2.2.3). However, the qualitative picture gives a useful insight into the formation of the reactive species.

The reaction path described for the switch of the covalent and ionic excited states of tryptophan is adiabatic and takes place along a conventional transition structure. However, in other cases the state switch can be nonadiabatic because the states are uncoupled at the intersection. In these cases the topology becomes, effectively, a seam of intersection rather than a double cone. This case has been identified in the intermolecular electronic energy transfer of a rigidly-linked biketone (Figure 2.11), a bicyclooctadione (BOD).[30] In this molecule there are two (n,π^*) excited states, with the excitation localized on either of the chromophores. The topology of the crossing can be rationalized using the general terms of

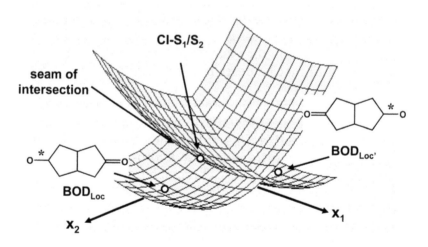

Figure 2.11 Potential energy surface for an $(n\text{-}1)$-dimensional seam of intersection in the intramolecular excitation-energy transfer of BOD.

the discussion of Figure 2.7. The states A and B are the linear combinations of the local excitations (i.e., states where the excitation is delocalized between the two chromophores). When the two carbonyl groups have the same bond length, the two states (S_1 and S_2) approach degeneracy. The minimum for this degeneracy is the conical intersection CI–S_1/S_2. Displacement from the conical intersection along the interstate coupling coordinate x_2 leads to the two minima BOD_{Loc} and $BOD_{Loc'}$, where the excitation is localized on one of the carbonyl groups. The interstate coupling coordinate is an antisymmetric stretching of the carbonyl bonds, and the excitation localizes on the carbonyl with the longer C–O bond.

In principle, for a peaked-intersection topology, along the gradient difference coordinate x_1 there should be two minima for the delocalized states (equivalent to Min_A and Min_B in Figure 2.7), and these minima should be the transition structures for the excitation energy transfer between the two carbonyl groups. However, the gradient difference coordinate at CI–S_1/S_2 has zero length, which means the degeneracy is not lifted along this coordinate; that is, the crossing becomes nonavoided. This occurs because the carbon framework that connects the two chromophores does not couple them electronically. The effect on the topology is that the double-cone shape changes to a seam of intersection, and the intersection space becomes $(n-1)$-dimensional instead of $(n-2)$-dimensional as in the normal case. CI–S_1/S_2 is the point of minimum energy on the seam. From the point of view of the reactivity of the system, there will be no conventional, adiabatic transition structure for the excitation energy transfer. Instead, the process will be nonadiabatic, because the energy transfer must take place through the crossing. Because there is no conventional transition structure, reaction-dynamics studies are required to gain further insight into this process.

We now turn our attention to singlet–triplet crossings. The theoretical treatment of crossings of different multiplicities is different from conical intersections between states of the same multiplicity, because at the crossing there is only one degeneracy-lifting coordinate, the gradient difference x_1. For singlet–triplet crossings, the interstate coupling vector x_2

is zero by definition, and the conical intersection becomes a seam of intersection. In principle, the optimization of singlet–triplet crossing points is no different from the optimization of conical intersections, and some details will be given in the following section. The probability of intersystem crossing between two surfaces of different multiplicity — in organic molecules, usually singlet and triplet — is given by the spin-orbit coupling, which depends on the structure of the molecule.[31] One subtle point is that the interstate coupling at conical intersections is a vector; that is, the splitting of the degeneracy is caused by a nuclear displacement along the vector. In contrast to this, the spin-orbit coupling lifts the degeneracy between the states without a displacement. This difference is reflected in how state crossings (i.e., the nonadiabatic events) are treated for these reactions. Another difference in the coupling between states of the same and different multiplicities is the timescale. Thus, at a point of singlet–singlet degeneracy, the crossing can take place in a single vibrational period (femtosecond timescale), whereas intersystem crossing is slower by several orders of magnitude (usually nanosecond scale or slower).

2.4.4 Potential Energy Surfaces for Ground-State Electron Transfer. Relation to Photochemistry: Nonadiabatic Chemistry

The ideas described in this chapter about the role of conical intersections in photochemistry can, in principle, be extended to other nonadiabatic reactions, where surface crossings are important. In this context, it has been postulated for radical cations that the reactivity is influenced by vibronic interactions between the ground state and the low-lying excited state.[9] Sastry and coworkers[32] have shown that the origin of these interactions in the case of the butadiene radical cation is a conical intersection region. We have found that this is also the case for the bicyclopentane radical cation.[33] More importantly, the ground-state intramolecular electron transfer (ET) in organic radical cations is also centered on a conical intersection region.[11,34] These reactions are typically described by Marcus theory (Figure 2.12). Analogous to the one-dimen-

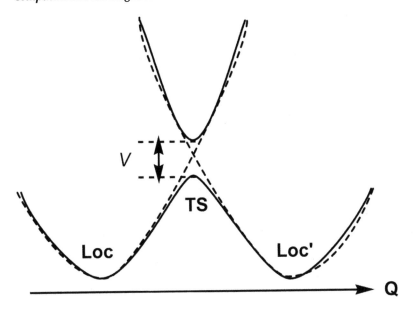

Figure 2.12   One-dimensional Marcus profile for electron transfer.

sional model described above for avoided crossings (Figure 2.9), Marcus theory uses two parabolas to represent the potential of the states, with the charge localized in one of the two donor or acceptor units (dashed parabolas in Figure 2.12). At the crossing point, the two states are coupled by the coupling element V. In cases where V is large enough, the ET becomes adiabatic and takes place smoothly on the lower surface (solid line). However, when V is small, the probability of ET at the avoided crossing becomes less than 1 and crossing to the upper surface is possible. Thus, Marcus theory provides a kinetic rate expression for the nonadiabatic case that depends on, among other things, the energy barrier to the crossing and the coupling element.[35,36]

In our calculated potential energy surfaces, we have shown that the avoided crossing, which is the transition state for ET in Marcus theory, is a region centered on a conical intersection. The topology is similar to the one shown for the model peaked intersection, and in general cases the reaction will take place on the lower surface of the double cone (Figure 2.13).

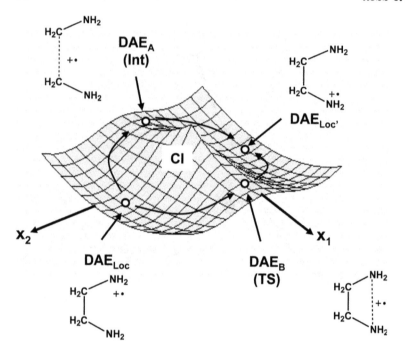

Figure 2.13 Potential energy surface for ground-state intramolecular electron transfer in diaminoethane (DAE)[11] around a conical intersection.

However, in some cases the avoided crossing path merges with the conical intersection, and the crossing becomes, effectively, a seam of intersection such as the one shown in Figure 2.11 for excitation energy transfer in a biketone. In these cases, the reaction becomes nonadiabatic and the reaction path goes through the intersection and crosses to the excited state. The study of the conical intersection region is conceptually interesting because it highlights the relation between our calculated surfaces and Marcus theory, which is a mathematical model of the reaction kinetics. In contrast to this, our surfaces are calculated *ab initio* and provide mechanistic insight about the structures and the reaction coordinate.

The general description of the conical intersections for ET goes along the lines used to describe the model peaked intersection in Figure 2.7 and the seam of intersection in

Figure 2.11. The crossing states come from different occupations of the frontier orbitals. In general, these are combinations of the charge-bearing orbitals of donor and acceptor. In our example (diaminoethane [DAE]; Figure 2.14a[11]), the donor and the acceptor are the nonbonding nitrogen orbitals. Their combinations give rise to delocalized orbitals, and the two states that cross at the conical intersection differ in the occupation of these delocalized orbitals.

The states are labeled A (orbital occupation $^1a^2b$) and B (orbital occupation $^2a^1b$), and the topology of the ground state around the intersection between the states is shown in Figure 2.13 (only the lower cone is shown). Thus, displacement from the conical intersection along the interstate coupling direction x_2 couples the states and localizes the charge on one of the nitrogen atoms (minima DAE_{Loc} and $DAE_{Loc'}$). The gradient difference coordinate x_1 leads to the minima for the two delocalized states. DAE_B is a transition structure for direct ET between the two nitrogen atoms. In contrast, DAE_A is an

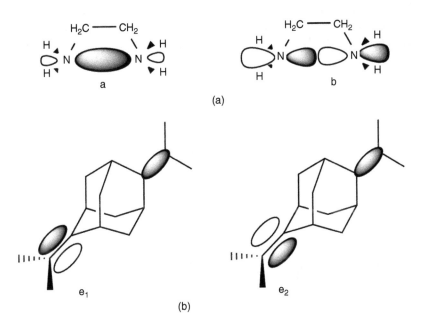

Figure 2.14 Charge-bearing orbitals (delocalized combinations) for (a) DAE and (b) BMA.

intermediate where the charge is localized on the carbon–carbon ethane bond (the bridge between donor and acceptor). Thus, there are two ET paths, which surround the cone along opposite sides (the arrows in Figure 2.13). In the case of DAE the delocalized structures lie about 10 kcal/mol below the intersection, so the ET will be adiabatic and there will be no crossing to the upper surface.

For other examples we have found the topology of a seam of intersection where we predict nonadiabatic reactivity. This is shown in Figure 2.15 for the bismethyleneadamantane radical cation (BMA). Here the delocalized orbitals formed from donor and acceptor (Figure 2.14b) are degenerate because the molecule has D_{2d} symmetry. The orbitals are labeled e_1 and e_2, and the two relevant states A and B have occupations $^2e_1^1e_2$ and $^2e_2^1e_1$. Because the orbitals are degenerate, the states will be degenerate at D_{2d} symmetry. The ET topology is therefore centered on a conical intersection, and the charge-localized minima lie along the interstate coupling coordinate. This coordinate is the antisymmetric stretching of the two π bonds, and the charge localizes on the longer bond. In principle, the minima of the delocalized states will lie along the gradient difference coordinate. These minima would be transition structures

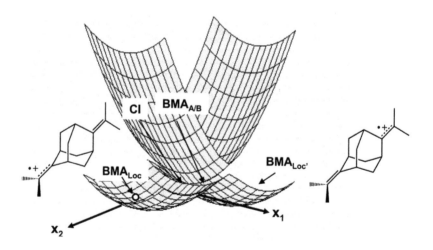

Figure 2.15 Potential energy surface for an (n-1)-dimensional seam of intersection in the intramolecular electron transfer of BMA.

or intermediates for ET (BMA_A and BMA_B). However, in the case of BMA the gradient difference vector (a skeletal deformation) has almost zero length. Because of this, the two formal transition structures lie very close to the intersection.

This gives rise to the topology given in Figure 2.15, where one of the formal transition structures is labeled $BMA_{A/B}$. The situation is similar to that described above for the model biketone in Figure 2.11. The ground- and excited-state surfaces are separated by a small energy gap of less than 1 kcal/mol at the nominal transition structures $BMA_{A/B}$, and the surface splitting along the gradient difference coordinate x_1 remains small. In the transition state and crossing region, the reacting molecules can either undergo ET to the opposite side of the seam or hop to the upper surface. The path taken will depend on the dynamics of the system. Also, vibrations of the carbon framework other than the coordinates x_1 and x_2 will affect the energy gap between the states, and therefore the dynamics have to include all degrees of freedom of the molecule.

2.4.5 Practical Issues: Optimization of Conical Intersection Minima

In this section we describe some of the practical issues associated with calculation of excited-state potential energy surfaces. We also describe the algorithms that have been developed by the London and Siena groups, some of them in collaboration with H. B. Schlegel from Wayne State University.

The optimization of a conical intersection[37] is a constrained optimization in the space orthogonal to the two degeneracy-lifting coordinates, x_1 and x_2. The optimization has an additional requirement, namely, the energy difference between the two states at the conical intersection is zero. Thus, the gradient used in our conical intersection optimization algorithm is the sum of two gradients:

$$\bar{g} = g + f \qquad (2.3)$$

The gradient f corresponds to the condition of minimizing the energy difference between the crossing states, whereas the gradient g corresponds to the constrained optimization in the $(n$-$2)$-dimensional space orthogonal to the plane $\mathbf{x}_1\mathbf{x}_2$ (where P is the projection of the gradient into this space):

$$f = 2(E_2 - E_1)\frac{x_1}{|x_1|} \tag{2.4}$$

$$g = P\frac{\partial E_2}{\partial q_\alpha} \tag{2.5}$$

The use of this technique gives a typical convergence behavior for a conical intersection optimization procedure, where in the first steps the energy difference is minimized at the cost of raising the energy of the states. Therefore, as one approaches the $(n$-$2)$-dimensional hyperline of degeneracy, the value of f becomes small. When the gradient f is sufficiently small, the remainder of the optimization is minimization of the gradient g (minimization of the energy in the intersection space). (For more details see [37].) For singlet–triplet crossings, the interstate coupling vector $\mathbf{x}_2$ is zero and the projection is done onto the $(n$-$1)$-dimensional space orthogonal to $\mathbf{x}_1$.

One of the crucial practical questions for the computational investigation of photochemical mechanisms remains to be discussed: Where is a good point to start a conical intersection optimization? One can say that peaked intersections are relatively lucky cases that can often be located approximately by a standard geometry optimization on an excited state. Here, peaked crossings may show up accidentally as regions where the energy will not go lower but for which the gradients do not go to zero.[38] The point where the optimization fails is a good point to start a conical intersection search using the algorithm described above. For sloped intersections, usually the best choice is to optimize the excited-state minimum and start the optimization from there. A particular difficulty can arise if the gradients of the crossing states are nearly parallel and of similar length. In that case the gradient difference $\mathbf{x}_1$ becomes very small, and this causes inaccuracies in the projection onto the constrained $(n$-$2)$-dimensional space. These cases require a restriction of the optimization stepsize, and eventually one has to settle for less strict convergence criteria than usual.[37]

In practice, the algorithm described in Bearpark et al.[37] (as implemented in Frisch et al.[39]) is quite efficient in finding crossings, provided you have an active space that correctly describes both of the states that will intersect (Section 2.2.1). For conjugated hydrocarbons, at least, we have had some success at guessing starting geometries for crossing searches by using prototypes such as the degeneracies of H_4 characterized by Gerhartz et al.[40]. (For an exhaustive application, see Bearpark et al.[41]). A VB model is useful here, because the crossings can be interpreted as points at which there are several weakly coupled electrons that can recouple in different ways, just as for the H_3 example described in Section 2.4.2 and Scheme 2.4.

2.5 COMPUTATION OF THE INITIAL RELAXATION DIRECTION

A photochemical reaction path can be unambiguously determined through the computation of two sequential steepest-descent lines in mass-weighted coordinates. The first line corresponds to the excited-state relaxation and reaction path (ESRP) and connects the FC point to a conical intersection (CI) where decay occurs. The second line describes the subsequent ground-state relaxation process (GSRP) and connects the conical intersection to a ground-state photoproduct minimum. In general a steepest descent line is defined by the direction of the gradient vector at the starting point: The first branch of the photochemical path may therefore be defined by the excited-state gradient at the FC region, and the second may be defined by the ground-state gradient computed at the conical intersection. These vectors define the initial relaxation direction (IRD), and in photochemical reaction-path computations they play the same role as the transition vector in thermal reaction path computations.

However, it has been shown that in realistic organic molecules,[42] the excited-state gradient vector at the FC point cannot be used for determining the IRD. The difficulty is illustrated in Figure 2.16 where we give a two-dimensional cross-section of excited- and ground-state potential energy surfaces connected through a conical intersection. Inspection

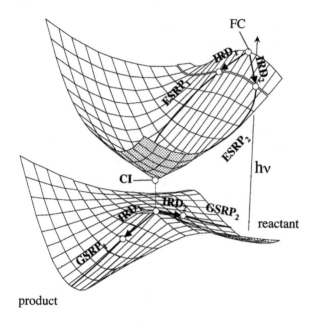

product

Figure 2.16 Construction of a photochemical reaction path. The full path is computed by joining two different steepest descent lines (e.g., $ESRP_1$ and $GSRP_1$), each providing information on a specific part of the excited- or ground-state potential-energy surface. (The excited- and ground-state energy surfaces are in contact at conical intersection, but in the figure this degeneracy is lifted for clarity). IRD_1 and IRD_2 are the vectors defining the initial direction of the ESRP and GSRP lines.

of the FC point reveals that, although the excited-state energy surface is smooth, its topography is such that two different reaction valleys develop away from the FC point. In other words, the FC point is located along a ridge rather than in a valley, and thus there must be two chemically relevant relaxation directions. Notice also that the gradient vector at the FC point does not lie along either of the two IRDs and thus cannot be used for the computations of steepest descent paths defining $ESRP_1$ and $ESRP_2$.

The situation is even more difficult if we attempt to use the ground-state gradient to determine the IRD defining the GSRP from the conical intersection. A conical intersection is a singularity point (a cusp) on the potential energy surface, and the gradient cannot be unambiguously defined. In practice, one can compute an infinite number of different gradient vectors as one moves along a small circle centered on the tip of the cusp. However, inspection of Figure 2.16 shows that only two different ground-state energy valleys develop from the conical intersection, and a chemist would want to be able to compute the IRDs defining the paths $GSRP_1$ and $GSRP_2$ running along these valleys.

The computation of excited- and ground-state IRDs can be carried out by the IRD method.[43] To illustrate the method, we start with a simple analogy: the computation of the familiar thermal reaction path. This path is constructed by computing steepest descent lines departing from the TS along the transition vector (i.e., the normal coordinate corresponding to the imaginary frequency of the TS). In other words, one takes a small step along this vector (shown as τ in Scheme 2.5) to points M_1 and M_2 and then follows the steepest descent paths connecting these points to the product or reactant wells. As shown in Figure 2.16, this procedure cannot be used to find the IRD for a photochemical reaction, because at the FC point or conical intersection there is no such unique direction τ describing the direction of the initial relaxation. Moreover, the TS is a stationary point, whereas the FC point and conical intersection are not.

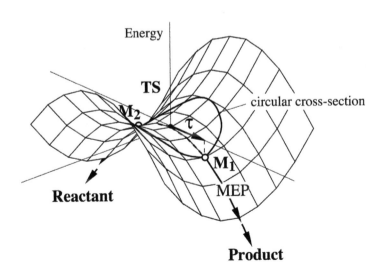

Scheme 2.5   IRD from a TS.

However, a simple alternative procedure can be devised for defining the IRD at a TS that does not explicitly use the transition vector. If one computes the energy of the system along a circular cross-section centered on the TS as illustrated in Scheme 2.5, then, provided that the radius of the circle is small enough, the energy minima M_1 and M_2 located on the circular cross-section provide an alternative but equivalent definition of the direction of relaxation toward the product and reactant. This procedure has two advantages. First, it can be carried out starting at stationary (TS) or nonstationary (FC, conical intersection) points, and second, it can locate any number of exit channels in the vicinity of the starting point.

2.5.1 IRD from the FC Geometry

In practice the IRD is computed by using a generalization of the strategy illustrated in Scheme 2.5 for a model two-dimensional potential-energy surface. We are therefore locating the energy minimum on a hyperspherical (i.e., $n-1$ dimensional) cross-section of the n-dimensional potential-energy surface centered on the starting point. The radius of this hypersphere is initially chosen to be small (typically ca. 0.25 to 0.5 au in mass-weighted coordinates) in order to locate the steepest direction in the vicinity of the starting point. Each IRD is then defined as the vector joining the starting point (i.e., the center of the hypersphere) to the energy minima found along the cross-section (e.g., IRD_1 and IRD_2 in Figure 2.16). Once the IRDs have been determined, the corresponding paths are computed as the steepest descent line in mass-weighted coordinates using the IRD vector to define the initial direction to follow. The resulting path is a minimum energy path (MEP) computed in mass-weighted coordinates. The procedure is summarized in Figure 2.17.

As an example of the use of this method, we will describe the computation of the ESRP for the photoisomerization of the minimal retinal protonated Schiff base model *tZt*-penta-3,5-dieniminium cation (cis-$C_5H_6NH_2{}^+$)[44] in which the proce-

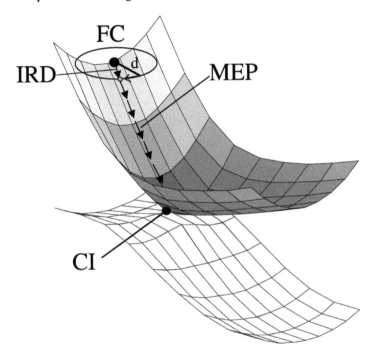

Figure 2.17   IRD search starting from the FC point.

dure to compute the initial direction of relaxation from the
FC structure has been successfully applied. The computed
S_1 path (ESRP) for cis-$C_5H_6NH_2^+$ is reported in Figure 2.18.
The initial steep part of the ESRP (region I) has no torsional
components and terminates at an inflection of the potential
energy along the ESRP. The subsequent steep energy
decrease toward the conical intersection (region II) is due to
the change in direction of the ESRP coordinate in the vicinity
of a bifurcation of the initial potential energy valley into two
mirror-image valleys. In other words, at a certain distance
from the FC region there are two energetically equivalent
ESRP valleys.

To characterize region I (Figure 2.18), a series of hypersphere calculations (radii 0.25, 0.50, 0.75, 1.00, and 1.50 au)
centered on the FC point were carried out. Scheme 2.6 shows
the result.[44]

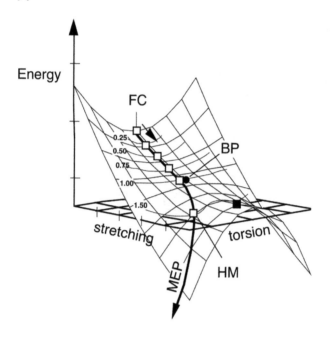

Scheme 2.6 Hyperspherical minima for a protonated Schiff base model.

The energy minima (open squares) on the (hyper)spherical cross-sections at 0.25 to 1.0 au are located at untwisted (i.e., symmetric) structures. This indicates that the potential energy surface has the shape of a valley in this region and is consistent with the fact that small torsional deformations in the FC region produce an energy increase. In contrast, the hypersphere energy minima at the 1.5 au cross-section correspond to about 12° twisted structures (HM), indicating that the potential energy surface at this distance has the shape of an unstable energy ridge. The results of (projected) analytical frequency computations at 0.25 and 0.75 au from the FC region and at HM confirm that a gradual change in surface shape occurs along the ESRP. In particular, the change from a valley shape to a ridge shape must be localized between 0.25 and 0.75 au along the path: The frequency corresponding to the torsional mode has a real value (212 cm^{-1}) at 0.25 au that becomes imaginary (239i cm^{-1}) at 0.75 au. The curvature

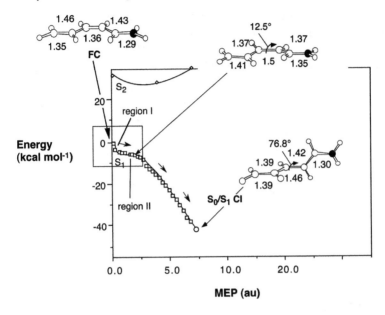

Figure 2.18 Energy profiles along the ESRP describing the excited-state relaxation from the FC point to the conical intersection decay point of *cis*-$C_5H_6NH_2^+$ (data from Garavelli et al.[44]). (Small square) the S_1 (1B$_u$-like) branch of the *cis*→*trans* photoisomerization path; (small diamond) the energy of the S_2 (2A$_g$-like) state along the same path; (small circle) the position of the conical intersection. (From Garavelli, M., Celani, P., Bernardi, F., Robb, M.A. and Olivucci, M., *J. Am. Chem. Soc.*, 119, 6891, 1997. With permission.)

becomes larger (frequency 264*i* cm^{-1}) at the stationary point located at 1.70 au, the transition structure with respect to the central bond torsion.

2.5.2   IRD from a Conical Intersection

We have already seen that a conical intersection forms a bottleneck separating the excited-state branch of a nonadiabatic photochemical reaction path from the ground-state branch. The nature of the products generated after decay at an intersection will depend on the corresponding ground-state valleys (reaction paths) that can be accessed from the conical intersection. The conical intersection can connect the same FC point to two or more ground-state energy minima (i.e., primary photoproducts; Section 2.4.1). If the base of the cone is circular (as in Figure 2.19), there is an infinite number of equivalent relaxation pathways on the ground-state surface. However, if the base of the cone is elliptic (as for the case illustrated in Figure 2.16), then there are two relaxation paths corresponding to the steepest sides of the cone (minor axis of the ellipse). More generally, a finite number of relaxation paths may exist.

In the following we will outline how the IRD method is used to locate IRDs and, consequently, GSRPs starting at the

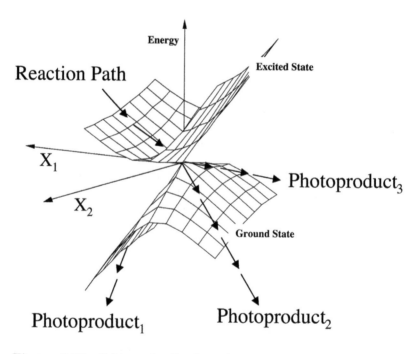

Figure 2.19 Schematic of a photochemical reaction path involving a conical intersection. The path starts on the excited-state potential energy surface and decays to the ground-state potential energy surface through a conical intersection, leading to the formation of different photoproducts 1, 2, and 3.

tip of a conical intersection. In Figure 2.20 we show the potential energy surface for a model "elliptic" conical intersection plotted in the x_1, x_2 branching plane. Because the cone is elliptic, there are two steep sides of the ground-state cone surface and two ridges. There are two preferential directions of downhill relaxation located on the steep sides of the ground-state cone surface. By analogy with the case of a TS, a simple procedure for defining these directions involves the computation of the energy profile along a circular cross-section of the branching plane centered on the vertex of the cone as illustrated in Figure 2.20a. This energy profile is given in Figure 2.20c as a function of the angle α and for a suitable choice of the radius d.

The energy profile contains two energy minima, M_1 and M_2. These minima uniquely define two IRDs at the vertex of the cone. The two associated steepest descent lines starting at M_1 and M_2 uniquely define two relaxation processes. Notice that there are also two energy maxima, TS_{21} and TS_{12}, in Figure 2.20c. These can be interpreted as the transition structures connecting M_1 and M_2 along the chosen circular cross-section. (The energy profile shown in Figure 2.20c is similar to the surfaces discussed in Section 2.4 [Figures 2.7, 2.10, and 2.13], and the structures M_1, M_2, TS_{12}, and TS_{21} can be characterized in terms of the intersecting electronic states as discussed there.)

In Figure 2.20b these transition structures locate the energy ridges that separate the IRD valleys located by M_1 and M_2. Thus, although there is no analogue of the transition vector for a conical intersection, the simple case of an elliptic cone shows that the IRDs are still uniquely defined in terms of M_1 and M_2.

The elliptic cone model of the potential energy surface at a conical intersection point discussed above is not general enough to give a correct description of the relaxation in realistic molecules. First, more than two possible IRDs may originate from the tip of the cone. Second, the first-order approximation (i.e., elliptic cone) may break down at larger distances, and some IRDs may lie out of the branching plane because the real

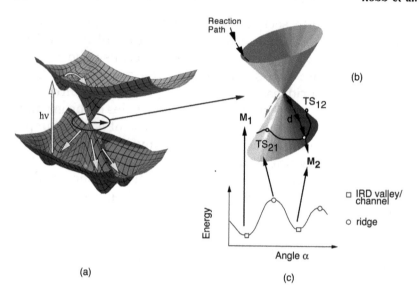

Figure 2.20 The general procedure used to locate the IRD toward the possible decay products. (a) General photochemical relaxation path leading (via conical intersection decay) to three final structures; (b) potential-energy surface for a model elliptic conical intersection plotted in the branching plane; (c) corresponding energy profile (as a function of the angle α) along a circular cross-section centered on the conical intersection point and with radius d.

(x_1, x_2) space is, in general, curved.[21] However, the ideas introduced above can be easily extended to search for IRDs in the full n-dimensional space surrounding a conical intersection point by replacing the circular cross-section with a hyperspherical cross-section centered at the vertex of the cone as shown in Figure 2.21. Thus, the search for energy minima in a one-dimensional circular cross-section (i.e., the circle in Figure 2.20a) is merely extended to an (n-1)-dimensional spherical cross-section (i.e., hypersphere) of the ground-state potential energy surface, as for the FC region.

We will now discuss an application of the strategy described above to the radiationless decay and competitive photoproduct formation process in the cyclohexadiene (CHD)/hexatriene (HT) system. A theoretical study of the first singlet excited state ($2A_1$) of cZc-hexa-1,3,5-triene (cZc-HT)

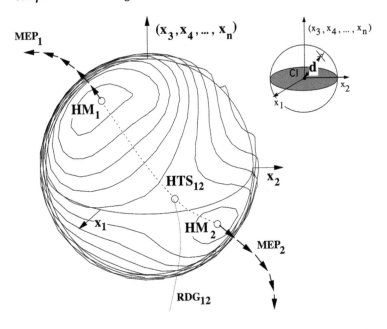

Figure 2.21 Spherical contour plot illustrating the behavior of the ground-state energy along a hyperspherical cross-section centered on a nonideal (i.e., nonelliptic) conical intersection point. The direction of the ideal tangent branching plane and tangent intersection space are also illustrated in the figure. Notice that the position of the hyperspherical energy minimum HM_1 lies well above the branching plane. HM_1 is connected to a second minimum HM_2 via the TS HTS_{12}. These minima on the hypersphere define two MEPs in the full coordinate space of the system that describe the two distinct relaxation paths on the ground-state potential-energy sheet. Similarly, HTS_{12} defines an energy ridge (RDG_{12}) that separates the two energy valleys where relaxation can occur.[49]

and CHD was reported in Celani et al.[45] These molecules are important active centers of many photochromic materials[46] and can be photochemically interconverted via direct irradiation experiments.[47,48] The conical intersection geometry optimized on the S_1 surface for the CHD/cZc-HT photochemical interconversion (CI_{CHD}) is shown in Figure 2.22. This structure has a characteristic -$(CH)_3$- kink. As shown in Figure 2.23a, this conical intersection is located slightly above an excited-state minimum.

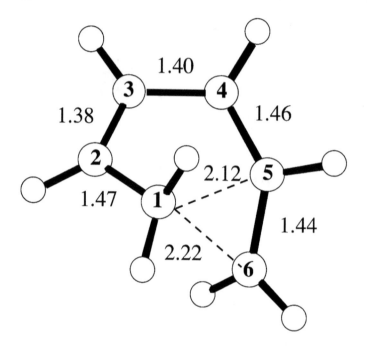

Figure 2.22 Top view of the *ab initio* optimized conical intersection structure CI_{CHD}. The relevant geometrical parameters[45] are given in Å.

Given the tetraradical nature of CI_{CHD} (see Figure 2.23b), the relaxation may, in principle, occur along three different routes. Each route is associated with a different bond-formation mode that is, in turn, driven by the recoupling of the four weakly interacting electrons (Section 2.4.2). Accordingly, route (a) leads to relaxation toward *cZc*-HT, route (b) leads to CHD, and route (c) leads to a methylenecyclopentene diradical (MCPD).

IRDs starting at CI_{CHD} have been located for each of the three routes proposed above. The MEPs (i.e., the GSRPs) computed starting from each IRD are reported in Figure 2.24. The photochemical reaction path bifurcates along two ground-state relaxation valleys leading to CHD and *cZc*-HT. A third relaxation valley leading to MCPD does not originate at CI_{CHD}, as the length of the corresponding IRD vector (i.e., the dis-

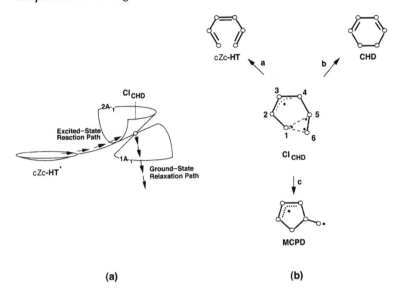

(a)   (b)

Figure 2.23 (a) The cyclohexadiene (CHD)/*cZc*-hexatriene (*cZc*-HT) photoconversion problem involves the formation of a common excited-state intermediate (*cZc*-HT*) and its decay via a conical intersection point (CI$_{CHD}$). (b) Because of the tetraradical electronic nature of CI$_{CHD}$, relaxation on S$_0$ may occur along three routes (a, b, and c) associated with different bond-formation modes and different recouplings. (From Garavelli, M., Celani, P., Fato, M., Bearpark, M.J., Smith, B.R., Olivucci, M. and Robb, M.A., *J. Phys. Chem. A*, 101, 2023, 1997. With permission.)

tance of the initial part of the valley from the tip of the cone) is large. The distances of the initial part of the CHD, *cZc*-HT, and MCPD valleys from the decay point and the magnitude of their slope provide qualitative information on the extent of the catchment region associated with a specific photoproduct. Figure 2.24 suggests that the size of the CHD and *cZc*-HT catchment regions must be similar. The MCPD product-formation path is initially inhibited because, in the immediate vicinity of the conical intersection, it corresponds to a ridge. For this reason, although the photolysis of either CHD or *cZc*-HT are predicted to have a similar outcome, MCPD will form only in limited amounts.

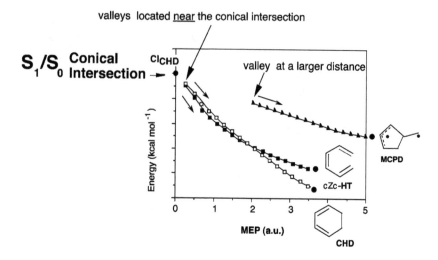

Figure 2.24 Energy profiles of the three valleys located around the conical intersection CI_{CHD}. (Adapted from Robb, M.A., Garavelli, M., Olivucci, M. and Bernardi, F., in *Reviews in Computational Chemistry*, Vol. 15. Lipkowitz, K.B. and Boyd, D.B., Eds., Wiley-VCH, New York, 2000.)

2.6 DYNAMICS

2.6.1 Introductory Examples

Of the methods available to the computational chemist, dynamics is the only one capable of yielding a reliable quantum yield or excited-state lifetime. However, dynamics can also be used as a subtle mechanistic probe, especially to interpret time-resolved spectroscopy in the femtosecond regime. We will focus on the mechanistic use of dynamics. Pump-probe experiments can now follow the time evolution of a photochemical event at a very detailed level; complementary trajectory calculations can supply mechanistic information to interpret such experiments. We will use two examples to illustrate this idea.

The Fuß group recently studied the photodissociation of $Cr(CO)_6$ with femtosecond pump-probe spectroscopy.[56] Fuß's results indicate that all fragment ions emerge from the photochemistry with an oscillation of the same period and phase, with a frequency of 96 cm.$^{-1}$ For a multidimensional potential energy surface, the vibrational energy should be partitioned equally among all vibrational modes (cf. Section 2.3) and one would not expect one signal to dominate over the others. To investigate this, dynamics computations (described in detail in Section 2.6.2) were recently carried out in our group.[57] Approximately 86 fs after the photon absorption, the $Cr(CO)_6$ molecule starts to dissociate. The departing CO ligand is left with some rotational excitation energy, whereas some (Coriolis type) coupling promotes a bending mode of two *trans* CO ligands of $Cr(CO)_5$ that fills the coordination hole left in the O_h molecular symmetry. After 100 fs, the first CO dissociation is complete and the $Cr(CO)_5$ fragment is essentially in trigonal bipyramidal C_{3v} coordination, oscillating in the moat surrounding a peaked conical intersection. Figure 2.25 shows the evolution of α, β, and γ (defined as the angles between the three equatorial ligands of a trigonal bipyramidal geometry) as a function of time during a single trajectory. At 130 fs, $Cr(CO)_5$ decays to the ground state, and the excitation energy is channeled into an apical-basal bending mode (in a way that is reminiscent of cytosine, discussed in Section 2.3 and Ismail et al.[20]). This results in a change of the molecular structure from trigonal C_{3v} to square planar pyramidal C_{4v} symmetry (when $\beta = \gamma$ in Figure 2.25). In the third phase of the trajectory, the molecule vibrates with a frequency of 98 cm^{-1} in a C_{4v} well. Both the computed frequency of the coherent oscillation and the delay time until this motion is observed are in good agreement with experimental results. Thus, our dynamics computations have provided a mechanistic explanation that could not be obtained in any other way. Experimentally, the total time from excitation until the first maximum of coherent oscillations is around 400 fs. Such timescales are typical for nonadiabatic events that do not involve a potential surface barrier. Fortunately, this timescale is just right for *ab initio* "on-the-fly" dynamics, so theory and experiment can be used in concert.

The second example of a successful joint experimental and theoretical investigation to unravel mechanistic features concerns the exceptional behavior of azulene. For most benzenoid aromatics, internal conversion and fluorescence emis-

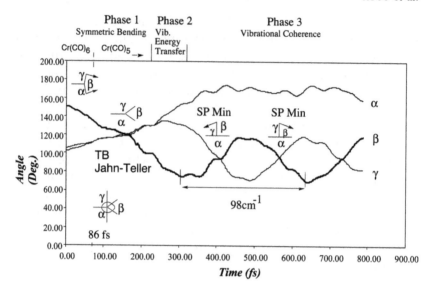

Figure 2.25 Variation in the three in-plane CO-Cr-CO angles along the entire $Cr(CO)_6$ and $Cr(CO)_5$ trajectory computed at the CASSCF/6-31G* quasi-classical/TSH level. In the first phase a symmetric bend is excited and the angles α and γ increase. The second phase corresponds to vibrational energy transfer from symmetric to antisymmetric bending coordinates. In the final phase, the molecule oscillates in a square planar minimum energy well with a frequency of 98 cm^{-1}. (Adapted from Paterson, M.J., Hunt, P.A. and Robb, M.A., *J. Phys. Chem. A,* 106, 10494–10504, 2002.)

sion both occur on comparable timescales (roughly from microseconds to nanoseconds).[58] This is consistent with a standard model in which the rate of internal conversion depends on the magnitude of the overlap between accidentally degenerate ground- and excited-state vibrational wave functions: the larger the overlap, the more efficiently the excited-state electronic energy is converted into ground-state vibrational energy. By contrast, the decay of azulene has been shown by vibrational and time-resolved spectroscopic methods to be too fast for the standard model (i.e., ultrafast).[59,60] The molecule displays an almost complete lack of fluorescence emission from its first excited S_1 state and is delivered to the ground state on the picosecond to femtosecond timescale before any vibration can be completed. More recently, Riedle and coworkers[61,62] showed that vibrationally cold azulene decays with a time constant close to 1.5 ps and that this value decreases as a function of the excess vibrational energy available on S_1. This strongly suggests that the S_1 state of azulene decays via a real crossing that it accesses after passing over a barrier thanks to its vibrational energy, as shown in Figure 2.5a. Riedle and coworkers could also identify two decay regimes. The first dominates at vibrational excess energies <1300 cm^{-1} and corresponds to monoexponential decay kinetics, with observed time constants ranging from 1.7 to 1.5 ps. For excess energies >1300 cm^{-1} the decay kinetics become more complex, with a change from monoexponential to nonexponential kinetics, indicating the superposition of two or more decay processes. The best fit of the experimental data suggests triexponential kinetics. In addition, Riedle's results suggest that one of these S_1/S_0 decay processes could occur in a stepwise manner.

Figure 2.26 summarizes the results of a full *ab initio* study that provided the initial evidence that the ultrafast decay of azulene occurs via a (sloped) conical intersection. The minimum of the S_0 surface was found to exhibit a single transannular bond, whereas the minimum on the S_1 surface has a double bond.[63] These two localized VB structures are degenerate at the conical intersection point. Consistent with the argument developed in Section 2.4.2, the gradient difference coordinate ($\mathbf{x}_1$) is calculated to have a strong component along the transannular bond. More recently, two sloped conical intersection points of C_s and C_{2v} symmetry could be located, lying a few kcal/mol^{-1} above the C_{2v} equilibrium structure of S_1 azulene. Recent molecular mechanics valence bond (MMVB)[72] dynamics calculations refined this picture.[64,65] For rather small excitation energy, monoexponential decay is predicted with the hop occurring preferentially through the C_{2v} or C_s real crossing depending on whether the system undertook vertical or 0-0 excitation. Above 1300 cm^{-1}, trajectories undergo two sorts of transition events. In the fast process (cf. one third of the trajectories), which is independent of the excess energy, the system hits the intersection space directly on the first oscillation along the S_1 surface (Figure 2.26) and

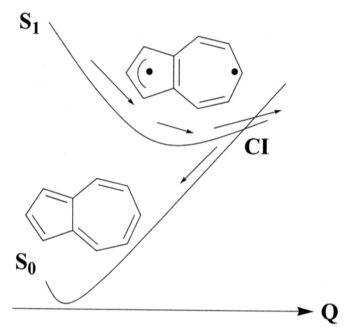

Figure 2.26 Reaction profile for the excited-state decay of azulene through a conical intersection.

takes a pure diabatic hop with instantaneous population transfer to S_0.

Figure 2.27a shows an example of the typical decay of the S_1 population for the slow process (the remaining two-thirds of the trajectories). Consistent with the experimental data, we see that at least one decay channel involves a population transfer occurring in several distinct and long-lived (>10 fs) steps. Interestingly, the hop from one population plateau to another coincides with a strong coupling of the S_0 and S_1 states (small S_1–S_0 gap; Figure 2.27b). Further analysis shows that the recoupling is quasi-periodic and controlled by the symmetric C_{2v} transannular bond-stretching mode. When the excess energy is greater than the energy difference between the C_{2v} S_1 minimum and the S_1/S_0 intersection space, the system can oscillate on the S_1 surface and periodically traverse the intersection space where the population transfer is possible.

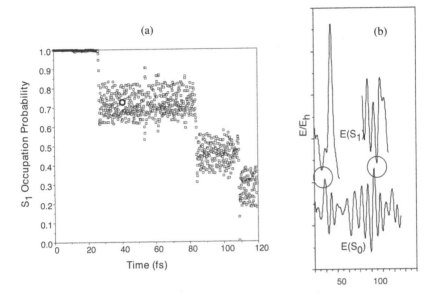

Figure 2.27 (a) State S_1 occupation probability as a function of time (b) energies of the S_1 and S_0 states (strong coupling regions are indicated by circles) for an MMVB quasi-classical/Ehrenfest trajectory on azulene. (Adapted from Klein, S., Bearpark, M.J., Smith, B.R., Robb, M.A., Olivucci, M. and Bernardi, F., *Chem. Phys. Lett.*, 292, 259–266, 1998.)

Calculations explain many aspects of the complex S_1/S_0 azulene decay. We have so far not been able to identify a third channel operating at low excess energies. This kind of disagreement is almost certainly related to the limitation of the MMVB potential and could be overcome if it was possible to carry out dynamics at a much higher level of theory.

There are many uses of dynamics calculations other than those described above. As a general rule, dynamics studies are necessary every time the chemical event does not take place mainly along the MEP. This echoes the use of dynamics as a potent tool to explore the topology of potential energy surface (PES), which has been initiated and popularized by the computational biochemistry community. Every time there is a bifurcation or crossing, the ratio can be predicted only by a dynamics study. Indeed, at any intersection, the branching of the reaction toward the products does not depend exclusively on the topology of the surface because it also depends on the distribution of the kinetic energy in the vibrational modes.

2.6.2 Theoretical Aspects of Dynamics

2.6.2.1 From Full-Quantum to Semi-Classical Dynamics Methods

Although the realistic photochemical problems presented in this review can be treated only by semi- or quasi-classical direct dynamics, it is useful to place this formalism into a broader perspective. Scheme 2.7 shows the basic require-ments of a dynamics calculation: the equations of motion to drive the dynamics (Newton, Ehrenfest, Car-Parinello, or time-dependent Hamiltonian formalism), a well-defined start-ing point (the initial conditions), and the potential energy surface gradients and Hessian along the trajectory. Depend-ing on how each of these requirements is addressed, it is possible to distinguish three categories of dynamics formal-isms as sketched in Table 2.1.

The full-quantum dynamics method[65,66] involves the approximate but explicit solution of the time-dependent Schrödinger equation by using wavepackets to represent the nuclear motion. Full-quantum dynamics requires an analyt-ical expression of the PES. In practice, a grid of points that covers the region to be studied is generated. The potential energy surface is then fitted to an analytic mathematical function, from which the gradient and Hessian needed for the wavepacket propagation of the nuclei can readily be calcu-lated. Analytic fitting is feasible only for a small number of degrees of freedom. For large systems, one must "freeze" the less relevant coordinates during the dynamics, an approxima-tion that cannot always be justified.[66] Thus, the full-quantum dynamics method remains limited to a small number of nuclear degrees of freedom.

The opposite extreme — adopted by biochemists studying extremely large systems such as proteins — is to carry out the dynamics relying on classical mechanics methods[67] only. The energy and first derivatives of the potential-energy sur-face (i.e., force constants) are computed on-the-fly using force-field methods (i.e., molecular mechanics [MM]) and are passed

TABLE 2.1   Fundamental Features of the Three Main Types of Dynamics Formalisms

Dynamics Method	Fully Quantum	Quasi-Classical Trajectory Surface Hop	Semi-Classical (Ehrenfest)	Molecular Mechanics
Number of atoms	1–3	3–100		>100, proteins
Equation of motion	Time-dependent Schrödinger	Newton		Newton
Integration of equation of motion	Numerical step-by-step	Numerical step-by-step		Numerical step-by-step
Hessian	Analytical	On-the-fly		On-the-fly
PES	Fitted analytical expression	On-the-fly		On-the-fly
Time-step	Large	Large	Small	Small
Gradients	Full nonadiabatic gradients	Pure state gradient	Mix of pure state gradients	Pure state gradient

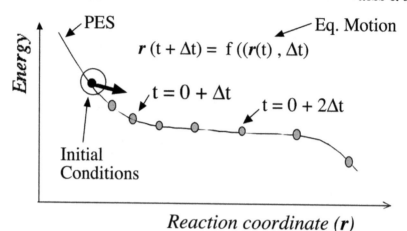

Scheme 2.7 Basic requirements of a dynamics calculation. The continuous trajectory is approximated by calculations of discrete points. The initial conditions consist of positioning the first point (in the circle) on the PES and providing a distribution of velocities (arrow).

to Newton's equations of motion (Equation 2.6) to propagate the trajectory.

$$F = ma = -m\frac{dV}{dq} \tag{2.6}$$

Often the integration steps have to be very small because it is not possible to evaluate the second derivative (Hessian) matrix of such systems. In such methods, the real nuclear (quantum) wavepacket must be emulated by a swarm of trajectories. Such trajectories are generated by sampling, which should be extensive enough (i.e., the swarm contains a sufficient variety of trajectories) to ensure that all relevant geometries involved in the chemical event have been explored.

Unfortunately, the MM formalism is not a method of general applicability. First (and despite progress in, for example, docking procedures and TS modeling[68]), force-field classical MM is fundamentally designed to address structural problems and cannot describe bond making, bond breaking, or charge transfer. Such problems require a (quantum) electronic structure treatment. In addition, MM cannot address photochemical problems because the force fields have been developed for the ground-state surface and not excited states. Even if it were technically feasible to fit a specific force field for each excited state of a molecule, there would still remain the problem of the crossing from excited to ground states, which requires a quantum description.

Between these two extreme approaches, quasi-classical dynamics[69,70] is an alternative suitable for medium-sized molecules. It is compatible with the Born–Oppenheimer approximation, because the electrons and the nuclei are treated differently. The electronic wave functions are computed by quantum mechanics as solutions of the Schrödinger equation, whereas the propagation of the nuclei is treated classically via Newton's equations of motion, using the surface computed in the electronic structure computation and the velocity. In addition to the force constants, the electronic structure computation produces the Hessian. With the gradient and Hessian at each step of the trajectory, the surrounding potential energy surface can be extrapolated using a quadratic approximation (the harmonic surface has parabolic shape along each normal mode), often complemented by (anharmonic) fifth-order corrections.[70] This locally-fitted surface is then used for the propagation of the nuclei in a much more efficient way than in techniques based on gradients only. Practically, it means that a much larger time step can confidently be taken. (In cases where the second derivative calculation is too expensive, updated techniques can be used to evaluate the Hessian rather than calculating it for every single step.)

In principle, any electronic structure method that can produce a gradient and Hessian can be used in quasi-classical dynamics. However, density functional theory (DFT)-based methods are, in general (at the time of writing), limited to ground-state surfaces. CASSCF-based dynamics calculations can be used for excited-state computations, and we will focus our discussion on this method. In this case, the size of the molecule that can be studied is limited by the size of the active space (Section 2.2.1): at present, no more than 10 active electrons or orbitals for a trajectory requiring the calculation of the second derivatives, without the possibility of using symmetry.

For dynamics in large systems where the electronic structure must be computed with quantum mechanics (QM), hybrid QM/MM methods can be used. In QM/MM, the active part of the molecule is computed at the expensive but general QM level, whereas the remaining atoms are treated by a cheaper MM force field. This is a particularly interesting approach, because even for very large molecules such as proteins it retains the main advantage of on-the-fly dynamics, namely, that all the degrees of freedom of the molecule are considered. There have been a number of implementations of direct hybrid and purely *ab initio* dynamics.[71a,b] In Siena and London, our groups developed a dynamics method based on the MMVB formalism, which combines a classical MM2 force field with a CASSCF/4-31G calculation emulated via the analytical construction of a parameterized Heisenberg Hamiltonian.[72]

2.6.2.2 Trajectory Surface Hop (TSH) Methods and Ehrenfest Dynamics[73]

A fundamental theoretical issue for computational photochemistry is the treatment of the hop (nonadiabatic) event. One needs to add the time propagation of the solutions of the time-dependent Schrödinger equation for electronic motion to the classical propagation of the nuclei, thus obtaining the populations of each adiabatic state. The time-dependent wave function for electronic motion is just a time-dependent configuration interaction vector:

$$a(t) = \begin{pmatrix} C_1(t) \\ \bullet \\ \bullet \\ \bullet \\ C_k(t) \\ \bullet \end{pmatrix} \quad (2.7)$$

where C_K is a complex coefficient giving the contribution of state K. The projection $(\langle a(t) | \Psi^k \rangle)$ of the $a(t)$ on the adiabatic basis states (i.e., the Ψ^k, eigenvectors of H) gives the "occupancy," or population, of these states. If these populations are used to determine a surface hop, we have the TSH method; if the gradient of the wave function in Equation 2.7 is used to propagate the nuclei, we have Ehrenfest dynamics. We now discuss each approach in more detail.

In the TSH method,[74] the molecule always "feels" the potential energy surface of only one of the two adiabatic states Ψ^k. (For a comparison of TSH with quantum wavepackets see, e.g., Worth et al.[66].) The populations $\langle a(t)|\Psi^k \rangle$ are monitored to decide which adiabatic state should be used to compute the gradient and to propagate the nuclei. When the occupation number of a state reaches a given threshold (e.g., 0.2), then one has a surface hop, and the wave function of the other state is now used to compute the energy and gradient. TSH techniques are particularly suitable in the case of diabatic (instantaneous population) transfer, that is when the trajectory stays in the region of degeneracy for an exceedingly short time (e.g., a strongly peaked conical intersection).

In Ehrenfest nuclear dynamics, the gradient is computed directly from the vector in Equation 2.7. Thus, the electronic wave function $a(t)$ corresponds to a mixture of adiabatic states. A mixed-state Ehrenfest trajectory drives the Newtonian mechanics fed with a mixed-state gradient, and consequently the trajectory feels several potential surfaces together with the nonadiabatic couplings while passing through the region of the surface hop. The distinguishing feature of Ehrenfest dynamics is the integration of the time-dependent Schrödinger equation for the electronic wave function in concert with nuclear propagation. This non-Born–Oppenheimer method is obviously closer to exact quantum methods than is the TSH approach. Accordingly, Ehrenfest dynamics is particularly suitable for treating strongly nonadiabatic reactions, such as when the system stays in the region of degeneracy for long times. This situation is common when the velocity and the gradient of the surface have opposite directions along the reaction coordinate, such as in a sloped conical intersection (when the system can go through multiple recrossings, e.g., azulene Figure 2.26) or in an upward hop from ground state to excited state (such as the electron transfer problems discussed earlier in Section 2.4.4).

The mixed-state Ehrenfest dynamics has problems after leaving the region of the nonadiabatic event. When the surfaces are sufficiently close in energy, the semi-classical Ehrenfest dynamics is switched on. Away from the degeneracy, the population on a single surface is recovered by reverting to the single-state quasi-classical dynamics.

2.6.2.3 Microcanonical Sampling[75]

In the full-quantum dynamics method, the distribution of nuclear positions is accounted for in nuclear wavepacket form, that is, by a function that defines the distribution of momenta of each atom and the distribution of the position in the space of each atom. In classical and semi-classical or quasi-classical dynamics methods, the wavepacket distribution is emulated by a swarm of trajectories. We now briefly discuss how sampling can generate this swarm.

For each member of the swarm of trajectories, some initial conditions are specified by sampling as shown in Scheme 2.8.

The starting conditions are generated by converting the zero-point vibrational energy into an initial geometry and velocity, where the partition is generated via a random number.

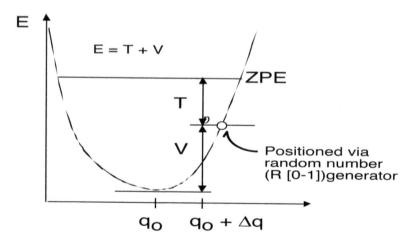

Scheme 2.8 Typical microcanonical sampling procedure.

One selects a snapshot of the system using a random location of the PES as illustrated on the left-hand side of Scheme 2.8. The kinetic energy (T) is related to the velocity of the atoms of the molecule, whereas the potential energy (V) arises from the distortion (Δq) of the sampled geometry from the minimum energy structure (q_o). The total energy (E) to distribute randomly is obviously the total energy of the molecule. At zero Kelvin this is the zero-point energy (ZPE). At higher temperatures, the total energy is augmented by the thermal factor.

Let us now consider the phase space $H(P,Q)$, defined as a space of dimension $6N$ ($3N$ dimensions for the coordinates q and $3N$ dimensions for the momenta p, where N is the total number of degrees of freedom) of all possible distributions so that the sum of their potential and kinetic energies is a constant. Microcanonical sampling[75] is a sampling procedure whose key idea is to generate a (classical) uniform distribution in this phase space. Practically, it is based on an analytical frequency calculation at the reference geometry and on the use of a random number R (see the right-hand side of Scheme 2.8). The total energy is formally split into elementary individual Q_i and P_i contributions that are related respectively to V and T.

In the case of a photochemical reaction, the species is instantaneously promoted toward the excited surface by the photon absorption. The species on the excited-state surface (generally S_1 or S_2) will initially have all the characteristics of the species on the S_0 surface. The typical sampling for a photochemical dynamics calculation should therefore be carried out on the S_0 surface. When studying a very rare photochemical event (e.g., one associated with a huge activation barrier), there is no guarantee that any of the resulting trajectories will ever follow the desired path. There are several approaches to address this issue.

In the most favorable cases, advantage can be taken of the fact that the equations of motion are reversible by computing the product→reactant (P→R) rather than the R→P trajectory. In this case, the sampling is conducted on the product. Very recent techniques that explicitly connect the product (P) to the reactant (R) are therefore of general interest for the study of rare events.[77] Less rigorously, we can resort to a variety of other computational tricks to solve the problem. For example, it is possible to edit the initial conditions intuitively in order to push the reaction in the right direction.[78] This is usually done by manually populating the mode that directs the reaction toward a desired channel (so that the energy in this mode is greater than the barrier to pass). Under the constraint of total energy conservation, the microcanonical sampling could then be carried out on the n-1 remaining vibrational modes. Unfortunately, this method will seldom succeed for large systems for which the complex reaction coordinate is the combination of many normal modes. A more effective alternative consists of starting the trajectory from the TS. Either sampling is carried out at the TS, or a very small amount of energy is provided to the system to escape the TS region and fall downhill.

2.6.3 Practical Issues

We will now illustrate the practical and technical points concerning running dynamics computations (the quality of the potential surface, the influence of sampling, and the effect of different surface hop algorithms, etc.). We will do this via some case-study dynamics calculations on the *cis→trans* isomerization of a model chromophore of a protonated Schiff base (PSB) of retinal. We have run all our trajectory calculations using the CASSCF and dynamics methods in Gaussian.[39] We begin with some background on the PSB example.

The PSB of retinal is the chromophore of rhodopsin proteins (Scheme 2.9). After photon absorption, the $S_1{\rightarrow}S_0$ radiationless decay through a conical intersection results in an 11-*cis→trans* isomerization, which, in turn, induces a conformational change in the protein. This is the first step of a long chain of chemical events behind the vision process in higher animals, for instance. Both the *tZt*-penta-3,5-dieniminium cation (*cis*-$C_5H_6NH_2{}^+$ **1**) and its methylated parent (*cis*-$C_{56}H_9NH_2{}^+$ **1m**) can be used as models for the natural retinal chromophore (Scheme 2.9).

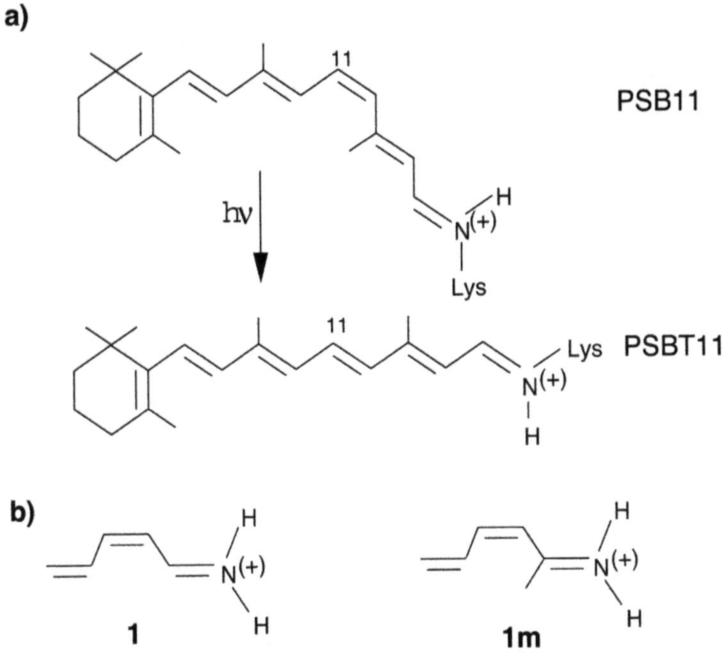

a)

PSB11

hv

Lys

11

PSBT11

Lys

b)

1

1m

Scheme 2.9 (a) 11 *cis→trans* isomerization of the natural chromophore and (b) **1** and **1m** model chromophores.

We will not be concerned with the photochemistry of the model chromophore *per se*; rather, we will use this molecule as a test system. We choose these species because a large body of experimental data is available,[76] and the model PSB has been extensively studied by our groups.[42,44,79,80] The key findings of our computational survey, especially those related to the topology of the S_1 surface, have already been presented in detail in Section 2.5. In addition, the families of *tZt*-penta-3,5-dieniminium cations are well adapted to our testing purpose, because their small size makes it possible to run a reasonable number of trajectories at relatively high levels of theory.

2.6.3.1 Quality of PES Topology

What is the impact of the quality of the computed PES on the predictions of dynamics calculations? To answer this question, we compare results for trajectories run with two basis sets: 3-21G and 6-31G*. We monitor two observables: the S_1 excited-state lifetime and the conversion rate to the *trans* configuration. The S_1 lifetime is defined from a computational point of view as the time between the start of the trajectory calculation on the S_1 surface and the instance of the surface hop. All trajectory calculations are started from a *cis* conformation. The conversion rate is defined as the number of trajectories for which the molecule, after decay, arrives in a *trans* conformation on the S_0 surface, relative to the total number of trajectories. For our comparison, we follow the evolution of the lifetime (τ) and conversion rate (ρ) average values as a function of the number of trajectories run from the swarm. These so-called instantaneous average values also allow us to check the convergence behavior of the swarm, that is, the number of trajectories necessary to obtain a representative, stable result.

The dynamics with both basis sets follow the MEP shown in Scheme 2.6 (Section 2.5) closely, and the reaction path is dominated by an increase of the twisting angle θ, as shown in Figure 2.18 and Figure 2.29. Figure 2.28 shows the accumulated instantaneous average value of the lifetime (τ) and conversion rate (ρ) computed using 3-21G and 6-31G* basis sets. Remarkably, both test observables appear to converge at

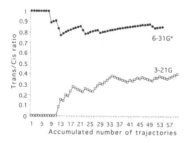
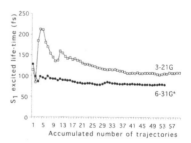

Figure 2.28 Instantaneous average values of the lifetime (τ) and *cis→trans* isomerisation ratio (ρ) of model chromophore **1** (Scheme 2.9) computed using 3-21G or 6-31G* basis sets (in combined quasi-classical CASSCF trajectories calculations).

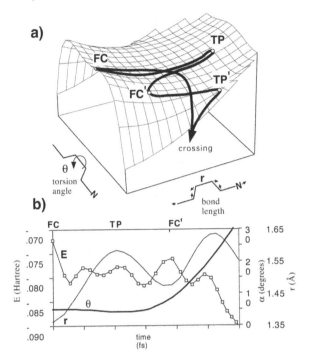

Figure 2.29 (a) Schematic structure of the S_1 energy surface of compound **1** (Scheme 2.9). (b) Potential energy (E, with -248 Hartree offset), central C=C twist (θ), and central length C=C (r) along the initial part of the trajectory.

around 50 trajectories. However, with the 3-21G basis set, the S_1-S_0 energy gap is too small, and there are premature hops at very small torsion angles (*cis* conformation). Thus, the *cis→trans* conversion rate is too small at the 3-21G level (around 40% with the 3-21G and 80% with the 6-31G* basis set — a large discrepancy). Similarly, τ is overestimated by 25 to 30 fs, although the two predicted S_1 excited-state lifetimes are of the same order of magnitude.

This example shows that the quality of the PES has a great impact on the quantitative outcome of a dynamics calculation. The 3-21G computations are certainly cheaper to run, but they are useful only for exploratory work.

2.6.3.2 Sampling without Initial Kinetic Energy (Reaction-Path Dynamics)

In the case where the PES can be computed reliably only with expensive calculations, it may not be possible to run many trajectories. One is thus forced to assume that the most interesting part of the PES is close to the MEP and start the dynamics at the FC structure with no initial kinetic energy at all. In the literature such a strategy is sometimes called reaction path dynamics.[81] It is also possible to start the dynamics with a geometry that is slightly distorted.

This trick has been used in our initial dynamics investigation of the decay mechanism of **1**. This initial theoretical survey relied on only four trajectory calculations at the CASSCF/6-31G* level of theory. The four trajectories differ from each other only by their initial torsion angle θ, which is 0 (the FC structure), 5, 10, and 20 degrees. As shown in Figure 2.29, the calculations demonstrated that the system relaxes from the FC point along a totally symmetric valley and performs a complete oscillation (from FC to FC') before significant twisting motion begins in the late stage of the trajectory. In other words, the isomerization is the direct consequence of an efficient intramolecular vibrational relaxation (IVR) to a torsion mode in the region near the oscillation turning point. This overall picture is perfectly consistent with what has been predicted earlier by reaction path calculations (IRD shown in Figure 2.18) but more complete.

2.6.3.3 Size of the Swarm and Sampling

Two key questions for the assessment of dynamics calculations are (a) are the initial conditions realistic? and (b) how many trajectories should be run before the swarm can be said to be of sufficient size? To investigate these two points, we ran three series of trajectories in a 3-21G basis on the potential surface for *cis→trans* conversion of model chromophore **1** and **1m** (Scheme 2.9). The first series used a genuine $(T + V)$ microcanonical sampling of the velocity and geometry simultaneously. In the second series (T sampling only), we used the same velocity as in the first series but did not allow any geometrical distortion: All the trajectories were started at exactly the FC geometry. In the last series, we used the same FC geometry but in combination with a velocity that has been computed at the 6-31G* level of theory.

Figure 2.30 shows the accumulated instantaneous excited state lifetime (τ) and *cis→trans* conversion ratio (ρ) for **1** and **1m** computed at the CASSCF/3-21G level of theory. For each individual series of calculations, the computed observables reach a limiting value after 50 trajectories. The swarm seems to be already of reasonable size to draw qualitative conclusions. For example, one has confidence in the conclusion that the methylation of the model chromophore reduces (by 20 to 60%) the efficiency of the conversion and extends (by about 40 to 50 fs) the lifetime of the S_1 excited state.

After about 50 extra trajectories, τ and ρ are quantitatively converged in each of the three different sampling methods. Remarkably, they predict similar lifetimes (127.2, 132.4, and 131.3 fs) and very similar conversion ratios (0.29, 0.33, and 0.25). This convergence might be interpreted as an indication that the nature of the sampling has little impact on

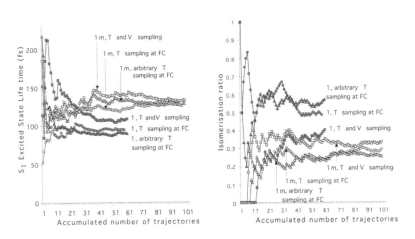

Figure 2.30   Instantaneous average values of the lifetime (τ) and *cis→trans* isomerization ratio (ρ) of the model chromophores **1** and **1m** (computed by quasi-classical CASSCF/3-21G trajectory calculations) for three types of sampling.

the final outcome of the calculations once the swarm has passed a certain critical size. Alternatively, the observed convergence might indicate that all relevant sections of the PES have been sampled. In fact, as we will now discuss, this is not the case. The quality of the sampling does matter (even if it does not directly affect observables such as the lifetime and the isomerization rate of our case study).

Figure 2.31 shows the corresponding distribution of energy at the hop for the three series of calculations on **1m**. We can see here that the trajectories started with lower potential energies hop at significantly lower energies than those whose initial geometries are deformed. The trajectories of the series 2 and 3 (which are run from the FC geometry) hop at an energy described by a Gaussian distribution centered at about −263.64 Hartrees, whereas the Gaussian distribution is centered at about −263.62 Hartrees in the case of series 1

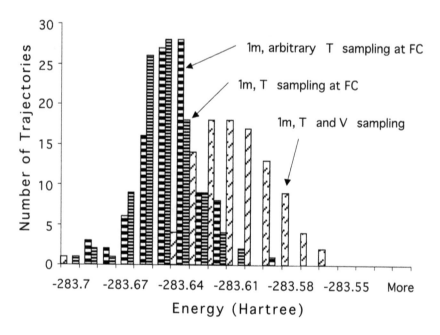

Figure 2.31 Distribution of the energy at the hop geometry of the model chromophore **1m** (computed by combined quasi-classical CASSCF/3-21G trajectory calculations) for the three series of calculations presented in Figure 2.30.

(full sampling). Moreover, the Gaussian function is much broader for series 1 (full sampling) than for the other two series. In other words, trajectories artificially started in the energy valley tend to stay confined to the low energy areas of the PES. The extent to which the details of the PES are properly accounted for depends clearly on the quality of the sampling.

To conclude this section, we should emphasize the reason for running dynamics computations at all, and why sampling is such a vital issue. In Figure 2.32 we show the distribution of the computed lifetime and torsional coordinates (in the 3-21G basis) for the S_1 to S_0 hop geometry of **1m**. Notice that each distribution is bimodal. The twisting angle of the hop geometry has maxima at about 30 and 100 degrees. The distribution of the S_1 excited lifetime has peaks at 70 and 250 fs. These two groups are related to each other. The short lifetimes and small torsional angles are associated with hops that have a large gap and give rise to a low *cis→trans* conversion rate. The long lifetime and large torsional angles correspond to situations where the transfer from excited state is almost diabatic, with a large *cis→trans* conversion rate. These computations are not quantitatively accurate. However, they show a physical effect that may be verifiable experimentally. The central point is that such an effect comes only from dynamics: A reaction path study can say nothing about such things. Further, it is clear that the generation of the initial

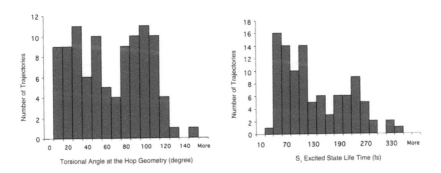

Figure 2.32 Distribution (for 100 CASSCF/3-21G combined quasi-classical trajectories) of the lifetimes (τ) and torsion angles (θ) of the hop geometry of model chromophore **1m**.

conditions by sampling is vital. One sees such effects only after running a statistically large number of trajectories.

REFERENCES

1. Molecular Quantum Mechanics: The Right Answer for the Right Reason, an International Conference in Honour of Prof. E. R. Davidson, July 21–26, 2001, University of Washington, Seattle.

2. Robb, M.A., Garavelli, M., Olivucci, M., and Bernardi, F., in *Reviews in Computational Chemistry*, Vol. 15. Lipkowitz, K.B. and Boyd, D.B., Eds., Wiley-VCH, New York, 2000, pp. 87–212; Bernardi, F., Olivucci, M., and Robb, M.A., *Chem. Soc. Rev.* 25, 321–328, 1996.

3. For a much more extensive discussion, see Schmidt, M.W. and Gordon, M.S., *Annu. Rev. Phys. Chem.*, 49, 233–266, 1998; Roos, B.O., *Adv. Chem. Phys.*, 69, 399, 1987.

4. Salem, L., *Electrons in Chemical Reactions: First Principles*, John Wiley and Sons, New York, 1982.

5. Mulliken, R.S., Nobel Lecture, December 12, 1966: Spectroscopy, molecular orbitals, and chemical bonding, in Nobel Lectures, Chemistry 1963–1970, World Scientific, Singapore, 1999.

6. Ashfold, M. N. R., Dixon, R. N., Kono, M., Mordaunt, D. H., and Reed, C. L., *Philos. Trans. R. Soc. Lond. A*, 355, 1659–1676, 1997; Langford, S. R., Orr-Ewing, A. J., Morgan, R. A., Western, C. M., Ashfold, M. N. R., Rijkenberg, A., Scheper, C. R., Buma, W. J., and de Lange, C. A., *J. Chem. Phys.*, 108, 6667–6680, 1998.

7. Hamilton, T. P. and Pulay, P., *J. Chem. Phys.*, 90, 4926, 1988.

8. Diffenderfer, R.N. and Yarkony, D.R., *J. Phys. Chem.*, 86, 5098–5105, 1982; Docken, K. and Hinze, J., *J. Chem. Phys.*, 47, 4928, 1972; Werner, H.J. and Knowles, P.J., *J. Chem. Phys.*, 82, 5053–5063, 1985.

9. Bally, T. and Borden, W.T., *Rev. Comput. Chem.*, 13, 1–97, 1998.

10. Davidson, E.R., and Borden, W.T., *J. Phys. Chem.*, 87, 4783–4790, 1983; Ayala, P.Y., and Schlegel, H.B., *J. Chem. Phys.*, 108, 7560–7567, 1983 and references cited therein.

11. Blancafort, L., Jolibois, F., Olivucci, M., and Robb, M.A., *J. Am. Chem. Soc.*, 123, 722–732, 2001.

12. Blancafort, L., Celani, P., Bearpark, M.J., and Robb, M.A., *Theor. Chem. Acc.,* 110, 92–99, 2003.

13. Boggio-Pasqua, M., Bearpark, M.J., Klene, M., and Robb, M.A., *J. Chem. Phys.,* 120, 7849–7860, 2004.

14. Roos, B.O., *Acc. Chem. Res.,* 32, 137–144, 1999; Roos, B.O., Andersson, K., Fuelscher, M.P., Malmqvist, P.-Å., Serrano-Andrés, L., Pierloot, K., and Merchán, M., *Adv. Chem. Phys.,* 93, 219–331, 1996.

15. Bauschlicher, C.W., Langhoff, S.R., and Taylor, P.R., *Adv. Chem. Phys.,* 77, 103, 1990.

16. Casida, M.E., Casida, K.C., and Salahub, D.R., *Int. J. Quant Chem.,* 70, 933–941, 1998; Furche, F. and Ahlrichs, R., *J. Chem. Phys.,* 117, 7433–7447, 2002; Tozer, D.J. and Handy, N.C., *Phys. Chem. Chem. Phys.,* 2, 2117–2121, 2000.

17. See for example Klessinger, M. and Michl, J., *Excited States and Photochemistry of Organic Molecules,* VCH, New York, 1995, p. 253.

18. Garavelli, M., Page, C.S., Celani, P., Olivucci, M., Schmid, W.E., Trushin, S.A., and Fuss, W., *J. Phys. Chem. A,* 105, 4458–4469, 2001.

19. Mann, D.J. and Hase, W.L., *J. Am. Chem. Soc.,* 124, 3208–3209, 2002; Nummela, J.A. and Carpenter, B.K., *J. Am. Chem. Soc.,* 124, 8512–8513, 2002; Doubleday, C., *J. Phys. Chem. A,* 105, 6333–6341, 2001; Carpenter, B.K., *Angew. Chem. Int. Ed.,* 37, 3340–3350, 1998; Carpenter, B.K., *J. Am. Chem. Soc.,* 117, 6336–6344, 1995; Carpenter, B.K., *J. Am. Chem. Soc.,* 118, 10329–10330, 1996; Hrovat, D.A., Fang, S., Borden, W.T., and Carpenter, B.K., *J. Am. Chem. Soc.,* 119, 5253–5254, 1997.

20. Ismail, N., Blancafort, L., Olivucci, M., Kohler, B., and Robb, M.A., *J. Am. Chem. Soc.,* 124, 6818–6819, 2002.

21. Atchity, G.J., Xantheas, S.S., and Ruedenberg, K., *J. Chem. Phys.,* 95, 1862–1876, 1991.

22. Sinicropi, A., Pogni, R., Basosi, R., Robb, M.A., Gramlich, G., Nau, W.M., and Olivucci, M., *Angew. Chem. Int. Ed.,* 40, 4185–4189, 2001.

23. Klessinger, M. and Michl, J., *Excited States and Photochemistry of Organic Molecules,* VCH, New York, 1995, pp. 183–184; Yarkony, D., *J. Phys. Chem. A,* 105, 6277–6293, 2001.

24. Deleted in proof

25. Garavelli, M., Bernardi, F., Olivucci, M., Bearpark, M.J., Klein, S., and Robb, M.A., *J. Phys. Chem. A*, 105, 11496–11504, 2001.

26a. Barckholtz, T.A. and Miller, T.A., *Int. Rev. Phys. Chem.*, 17, 435–524, 1998.

26b. Köppel, H., Domcke, W., and Cederbaum, L.S., *Adv. Chem. Phys.*, 57, 59–246, 1984.

27. Longuet-Higgins, H.C., *Adv. Spectr.*, 2, 429–472, 1961; Frey, R.F. and Davidson, E.R., *Adv. Molec. Electr. Struct. Theor.*, 1, 213–262, 1990; Bersuker, I.B., *Chem. Rev.*, 101, 1067–1114, 2001; Kuppermann, A., and Wu, Y.S.M., *Chem. Phys. Lett.*, 205, 577–586, 1993; Wu, Y.S.M., Kuppermann, A., and Anderson, J.B., *Phys. Chem. Chem. Phys.*, 1, 929–937, 1999; Herzberg, G., *Molecular Spectra and Molecular Structure. III. Electronic Spectra and Electronic Structure of Polyatomic Molecules*, Van Nostrand, Princeton, NJ, 1966.

28. Blancafort, L., González, D., Olivucci, M., and Robb, M.A., *J. Am. Chem. Soc.*, 124, 6398–6406, 2002.

29. Platt, J.R., *J. Chem. Phys.*, 17, 489, 1949.

30. Jolibois, F., Bearpark, M.J., and Robb, M.A., *J. Phys. Chem. A*, 106, 4358–4367, 2002.

31. Salem, L. and Rowland, C., *Angew. Chem. Int. Ed.*, 11, 92, 1972; Yarkony, D.R., *Int. Rev. Phys. Chem.*, 11, 195–242, 1992.

32. Sastry, G.N., Bally, T., Hrouda, V., and Cársky, P. *J. Am. Chem. Soc.*, 120, 9323–9334, 1998.

33. Blancafort, L., Adam, W., González, D., Olivucci, M., Vreven, T., and Robb, M.A., *J. Am. Chem. Soc.* 121, 10583–10590, 1999.

34. Fernández, E., Blancafort, L., Olivucci, M., and Robb, M.A., *J. Am. Chem. Soc.*, 122, 7528–7533, 2000.

35. Newton, M.D., *Chem. Rev.*, 91, 767–792, 1991; Sutin, N., *Progr. Inorg. Chem.*, 30, 441–498, 1983.

36. Marcus, R.A., *Annu. Rev. Phys. Chem.*, 15, 155–196, 1964.

37. Bearpark, M.J., Robb, M.A., and Schlegel, H.B., *Chem. Phys. Lett.*, 223, 269–274, 1994.

38. Bernardi, F., De, S., Olivucci, M., and Robb, M.A., *J. Am. Chem. Soc.*, 112, 1737–1744, 1990.

39. Frisch, M.J., Trucks, G.W., Schlegel, H.B., Scuseria, G.E., Robb, M.A., Cheeseman, J.R., Zakrzewski, V.G., Montgomery, J.A., Jr., Stratmann, R.E., Burant, J.C., Dapprich, S., Millam, J.M., Daniels, A.D., Kudin, K.N., Strain, M.C., Farkas, O., Tomasi, J., Barone, V., Cossi, M., Cammi, R., Mennucci, B., Pomelli, C., Adamo, C., Clifford, S., Ochterski, J., Petersson, G.A., Ayala, P.Y., Cui, Q., Morokuma, K., Salvador, P., Dannenberg, J.J., Malick, D.K., Rabuck, A.D., Raghavachari, K., Foresman, J.B., Cioslowski, J., Ortiz, J.V., Baboul, A.G., Stefanov, B.B., Liu, G., Liashenko, A., Piskorz, P., Komaromi, I., Gomperts, R., Martin, R.L., Fox, D.J., Keith, T., Al-Laham, M.A., Peng, C.Y., Nanay-akkara, A., Challacombe, M., Gill, P.M.W., Johnson, B., Chen, W., Wong, M.W., Andres, J.L., Gonzalez, C., Head-Gordon, M., Replogle, E.S., and Pople, J.A., *Gaussian 98 (Revision A.1x)*, Gaussian, Inc., Pittsburgh, PA, 2001.

40. Gerhartz, W., Poshusta, R.D., and Michl, J., *J. Am. Chem. Soc.*, 98, 6427–6443, 1976; Gerhartz, W., Poshusta, R.D., and Michl, J., *J. Am. Chem. Soc.*, 99, 4263–4271, 1977.

41. Bearpark, M.J., Deumal, M., Robb, M.A., Vreven, T., Yamamoto, N., Olivucci, M., and Bernardi, F., *J. Am. Chem. Soc.*, 119, 709, 1997.

42. Garavelli, M., Bernardi, F., Olivucci, M., Vreven, T., Klein, S., Celani, P., and Robb, M.A., *Faraday Discuss.*, 110, 51–70, 1998.

43. Celani, P., Robb, M.A., Garavelli, M., Bernardi, F., and Olivucci, M., *Chem. Phys. Lett.*, 243, 1, 1995.

44. Garavelli, M., Celani, P., Bernardi, F., Robb, M.A., and Olivucci, M., *J. Am. Chem. Soc.*, 119, 6891, 1997.

45. Celani, P., Ottani, S., Olivucci, M., Bernardi, F., and Robb, M.A., *J. Am. Chem. Soc.*, 116, 10141, 1994.

46. Heller, H.G., Elliot, C.C., Koh, K., Al-Shihry, S., and Whittall, J., in *Photochemistry and Polymeric Systems*, Kelly, J.M., Ardle, C.B.M., and de F. Maunder, M.J., Eds., Royal Society of Chemistry, London, 1993, p. 156.

47. Jacobs, H.J.C. and Havinga, E., in *Advances in Photochemistry*, Vol. 11, Pitts, J.N. Jr., Hammond, G.S., and Gollnick, K., Eds., John Wiley & Sons, New York, 1979, p. 305.

48. *Photochemical Rearrangements in Trienes.* Dauben, W.G., McInnis, E.L., and Michno, D.M., *In Rearrangements in Ground and Excited States*; de Mayo, P., Academic Press: New York, 1980; Vol III, p91.

49. Garavelli, M., Celani, P., Fato, M., Bearpark, M.J., Smith, B.R., Olivucci, M., and Robb, M.A., *J. Phys. Chem. A,* 101, 2023, 1997.

50. Zilberg, S. and Haas, Y., *Chem. Phys.,* 259, 249–261, 2000.

51. Applegate, B.E., Barckholtz, T.A. and Miller, T.A., *Chem. Soc. Rev.,* 32, 38–49, 2003.

52. For a more detailed discussion, see Malrieu, J.P., *Theoret. Chim. Acta,* 59, 251–279, 1981.

53. Be careful not to confuse CI = configuration interaction in this section with CI = conical intersection occasionally in the rest of this chapter (for example Figure 2.10), and elsewhere in the literature. The coincidence is unfortunate.

54. Cooper, D.L., Ponec, R., Thorsteinsson, T., and Rao, G., *Int. J. Quant. Chem.,* 57, 501–518, 1996; Karadakov, P.B., Cooper, D.L., and Gerratt, J., *J. Am. Chem. Soc.,* 120, 3975–3981, 1998; Shaik, S. and Shurki, A., *Angew. Chem. Int. Ed.,* 38, 586–625, 1999.

55. Page, C.S. and Olivucci, M., *J. Comput. Chem.,* 24, 298–309, 2003.

56. Trushin, S.A., Fuβ, W., Schmid, W.E., and Kompa, K.L., *J. Phys. Chem.,* 102, 4129–4137, 1998; Trushin, S.A., Fuβ, W., and Schmid, W.E., *Chem. Phys.,* 259, 313–330, 2000; Fuβ, W., Trushin, S.A., and Schmid, W.E., *Res. Chem. Intermed.,* 27, 447–457, 2001.

57. Paterson, M.J., Hunt, P.A., and Robb, M.A., *J. Phys. Chem. A,* 106, 10494–10504, 2002.

58. Birks, J.B., *Photophysics of Aromatic Molecules,* Wiley-Interscience, London, 1970.

59. Beer and Longuet-Higgins (Beer, M., Longuet-Higgins, H.C., *J. Chem. Phys.,* 23, 1390–1391, 1955) were the first to observe that the S_1 state of azulene is nonfluorescent and suggest that this was due to a nonradiative decay at a real crossing.

60. For femtosecond laser experiment see, for instance, Tittelbach-Helmrich, D. and Steer, R.P., *Chem. Phys.*, 197, 99–106, 1995; Schwarzer, D., Troe, J., and Schroeder, J., *Ber Bunsen-Ges. Phys.* 95, 933–934, 1991 and references therein. Diau, E.W.G., De Feyter, S., and Zewail, A.H., *J. Chem. Phys.*, 110, 9785–9788, 1999.

61. Wurzer, A.J., PhD dissertation, Universitat München, München, 1997.

62. Wilhem, T., Wurzer, A.J., Piel, J., and Riedle, E., personal communication, 1998.

63. Bearpark, M.J., Bernardi, F., Clifford, S., Olivucci, M., Robb, M.A., Smith, B.R., and Vreven, T., *J. Am. Chem. Soc.*, 118, 169–175, 1996.

64. Klein, S., Bearpark, M.J., Smith, B.R., Robb, M.A., Olivucci, M., and Bernardi, F., *Chem. Phys. Lett.*, 292, 259–266, 1998.

65. Worth, G.A. and Robb, M.A., *Adv. Chem. Phys.*, 124, 355–341, 2002; Marx, D. and Hutter, J., *Modern Methods and Algorithms of Quantum Chemistry*, Proceedings, second Edition, J. Grotendorst (Ed.), John Von Neumann Institute for Computing, Jülich, NIC Series, Vol. 3, pp.329–477, 2000.

66. For an example of full-quantum dynamics in subspace see Worth, G.A., Hunt, P., and Robb, MA., *J. Phys. Chem.*, 107, 621–631,.

67. Brooks, C.L. III, Karplus, M., and Pettitt, B.M., *Proteins: A Theoretical Perspective of Dynamics, Structure, and Thermodynamics*, John Wiley and Sons, New York, 1988.

68. Houk, K.N., Tucker, J.A., and Dorigo, A.E., *Acc. Chem. Res.*, 23, 107, 1990.

69. Hunt, P. and Robb, M.A., in preparation.

70. Millam, J.M., Bakken, V., Chen, W., Hase, W.L., and Schlegel, H.B., *J. Chem. Phys.*, 111, 3800–3805, 1999.

71a.For an example of the implementation of direct *ab initio* dynamics see, e.g., Schlegel, H.B., Iyengar, S.S., Li, X., Millam, J.M., Voth, G.A., Scuseria, G.E., and Frisch, M.J., *J. Chem. Phys.*, 117, 8694–8704, 2002 and the series of references to articles by the same authors therein.

71b. For an example of application see Li, X., Millam, J.M., and Schlegel, H.B., *J. Chem. Phys.,* 113, 10062–10067, 2002.

71c. Helgaker, T., Uggerud, E., and Jensen, H.J.A., *Chem. Phys. Lett.,* 173, 145–150, 1990.

72. The MMVB formalism is presented in Bernardi, F., Olivucci, M., and Robb, M.A., *J. Am. Chem. Soc.,* 114, 1606–1616, 1992. For recent examples of a study based on MMVB dynamics, see, e.g., [25].

73. Hack, M.D. and Truhlar, D.G., *J. Phys. Chem. A,* 104, 7917–7926, 2000.

74. Tully, J.C. and Preston, R.K., *J. Chem. Phys.,* 55, 562–572, 1971.

75. Hase, W.L., *Encyclopaedia of Computational Chemistry,* John Willey and Sons, New York, 1998, pp. 402–407.

76. See for instance Kochendoerfer, G.G. and Mathies, R.A., *J. Phys. Chem.,* 100, 14526–14532, 1996; Kandori, H., Sasabe, H., Nakanishi, K., Yoshizawa, T., Mizukami, T., and Schichida, U., *J. Am. Chem. Soc.,* 118, 1002–1005, 1996; Wang, Q., Schoenlein, R.W., Peteanu, L.A., Mathies, R.A., and Shank, C.V., *Science,* 266, 422–424, 1994.

77. Passerone, D. and Parrinello, M., *Phys. Rev. Lett.,* 87, art no. 108302, 2001.

78. Yoshizawa, K., Kamachi, T., and Shiota, Y., *J. Am. Chem. Soc.,* 123, 9806–9816, 2001.

79. Garavelli, M., Vreven, T., Celani, P., Bernardi, F., Robb, M.A., and Olivucci, M., *J. Am. Chem. Soc.,* 120, 1285–1288, 1998; Migani, A., Robb, M.A., and Olivucci, M., *J. Am. Chem. Soc.,* 125, 2804–2808, 2003.

80. Vreven, T., Bernardi, F., Garavelli, M., Olivucci, M., Robb, M.A., Schlegel, H.B., and Olivucci, M., *J. Am. Chem. Soc.,* 119, 12687–12688, 1997.

81. Natanson, G.A., Garrett, B.C., Truong, T.N., Joseph, T., and Truhlar, D.G., *J. Chem. Phys.,* 94, 7875–7892, 1991.

3

Spin-Orbit Coupling

ZDENĚK HAVLAS*, MOJMÍR KÝVALA*,
AND JOSEF MICHL**

*Institute of Organic Chemistry
and Biochemistry,
Academy of Sciences of the Czech Republic
and Center for Complex Molecular Systems
and Biomolecules,
Prague, Czech Republic

**Department of Chemistry and Biochemistry,
University of Colorado,
Boulder, CO

CONTENTS

3.1. INTRODUCTION

For molecules composed of light elements, and particularly organic molecules, spin-orbit coupling (H^{SO}) usually is the most important of the small terms in the molecular Hamiltonian operator that mix states of different zero-order spin multiplicity. Hyperfine coupling with magnetic nuclei (H^{HF}) is another, and spin–spin dipolar coupling (H^{SS}) a distant third. The most commonly encountered case is the mixing of zero-order singlets and triplets, and only rarely does one need to consider the mixing of doublets and quartets. The primary importance of such mixing is that it induces transitions between zero-order states of different spin multiplicity. Radiative transitions of this type are referred to as phosphorescence and spin-forbidden (usually singlet-triplet) absorption, and they are important in spectroscopy and photophysics. Nonradiative transitions of this type are known as intersystem crossing, and they are of particular interest to photophysicists and photochemists.

Another phenomenon for which the presence of the spin-orbit coupling term in the Hamiltonian is important is zero-field splitting, the removal of degeneracy of the individual components of spin states of multiplicity higher than doublet. In organic molecules, spin-orbit coupling plays only a subordinate role, while spin-spin dipolar coupling usually dominates. Only when molecular symmetry is high, and especially when it forces the spin-spin dipolar contributions to vanish, is there a good opportunity for the contribution from spin–orbit coupling to play a significant role. This is more likely to occur in inorganic molecules, such as transition metal complexes. Zero-field splitting is particularly important in electron paramagnetic resonance spectroscopy. Because the H^{SO} and the H^{SS} terms in the Hamiltonian play related roles, we will mention both of them in this chapter, but our primary focus is the former.

For molecules containing atoms of very heavy elements, the spin-orbit term in the Hamiltonian can no longer be viewed as a small perturbation; the importance of spin-orbit coupling may be comparable with that of electron repulsion, and it is no longer useful to refer to zero-order states of pure spin multiplicity. Under such circumstances, relativistic terms other than spin-orbit coupling normally also have to be considered, and computational treatment of such molecules needs to be more elaborate than those that are adequate for organic molecules.

We will deal only with computational procedures that are normal for molecules encountered in organic photochemistry. These methods depend heavily on the assumption that the spin-orbit coupling term H^{SO} is only a minor perturbation. In such an instance, it is common to not include small relativistic terms such as spin-orbit coupling in the Hamiltonian from the beginning but to include them as an afterthought after an ordinary nonrelativistic calculation. This is usually done using perturbation theory or response theory.

The chapter is organized as follows. In Section 3.2, we describe the Breit–Pauli Hamiltonian, which forms the historical basis for most spin-orbit calculations for organic molecules. We mention only the H^{SO} and H^{SS} terms explicitly and comment briefly on the existence of less approximate relativistic Hamiltonians. We then describe simplified expressions for H^{SO} that are used in numerical computations. We attempt to comment on the strengths and weaknesses of the various Hamiltonians. Unfortunately, it has been only in the past decade that spin-orbit calculations for organic molecules have received much attention, and too little is known at present to formulate definitive statements on their performance for the purposes of organic photophysics and photochemistry. The less approximate methods perform better but are computationally more expensive to apply and can therefore be used only for smaller molecules. In Section 3.3, we consider the various choices of initial nonrelativistic wave functions that have been used as the starting point for the evaluation of spin-orbit coupling and try to evaluate their relative merits. Unfortunately, once again, it is too early for definitive statements and recommendations.

In Section 3.4, we focus on the case of spin-orbit coupling between singlets and triplets, which is by far the most important in organic photochemistry. We describe the effect of spin-orbit coupling on the zero-field splitting, define the spin-orbit coupling vector, consider the use of molecular symmetry for the derivation of selection rules for its components, and point out their relation to the intersystem crossing rates between the singlet state and the individual sublevels of the triplet. We also mention the analysis of the structural origin of spin-orbit coupling in terms of contributions from pairs of natural hybrid orbitals. In Section 3.5; we describe the case of spin-orbit coupling in biradicals and biradicaloids, which are of special importance as intermediates in photochemical reactions.

In the remaining sections, we provide a series of illustrative examples of spin-orbit calculations for organic molecules selected from recent literature. These are of two types. Some are for stable molecules at their equilibrium geometries and are of particular interest for spectroscopy and photophysics, especially the understanding of phosphorescence rate constants and intensities of spin-forbidden absorption transitions, as well as intersystem crossing rates in vertically excited molecules. In Section 3.6, we deal with nonconjugated π systems; in Section 3.7, with linearly conjugated π systems; and in Section 3.8, with cyclic conjugated π systems.

Other calculations are for reactive intermediates suspected to occur along photochemical reaction paths and are useful for unraveling photochemical reaction mechanisms. In this activity, the understanding of spin-orbit coupling in biradicals is particularly important and is treated in Section 3.9. Finally, in Section 3,10, we summarize the most important points.

Throughout, we have avoided mathematical and theoretical detail. For this and for additional selected applications, the reader is referred to several review articles.[1-7]

3.2 THE SPIN-ORBIT COUPLING HAMILTONIAN

Einstein's special theory of relativity postulates that the finite velocity of light in vacuum and the laws of nature are the same in all inertial frames of reference. The consequences unique to this theory, relative to classical quantum theory, are usually called relativistic effects. The couplings responsible for the molecular spectral fine structures and spin-forbidden transitions fall into this category. A complete theory of molecular properties, including interactions with external and internal electric and magnetic fields, requires a combination of the theory of relativity, electrodynamics, and quantum theory. For organic molecules containing no heavy atoms, the speed of electrons is low compared with the speed of light, and relatively rough approximations can therefore still give useful results.

3.2.1 The Breit–Pauli Hamiltonian

Most applications to molecules that do not contain heavy atoms (beyond Br in the periodic table) have been based on the Breit–Pauli Hamiltonian (H^{BP}), which consists of six terms and operates on ordinary nonrelativistic wave functions. In the Breit–Pauli approximation, the total Hamiltonian is defined as

$$H = H^0 + H^{BP} \tag{3.1}$$

$$H^{BP} = H^{SO} + H^{SS} + H^{FC} + H^{OO} + H^D + H^{MV} \tag{3.2}$$

where H^0 is the nonrelativistic electronic Hamiltonian, H^{SO} is the spin-orbit coupling term, H^{SS} is the dipolar spin–spin term, H^{FC} is the Fermi contact term, H^{OO} is the orbit–orbit term, H^D is the Darwin term, and H^{MV} is the mass-velocity term.

For our purposes, only the first two terms are important. They are defined by the following equations:

$$H^{SO} = H_1^{SO} + H_2^{SO} = \frac{e^2}{8\pi\varepsilon_0 m^2 c^2}$$

$$\left[\sum_{i=1}^{n}\sum_{a=1}^{N} Z_a r_{ia}^{-3} \left(\mathbf{r}_{ia} \times \mathbf{p}_i \right) \mathbf{s}_i - \sum_{i=1}^{n}\sum_{j\neq i}^{n} r_{ij}^{-3} \left(\mathbf{r}_{ij} \times \mathbf{p}_i \right) \left(\mathbf{s}_i + 2\mathbf{s}_j \right) \right] \tag{3.3}$$

$$H^{ss} = \frac{e^2}{4\pi\varepsilon_0 m^2 c^2} \sum_{i=1}^{n}\sum_{j>i}^{n} \left[r_{ij}^{-3} \mathbf{s}_i \mathbf{s}_j - 3 r_{ij}^{-5} \left(\mathbf{r}_{ij}\mathbf{s}_i \right) \left(\mathbf{r}_{ij}\mathbf{s}_j \right) \right]. \tag{3.4}$$

The vector $r_{i\alpha}$ points from nucleus α to electron i, and the vector r_{ij} points from electron j to electron i; p_i is the linear momentum vector operator of electron i, and s_i is the spin angular momentum vector operator of electron i. The presence of electron spin operators in H^{SO} allows it to change the spin part of an electron wave function on which it operates, and hence the operator is able to mix wave functions of different spin multiplicities. The spin-orbit coupling term contains a one-electron part $H_1{}^{SO}$ and a two-electron part $H_2{}^{SO}$, and its mathematical form accounts for the origin of the term *spin–orbit coupling.*

The one-electron part describes the interaction of the magnetic dipole moment associated with the spin of each electron i, represented (in SI units) by the operator $(-e/m)s_i$, with the magnetic field generated by the effect of a nucleus α of positive charge eZ_α, located at a distance $r_{i\alpha}$, on the motion of the same electron. The presence of the orbital magnetic moment operator $(eZ_\alpha/2m)(r_{i\alpha} \times p_i)$, formally acting at the electron but effectively describing the relative rotation of the nucleus around the electron, ensures that the effect vanishes when the motion of the electron is described by an atomic s orbital, that is, when it is directed straight at or away from the nucleus, because then no curvature and hence no magnetic moment results.

The two-electron part is analogous to the one-electron part, except that now the bending of the electron path of the electron i is due to repulsion by another electron j.

The spin–spin dipolar interaction operator has the form expected for the interaction of two magnetic dipoles described by the operators $(e/m)s_i$ and separated by r_{ij}.

Spin-orbit coupling in a molecule can be expressed approximately as a sum of atomic contributions. The magnitude of the contribution from a particular atom is often stated to increase approximately with the fourth power of its atomic number Z, and the introduction of even only one atom from a second or lower full row of the periodic table into an organic molecule can increase the strength of spin-orbit coupling dramatically. This is frequently referred to as the "heavy atom effect," although it would be more correctly called the "atomic number effect".[8] In addition to the intramolecular (internal) heavy atom effect, an external effect is also known in which the enhancement of spin-orbit coupling in a molecule is due to the presence of high-Z atoms in the solvent.

The heavy atom effect can be comprehended qualitatively on the example of a hydrogen-like atom. For a constant principal quantum number n, in such an atom the distance between the nucleus and the electron scales as Z^{-1}, and the factor $Z_\alpha r_{i\alpha}^{-3}$ therefore scales as Z^4. In multielectron systems the situation is more complicated, because the principal quantum number of the valence shell changes as Z changes, and the other electrons effectively screen the nucleus, such that a single electron experiences only a reduced nuclear charge. In practice, the one-electron contributions to the spin-orbit coupling in multielectron systems grow at most quadratically, and the compensating two-electron contributions grow only linearly, with respect to the nuclear charge (in the series of molecules[9] XH_2, X = C, Si, Ge, Sn, Pb, and X_2^+, X = O, S, Se, Te, and atoms[10] F, Cl, Br). As chemical bonding results in a withdrawal of electrons away from nuclei, formation of molecules decreases the spin-orbit coupling compared with atoms. This diminution is less pronounced for heavy elements whose atoms contain a large number of inner shell electrons that are almost unaffected by chemical bonding.

The magnitude of the effect varies widely, from very strong to almost none. In exceptional cases, an "inverse heavy atom effect" can occur, in which the introduction of an atom with a large atomic number into a molecule actually reduces spin-orbit coupling.[11] The way in which molecular structure dictates the operation of the heavy atom effect will be discussed below.

3.2.2 An Improved Spin-Orbit Hamiltonian

The Breit–Pauli spin-orbit Hamiltonian is very useful for organic molecules when the matrix elements of H^{SO} are computed from the nonrelativistic wave functions using perturbation theory or response theory, but it often overestimates the magnitude of spin-orbit splitting. It also suffers from serious formal deficiencies. The total Hamiltonian, $H = H^0 + H^{BP}$, cannot be used in a direct variational evaluation of the relativistically corrected wave function and energy because it is not bounded from below and is not variationally stable. An attempt at full variational optimization will bring the total energy to minus infinity. However, even though it is not theoretically justified, H^{BP} can be used in variational calculations, using a limited expansion of configurations of moderate energies and leaving no room for variational collapse.[5,12]

A variationally stable modification of H^{SO} has been suggested by Hess[13] and Samzow et al.[14] in the framework of relativistic no-pair theory:

$$H^{so} = \frac{e^2 c^2}{2\pi\varepsilon_0} \left[\sum_i \sum_\alpha Z_\alpha \frac{A_i}{E_i + mc^2} r_{i\alpha}^{-3} \left(\mathbf{r}_{i\alpha} \times \mathbf{p}_i \right) \mathbf{s}_i \frac{A_i}{E_i + mc^2} \right.$$

$$\left. - \sum_i \sum_{j \neq i} \frac{A_i A_j}{E_i + mc^2} r_{ij}^{-3} \left(\mathbf{r}_{ij} \times \mathbf{p}_i \right) \cdot \left(\mathbf{s}_i + 2\mathbf{s}_j \right) \frac{A_i A_j}{E_i + mc^2} \right]$$

(3.5)

with the operators E_i and A_i defined as

$$E_i = \sqrt{p_i^2 c^2 + m^2 c^4}$$

$$A_i = \sqrt{\frac{E_i + mc^2}{2E_i}}.$$

(3.6)

This so-called no-pair or first-order Douglas–Kroll-transformed spin-orbit Hamiltonian is often used in computations on systems containing atoms of heavier elements.[5] The factors bracketing both the one- and two-electron terms of the no-pair spin-orbit Hamiltonian are clearly intended to damp the singularities of the included spatial operators for $r_{i\alpha} = 0$ and $r_{ij} = 0$, respectively. The Breit–Pauli Hamiltonian can be recovered by expanding the factors into a Taylor series and taking only the lowest-order terms. The neglect of the higher-order terms leads to an overestimation of spin-orbit coupling in the Breit–Pauli Hamiltonian.

The attribute "no-pair" originates in four-component relativistic theories that divide the four-component Hamiltonian into two parts. The no-pair partcommutes with the number operators for electrons and positrons, and the pair part allows for the creation and annihilation of electron-positron pairs. This seems to be a reasonable simplification because only the part conserving the number of electrons is usually needed in electronic structure calculations. Two-component operators, with the positronic degrees of freedom decoupled, need not be protected against the implications of pair creation and annihilation. Although the so-called no-pair spin–orbit Hamiltonian was initially derived by means of the first-order Douglas–Kroll transformation starting from the no-pair part of the four-component Dirac–Coulomb–Breit Hamiltonian, exactly the same operations applied to the full four-component Dirac–Coulomb–Breit Hamiltonian would lead to exactly the same spin-orbit Hamiltonian in any event.[5] The expression *no-pair* has no real meaning in connection with two-component operators, since all of them, including the Breit–Pauli Hamiltonian, are no-pair by nature. It is traditionally used only for identification purposes.

3.2.3 Computations with Spin-Orbit Operators

The presence of two-electron operators in the Breit–Pauli and similar expressions for H^{SO} makes their use computationally quite demanding, because such operators have nonvanishing matrix elements even between Slater determinants that differ in two spin-orbital occupancies and because there are many two-electron integrals. This is especially true in studies of photochemical reaction paths, where information about spin-orbit coupling is needed at many geometries. Several simplifications have been quite popular.

3.2.3.1 The Mean-Field Approximation

At present, this is the most important of the simplified methods.[15,16] The two-electron integrals are still calculated, but they are combined with the one-electron integrals in a way that produces an effective one-electron operator, whose matrix elements in the basis of atomic orbitals are:

$$\langle\mu|o|\nu\rangle^{MF} = \langle\mu|o|\nu\rangle + \frac{1}{2}\sum_{\lambda=1}^{AO}\sum_{\rho=1}^{AO} d_{\lambda\rho}\Big[2(\mu\nu|\lambda\rho) - 3(\mu\rho|\lambda\nu) - 3(\lambda\nu|\mu\rho)\Big]$$

$$d_{\lambda\rho} = \frac{1}{2}\sum_{i=1}^{MO} n_i c_{\lambda i}^* c_{\rho i}.$$

(3.7)

The mean-field approximation requires the proper occupation numbers n_i of the valence orbitals ($n_i = 2$ for core and $n_i = 0$ for virtual molecular orbitals). These should be taken from a nonrelativistic *ab initio* calculation. Nevertheless, all reasonable choices proved to be suitable for practical purposes because they cause only negligible fluctuations in the calculated energy of spin-orbit coupling.

As a result of the mean-field approximation, those pairs of Slater determinants that differ by more than one spin-orbital no longer contribute. This approximation is based on an idea similar to the conversion of the ordinary full electronic Hamiltonian into the one-electron Hartree–Fock operator and can be interpreted as describing electronic motion in an averaged field of the other electrons.

A simplified procedure results when it is noted that the magnitude of the spin-orbit coupling matrix elements falls with the third power of the electron-nucleus and electron-electron distance. This suggests that a neglect of all multicenter two-electron integrals might be reasonable. It turns out that this is acceptable only for molecules containing heavy atoms, in which the two-electron terms play a subordinate role, while molecules composed of light atoms have to be treated with greater care.[7,17]

3.2.3.2 The Atomic Mean-Field Approximation

It is better to go one step further and also neglect the multicenter one-electron integrals. The resulting atomic mean-field approximation seems to be good even for molecules composed of light atoms, because it appears that the one- and two-electron multicenter contributions to spin-orbit coupling partially compensate in a systematic manner.[17] This approximation is adopted in the atomic mean-field integral (AMFI) computer code developed by B. Schimmelpfennig at the University of Stockholm, which is often incorporated into programs for the calculation of spin-orbit coupling in molecules using the Breit–Pauli and no-pair spin-orbit Hamiltonians. In spite of its attractive features, AMFI should probably not be used with molecules that do not contain heavy atoms if high accuracy is required.

3.2.3.3 The Z_{eff} Approximation

An even simpler but less well-justified approximation avoids the calculation of the matrix elements of the two-electron part of the H^{SO} operator altogether. Only the matrix elements of the one-electron part of H_1^{SO} are computed, and in the sum over nuclei α in Equation 3.3, contributions from each atom are not multiplied by Z_α but by Z_{eff}, the effective spin–orbit coupling nuclear charge of atom α, which has been optimized empirically to represent the partial compensation of the one-electron part by the two-electron part of the H^{SO} operator. Recommended[18,19] values of Z_{eff} for atoms of main-group and transition metal elements are listed in Table 3.1. This method is generally acceptable in molecules containing heavy atoms but is not very accurate in those composed of light atoms only.

$$H^{SO} = \frac{e^2}{8\pi\varepsilon_0 m^2 c^2} \sum_{i=1}^{n} \sum_{\alpha=1}^{N} Z_{eff}(\alpha) r_{i\alpha}^{-3} (\mathbf{r}_{i\alpha} \times \mathbf{p}_i) \mathbf{s}_i \qquad (3.8)$$

3.2.3.4 Zero-Field Splitting

Except in the highest symmetry cases, *ab initio* calculation of zero-field splitting in organic molecules requires the use of H^{SS}, an operator that has only a two-electron part. Then the heavy computation involving two-electron terms cannot be avoided regardless of what spin-orbit Hamiltonian is used. These calculations are difficult because the correlation of the electrons has to be described very well before the zero-field splitting parameters are calculated accurately. Of the

TABLE 3.1 Effective Nuclear Charge (Scaling Parameter) Z_{eff} for Approximate Spin-Orbit Interaction Calculations Using the One-Electron Term in the Breit–Pauli Hamiltonian (Equation 3.8, developed by Koseki et al.[18,19a])

Elements	f (ECP)[b]	f (all electron)[c]
Li-F	$0.45 + 0.05n$	$0.045 + 0.05n$
Na-Cl	12	$0.98 - 0.01n$
K,Ca,Ga-Br	41	$1.21 - 0.03n$
(for Ga)	(11)	
Rb,Sr,In-I	110	1.24
(for In)	(33)	
Transition metals:		
First row	$0.385 + 0.025(m\text{-}2)$	
Second row	$4.680 + 0.060(m\text{-}2)$	—
Third row	$13.960 + 0.140(m\text{-}2)$	—

[a] Z_{eff} is defined as $Z_{eff} = f Z$, where Z is the atomic number. In the definition of f, n is the number of valence electrons of a neutral main group atom and m is the sum of d and sp valence electrons of a neutral transition metal atom. The functions f were fitted to experimental results for the fine structure splittings in ^2A states of diatomic hydrides (main group elements) or in atomic terms (transition metals).
[b] MCSCF/SBK(d,p) for main group elements and MCSCF/SBKJC(f) for transition metals wave functions with valence active space.
[c] MCSCF/3-21G(d,p) wave functions with valence active space.

two parameters typically deduced from EPR spectra, D is a measure of the average distance of the two unpaired electrons with parallel spins and is easier to compute accurately than E, which is a measure of the anisotropy of their distribution around the line connecting their average locations. These simple interpretations apply when the effect of H^{SO} on the zero-field splitting is negligible, as is often the case in organic molecules.

3.3 SOLUTION OF SCHRÖDINGER EQUATION WITH H^{SO} INCLUDED

The methods for solving the Schrödinger equation discussed in this section are in principle applicable to any of the spin-orbit Hamiltonians discussed above. Nevertheless, their computer implementations are often confined to just one of the approximate one-electron spin–orbit coupling operators, whereas the H^{SS} operator is usually not included at all. The reason is that a majority of the programs developed so far are mainly intended for calculations on heavy atoms and small molecules that contain heavy atoms, for which the scalar relativistic and one-electron spin-orbit contributions clearly dominate. Only a few of them are suitable for treating medium-size molecules composed of light atoms. Also, some of the methods for solving the Schrödinger equation are so demanding that they cannot be applied to medium-size organic molecules at present, and it would thus be useless to supply the programs in question with the full two-electron spin-orbit Hamiltonian.

3.3.1 Perturbational Calculations

The most straightforward, least demanding, and over a long period in the past the most common method for the calculation of the approximate eigenvalues and eigenfunctions of the relativistic Hamiltonian was the Rayleigh–Schrödinger perturbation theory. When we refer to zero-field splittings of spin-degenerate states or spin-forbidden transitions between two states of different spin multiplicities, we always mean zero-order nonrelativistic states and thus implicitly presume that perturbation theory is appropriate for the description of the effects. As noted in the introduction, perturbation theory is in principle applicable only to molecules that do not contain very heavy atoms. For computational work, its pure form is obsolete now, and in the past decade it has been dislodged by the more elegant response theory and the more general quasi-degenerate perturbation theory. It is still used for conceptual work and semiquantitative discussions.

In Rayleigh–Schrödinger perturbation theory one starts with a nonrelativistic solution for the state wave functions Ψ_i^0 and their energies E_i^0,

$$H^0\Psi_i^0 = E_i^0\Psi_i^0. \tag{3.9}$$

This calculation is typically performed in some form of a restricted configuration interaction (CI) expansion (CASSCF [complete active space self consistent field], MRCI [multireference configuration interaction]). The perturbation V is represented by the operators H^{SO} and H^{SS}. The perturbed wave function Ψ and energy E satisfy the equation

$$(H^0 + V)\Psi = E\Psi. \tag{3.10}$$

Two variants of Rayleigh–Schrödinger perturbation theory have to be distinguished. One applies to nondegenerate reference states and the other to degenerate reference states. For a nondegenerate energy level, explicit relations for wave functions up to the first-order correction and for energy up to the second-order correction are

$$\Psi = \Psi_i^0 + \sum_{j \neq i}^{\infty} \frac{\left\langle \Psi_j^0 \left| V \right| \Psi_i^0 \right\rangle}{E_i^0 - E_j^0} \Psi_j^0 \tag{3.11}$$

$$E = E_i^0 + \left\langle \Psi_i^0 \left| V \right| \Psi_i^0 \right\rangle + \sum_{j \neq i}^{\infty} \frac{\left| \left\langle \Psi_j^0 \left| V \right| \Psi_i^0 \right\rangle \right|^2}{E_i^0 - E_j^0}. \tag{3.12}$$

For a p-fold degenerate energy level, the eigenvalue problem of the perturbation operator V is first solved in the basis of the eigenfunctions of the unperturbed Hamiltonian H^0 that span the degenerate space associated with the energy level E_i^0

$$\mathbf{Vc} = (E - E_i^0)\mathbf{c} \tag{3.13}$$

where the elements of the matrix $\mathbf{V}$ are $V_{ij} = V_{ji} = \left\langle \Psi_i^0 \left| V \right| \Psi_j^0 \right\rangle$.

The first-order corrections to energy and the zero-order wave functions, with expansion coefficients $\mathbf{c}$, are obtained as the eigenvalues and eigenfunctions. Because the matrix $\mathbf{V}$ generally has p different eigenvalues, the degeneracy of the reference state is fully or at least partially removed. Second-order corrections to energy and the corresponding first-order wave functions are then obtained by solving the eigenvalue problem of the matrix $\mathbf{V}$ with elements V_{in} (the degenerate states are ordered as the first p states).

$$V_{in} = \left\langle \Psi_i^0 \left| V \right| \Psi_n^0 \right\rangle + \sum_{j>p}^{\infty} \frac{\left\langle \Psi_i^0 \left| V \right| \Psi_j^0 \right\rangle \left\langle \Psi_j^0 \left| V \right| \Psi_n^0 \right\rangle}{E_i^0 - E_j^0} \qquad (3.14)$$

$$\Psi = \sum_{i=1}^{p} c_i \left(\Psi_i^0 + \sum_{j>p}^{\infty} \frac{\left\langle \Psi_j^0 \left| V \right| \Psi_i^0 \right\rangle}{E_i^0 - E_j^0} \Psi_j^0 \right) \qquad (3.15)$$

Degenerate perturbation theory is implemented in the computer programs MELD[20] and DALTON.[21] Both use the Breit–Pauli spin-spin Hamiltonian H^{SS} with ROHF and MCSCF wave functions; MELD also allows the use of CI wave functions.

3.3.2 More General Perturbation Calculations

The principal disadvantage of the summation over all excited states is that it is impossible to determine more than several hundred eigenfunctions of the unperturbed Hamiltonian. This is a problem if the eigenstates are selected according to their nonrelativistic energies. Important contributions may be omitted from states for which both $\left| E_i - E_j \right|$ and $\left| \left\langle \Psi_j^0 \left| V \right| \Psi_i^0 \right\rangle \right|$,

are large. This is especially critical for the calculation of spin-forbidden radiative transition probabilities because some of the transition matrix elements between the dominant component of one perturbed state and a neglected minor component of another perturbed state may still be very large. The problem is less pressing in calculations of intersystem crossing rate between two zero-order states, which is of particular interest to photochemists.

So far, we have assumed that the basis Ψ^0 is represented by the orthonormal eigenfunctions of the unperturbed Hamiltonian. We can use a somewhat weaker assumption that this is true only for the zero-order wave function of the reference state i, Ψ_i^0, whereas the higher-order corrections are evaluated in some orthonormal basis Θ, required to be orthogonal only to Ψ_i^0:

$$\left\langle \Psi_i^0 \middle| \Theta_\mu \right\rangle = 0, \quad \mu = 1, \cdots, \infty. \tag{3.16}$$

The basis Θ need not be represented by the remaining eigenfunctions of the unperturbed Hamiltonian, which is fairly advantageous. As a result, the summation over an infinite number of states (eigenfunctions of the unperturbed Hamiltonian) is replaced with a solution of a system of an infinite number of linear equations. The energy of the perturbed state up to the second-order correction

$$E = E_i + \left\langle \Psi_i^0 \middle| V \middle| \Psi_i^0 \right\rangle + \mathbf{s}^\dagger \mathbf{t} \tag{3.17}$$

is obtained after solving the system of linear equations

$$(E_i \mathbf{1} - \mathbf{H})\mathbf{t} = \mathbf{s}, \tag{3.18}$$

where

$$\begin{aligned} H_{\mu\nu} &= \left\langle \Theta_\mu \middle| H^0 \middle| \Theta_\nu \right\rangle \\ s_\mu &= \left\langle \Theta_\mu \middle| V \middle| \Psi_i^0 \right\rangle. \end{aligned} \tag{3.19}$$

The wave function of the perturbed state up to the first-order correction is then

$$\Psi = \Psi_i^0 + \sum_{\mu=1}^{\infty} t_\mu \Theta_\mu. \tag{3.20}$$

Actual calculations involve a truncated basis set of a finite number of atomic basis functions, which limits the dimensionality of Ψ^0, Θ, and the system of linear equations.

It is easy to generate elements of the orthonormal basis Θ, as they are only required to be orthogonal to Ψ_i^0. Unfortunately, the Hermitian matrix $(E_i \mathbf{1} - \mathbf{H})$ is large and should be sparse, otherwise it might be difficult to solve the corresponding system of linear equations. The necessity to build up and invert the matrix $(E_i 1 - \mathbf{H})$ is the price paid for avoiding the computation of a huge number of eigenstates of the unperturbed Hamiltonian.

As an example, let us consider nonrelativistic singlet states. Singlets have nonzero spin-orbit matrix elements only with triplets. Because the bases of singlet and triplet configuration state functions (CSFs) are mutually orthogonal, the first-order correction to any given singlet eigenfunction Ψ_i^0 of the unperturbed Hamiltonian can be calculated using the basis Θ of triplet CSFs. For the calculation of the second-order correction to Ψ_i^0, functions built up from singlet CSFs and orthogonal to the reference state Ψ_i^0 would have to be included in the basis Θ. An analogous situation arises if the first-order correction to a nonrelativistic triplet eigenfunction is to be computed. Because triplets have nonzero spin–orbit matrix elements with singlets, triplets, and quintets, functions built up from triplet CSFs and orthogonal to the reference triplet must, in turn, be included in the basis. This is easily done for symmetric molecules provided the spin-orbit matrix elements between nonrelativistic states of the same spatial symmetry vanish. Then, only triplet CSFs that do not transform according to the same irreducible representation of the molecular point group as the reference triplet (and are thus orthogonal to it) are involved together with all the necessary singlet and quintet CSFs.

This version of perturbation theory without explicit consideration of the excited state wave functions was implemented by Yarkony et al. using the full Breit–Pauli spin–orbit[22,23] and spin–spin[24] Hamiltomians H^{SO} and H^{SS}.

3.3.3 Response Theory

Formalism of the so-called response theory is another, quite universal language for the description of the more general approach to Rayleigh–Schrödinger perturbation theory suggested above, in which the summation over excited states is effectively replaced with solving a large system of linear equations. Response theory has been developed in the framework of the time-dependent perturbation theory and describes how a property of a system responds to a time-dependent perturbation. This reformulation of the problem permits exploitation of the power and elegance of the well-developed theory of Green's functions. In its time-independent limit, response theory evaluates the required spin-orbit matrix elements and spin-forbidden radiative transition probabilities as the residues of the appropriately defined linear and quadratic response functions, respectively.[25]

The computer program DALTON implements[26,27] MCSCF response theory within the full Breit–Pauli H^{SO}. The coupled-cluster (CC) response theory is implemented[28] in the program ACES II. It calls DALTON if the full Breit–Pauli H^{SO} is to be used or the program AMFI for the mean-field approximation.

3.3.4 Quasi-Degenerate Perturbation Theory

Treating the spin-dependent part of the relativistic Hamiltonian by means of perturbation theory is reasonable only for molecules that are composed of light atoms. Otherwise, there is a risk that the energy of spin-orbit coupling is not a small perturbation with respect to the splitting caused by electron–electron repulsion. A similar situation may occur even for molecules composed of light atoms at geometries at which two nonrelativistic electronic states with different spin multiplicities accidentally lie close to each other. In such cases the simple perturbation theory described above should be replaced with some more universal approach treating both spin-orbit coupling and electron-electron repulsion on almost equal footing. The quasi-degenerate perturbation theory and variational theory represent two of the methods currently in use for this purpose.

The quasi-degenerate perturbation theory is an extension of the degenerate perturbation theory, in which not only the strictly degenerate eigenfunctions of the unperturbed Hamiltonian but also a number of other eigenfunctions that are nearly degenerate with them are included in initial diagonalization. The following eigenvalue problem, formulated in a more general form than Equation 3.13, which applies to both the quasi-degenerate and degenerate theories, is solved

$$\left(\mathbf{H}+\mathbf{V}\right)\mathbf{c}=E\mathbf{c}$$

$$H_{ij}=\left\langle\Psi_i^0\middle|H\middle|\Psi_j^0\right\rangle=E_i\delta_{ij}. \tag{3.21}$$

The basis set Ψ^0 in which the elements of matrices $\mathbf{H}$ and $\mathbf{V}$ are evaluated need not even be composed of the orthonormal eigenfunctions of the unperturbed Hamiltonian if we apply the quasi-degenerate perturbation theory only up to the first-order correction to energy. The zero-order eigenfunctions of the perturbed Hamiltonian may be evaluated directly in a nonorthogonal basis by introducing the Hermitian positive definite overlap matrix $\mathbf{S}$ of the nonorthogonal basis and solving

$$(\mathbf{H}+\mathbf{V})\mathbf{c}=E\mathbf{S}\mathbf{c}. \tag{3.22}$$

If the basis Ψ^0 is composed of the approximate nonorthogonal eigenfunctions of H, it is sometimes assumed that

$$H_{ij}\approx\frac{1}{2}\left(E_i+E_j\right)S_{ij} \tag{3.23}$$

where

$$E_i=\left\langle\Psi_i^0\middle|H\middle|\Psi_i^0\right\rangle,\ S_{ij}=\left\langle\Psi_i^0\middle|\Psi_j^0\right\rangle. \tag{3.24}$$

The solution by the quasi-degenerate perturbation theory up to the first-order correction to energy is equal to the variational solution to the perturbed Hamiltonian eigenvalue problem in a basis of selected eigenfunctions of the nonrelativistic Hamiltonian. The selection is usually driven by energy, but this is not a strict rule. Attempts have been made to select those states that are strongly spin-orbit coupled rather than simply close-lying in energy. The main argument is the very different nature of the operators of spin-orbit coupling and electron-electron Coulombic repulsion. For instance, single excitations give rise to much larger contributions to the energy of spin-orbit coupling than to the energy of electron-electron repulsion.

Quasi-degenerate perturbation theory is implemented in several program packages. The program GAMESS utilizes[9,29] MCSCF wave functions and the full Breit–Pauli Hamiltonian H^{SO}. In the same program, dynamic electron correlation and spin-orbit coupling are taken into account simultaneously by means of the spin-orbit multiconfiguration quasi-degenerate perturbation theory[30] (SO-MCQDPT) up to the second-order correction to energy. A similar method is used in the program CIPSO.[31] In the MOLCAS program, RASSCF wave functions are used to evaluate the spin-orbit coupling based on the one-center mean-field approximation by calling the program AMFI. The BNSOC program suite, developed at the University of Bonn by B. A. Hess and C. M. Marian with contributions from P. Chandra, S. Hutter, M. Klainschmidt, F. Rakowitz, R. Samzow, B. Schimmelpfennig, and J. Tatchen, is capable of using the MRCISD wave functions with the full Breit–Pauli H^{SO} or the variationally stable no-pair spin-orbit Hamiltonian and includes the AMFI program. The program MOLPRO[32] uses MCSCF wave functions in combination with the full Breit–Pauli H^{SO} and internally contracted MRCI wave functions with the same spin-orbit Hamiltonian for matrix elements involving only internal configurations and with a special kind of mean-field approximation for the others. The program SPOCK[33] uses MRCI or DFT/MRCI wave functions available in the TURBOMOLE program in combination with the AMFI program for the calculation of the mean-field spin-orbit integrals.

3.3.5 Variational Calculations

Another possibility is to include the relativistic operators (H^{SO} and H^{SS}) in the Hamiltonian during the variational determination of wave function. There are two entry points at which a variationally stable spin-orbit operator can be added to the nonrelativistic Hamiltonian.

First, the operator of spin-orbit coupling is inserted at the first step in which determination of the optimal single-configuration wave function is carried out. This method, called spin-orbit SCF[34] provides in general complex relativistic molecular spin-orbitals that are neither pure α nor pure β functions of the spin variable. In principle, the spin-orbit SCF can be followed with variational (MCSCF, MRCI) and nonvariational (MP2, CC) steps. All these approaches are commonly dubbed *jj*-coupling methods (in contrast to the approaches dealing with nonrelativistic molecular orbitals, called *LS*-coupling methods) and are, due to their complexity, rarely applied to molecules composed of more than two atoms.

Second, the operator of spin-orbit coupling is inserted at the last step defining the so-called spin-orbit configuration interaction (SOCI). The SOCI is therefore the variational solution to the relativistic Hamiltonian eigenvalue problem in a large basis of Slater determinants built up from nonrelativistic molecular orbitals. Although it is not as complex as a typical *jj*-coupling method, it is still much more demanding than a typical perturbational calculation and thus it is mostly used for small molecules, often those containing heavy atoms.

The fact that the relativistic Hamiltonian matrix is not evaluated in the basis of the eigenfunctions of the nonrelativistic Hamiltonian is the inherent difference between the SOCI and the quasi-degenerate perturbation theory up to the first-order correction to energy. Whereas the perturbation calculation is necessarily a two-step method, the SOCI calculations are carried out in either one or two steps. The first, nonrelativistic step of the SOCI is typically performed to obtain information on the importance of individual Slater determinants, which allows reduction of the basis for the final SOCI step. Because neither the molecular point-group symmetry operators nor the total spin angular momentum operators S^2 and S_z commute with the relativistic Hamiltonian, the SOCI matrix cannot be factored into separate diagonal blocks pertaining to individual spin multiplicities and irreducible representations of the molecular point group. Instead, the double-group and time-reversal symmetries are exploited to make the SOCI matrix block diagonal.

The program LUCIA implements nonrelativistic RASSCF for the evaluation of the molecular orbitals and initial estimates of relativistic energies and direct SO RASCI[12] in a basis of all Slater determinants. The program was enhanced[35,36] with the full no-pair spin-orbit Hamiltonian and coupled with the program AMFI. The program LUCIAREL[37] uses the AMFI program for direct SO GASCI. Most of the programs for SOCI do not use all-electron methods. Instead, they approximate the inner electrons by effective core potentials and use an empirical one-electron spin–orbit Hamiltonian.

3.4 SINGLET-TRIPLET COUPLING IN ORGANIC MOLECULES

In organic photochemistry, by far the most important case of spin-orbit coupling is that between a singlet and a triplet state. Specifically, the coupling of the lowest triplet T_1 to the lowest (S_0) or first excited (S_1) singlet state is of paramount importance, although sometimes the coupling of S_1 to the higher triplets T_2 or T_3 is important as well. In this section, we focus our discussion of spin-orbit coupling to this case, and refer to the triplet state involved as T and to the singlet state as S.

A triplet state consists of three sublevels. In the absence of an outside magnetic field these are degenerate if relativistic effects are ignored. They are split slightly (zero-field splitting), and in the Breit–Pauli description this is due to the spin–spin dipolar operator H^{SS} in the first order and the spin-orbit interaction operator H^{SO} in the second order of perturbation energy (the matrix elements of H^{SO} in the basis of the three sublevels of a spatially nondegenerate triplet state all vanish). In almost all organic molecules, the effect of H^{SS} on zero-field splitting of a triplet dominates by far over the effect of H^{SO}, but the rate of intersystem crossing between singlets and triplets is normally dominated by H^{SO}. For qualitative discussions of the ways in which spin-orbit coupling induces intersystem crossing, and even for semiquantitative calculations, it is reasonable to use H^{SS} alone to calculate the first-order approximation to the splitting energies in the spirit of degenerate perturbation theory (compare Equation 3.13 with $V = H^{SS}$) and to adapt the triplet spin eigenfunctions to H^{SS}. The total electronic wave function depends on the coordinates of all electrons and, in general, is a linear combination of products of spin functions with space functions. Below, we consider only the two unpaired electrons of the open shell as the active space and treat the rest implicitly as an inert core.

As defined in Equation 3.4, the spin part of H^{SS} is a traceless symmetric tensor operator that can have nonvanishing matrix elements between states of spin multiplicity different by up to four, such as singlet and quintet or triplet and septet, etc., but in first order we now consider only its elements in the basis of the three sublevels of the T state. H^{SS} is diagonalized in this subspace by rotation of the coordinate system into the principal system of axes x, y, and z in the molecular frame. In principle, the molecular magnetic axes defined in this fashion are different for every triplet state of the molecule. In the presence of symmetry, some or all of them are constrained to the molecular symmetry axes.

The H^{SS} eigenvalues $-D_x$, $-D_y$, and $-D_z$ for the resulting substates, T_x, T_y, and T_z, respectively, add up to zero. They are traditionally combined into two zero-field splitting parameters, $D = 3D_z/2$ and $E = (D_x - D_y)/2$. In terms of these parameters, the three eigenvalues are $D/3 - E$, $D/3 + E$, and $-2D/3$, respectively. In the usual convention, the labels of the axes are chosen to make $|D| \geq 3|E|$ and $DE < 0$. Then T_z is the lowest and T_x the highest in energy if D is positive, and the level order is the opposite if D is negative. In axially symmetrical molecules, $E = 0$ and the levels T_x and T_y are degenerate. In molecules of even higher symmetry, such as cubic, D vanishes as well and all three levels remain degenerate. This is not common in organic molecules, but it happens frequently for coordination compounds, and then the second-order effect of H^{SO} remains as a sole source of zero-field splitting.

The spin eigenfunction $\Theta[u]$ associated with an axis u ($u = x$, y, or z) has a zero projection of spin angular momentum into this axis. In terms of the usual electron spin functions α

and β, symmetry adapted to the z axis, the spin eigenfunctions for the two unpaired electrons of the triplet state are

$$\Theta[x] = -2^{-1/2}[\alpha(1)\alpha(2) - \beta(1)\beta(2)]$$

$$\Theta[y] = 2^{-1/2}i[\alpha(1)\alpha(2) + \beta(1)\beta(2)] \qquad (3.25)$$

$$\Theta[z] = 2^{-1/2}[\alpha(1)\beta(2) + \beta(1)\alpha(2)]$$

where the phase factors in front of the right-hand sides ensure cyclic permutation properties with regard to x, y, and z. The singlet spin function is

$$\Sigma = 2^{-1/2}[\alpha(1)\beta(2) - \beta(1)\alpha(2)]. \qquad (3.26)$$

Next, we consider the even weaker second-order perturbation of the three sublevels of T that is due to H^{SO}. Recall that the analogous second-order contributions from H^{SS} are neglected in this approach, as they are generally believed to be small, and the first-order contributions have already been included. It is seen from inspection of Equation 3.3 that the operator H^{SO} can mix each of these three triplet wave functions with singlet, other triplet, and quintet wave functions. This interaction has no first-order effect on the energies $-D_x$, $-D_y$, and $-D_z$ of the three substates, T_x, T_y, and T_z, respectively, but in second order, Equation 3.14 with $V = H^{SO}$, they will be affected somewhat and become $-D'_x$, $-D'_y$, and $-D'_z$. If we continue to define the zero-field splitting parameters by $D' = 3D'_z/2$ and $E' = (D'_x - D'_y)/2$, they can be compared with the values observed. In Section 3.3, we noted the difficulties involved in attempts to evaluate these values accurately in this fashion, due to the very large number of states over which the summation in Equation 3.14 is necessary, and we commented on alternative methods of evaluation such as response theory.

For triplet states of organic molecules, the effect of H^{SO} on the zero-field parameters is normally so small that it is often assumed to be unnecessary to distinguish between D' and D and between E' and E, and the symbols D and E are used both for the observed parameters and those obtained from computations or estimates based on H^{SS} alone. This is unfortunate, since already the presence of a single atom from

the second complete row of the periodic table, such as silicon
or sulfur, can make the unprimed and primed values quite
different from each other. For our purposes, a discussion of
zero-field splitting is not central, and in the remainder of the
text, the symbols D and E, when used at all, will have the
usual meaning of the observed zero-field splitting parameters
and the computed approximations to them, regardless of
whether the computation used only H^{SS}, only H^{SO}, or both.

The matter that is central to our photochemical interests
is the magnitude of the three matrix elements $\langle T_u | H^{SO} | S \rangle$,
$u = x, y, z$, that connect each of the three sublevels of T with
the singlet level S, because in the Fermi golden rule approx-
imation $|\langle T_u | H^{SO} | S \rangle|^2$ is proportional to the rate at which
molecules interconvert between T_u and S. The three numbers,
$\langle T_x | H^{SO} | S \rangle$, $\langle T_y | H^{SO} | S \rangle$, and $\langle T_z | H^{SO} | S \rangle$, can be viewed as
the x, y, and z coordinates of the "spin–orbit coupling vector"
(**SOC**), respectively. Under many experimental conditions,
molecules equilibrate among the three T sublevels much
faster than intersystem crossing to S takes place. In such a
case, the intersystem crossing rate is proportional to

$$|\mathbf{SOC}|^2 = \left|\langle T_x | H^{SO} | S \rangle\right|^2 + \left|\langle T_y | H^{SO} | S \rangle\right|^2 + \left|\langle T_2 | H^{SO} | S \rangle\right|^2 \quad (3.27)$$

The length $|\mathbf{SOC}|$ of the **SOC** vector is the only quantity
ordinarily reported in calculations of spin-orbit coupling
between a triplet and a singlet state in molecules and is
referred to as the "size of spin-orbit coupling." This is unfortunate for two reasons. First, in experiments at very low
temperatures or very short times, the equilibration of molecules among the three triplet sublevels is incomplete or
absent, and the individually observable rates are related to
the squares of the coordinates of the **SOC** vector and not to
its overall length. Second, if one wishes to rationalize the
origin of the computed numbers in terms of molecular structure and not merely accept them as statements from a wise
oracle, it is much easier to do so for the three coordinates
individually than for the overall length of the **SOC** vector
alone. This is important for endeavors such as understanding
the mechanism by which the heavy atom effect operates.

Although the direction of the **SOC** vector for an S–T pair of states in the molecular frame is well defined, its sense is not, because the wave function of the triplet or the singlet state can always be multiplied by –1 or another complex unity if desired. In this regard, the **SOC** vector is similar to the familiar electric dipole transition moment vector **M** between two states. In both cases, the ambiguity is immaterial because the values of observable properties are proportional to the absolute value square, the rate of intersystem crossing to $|\mathbf{SOC}|^2$, and the transition probability to $|\mathbf{M}|^2$.

The analogy between transition moment vectors and spin-orbit coupling vectors extends to the effects of symmetry. In molecules that belong to point groups with symmetry operations, the coordinate $\langle T_u | H^{SO} | S \rangle$ of the **SOC** vector connecting an S and a T state will vanish if the product of the irreducible representations to which T_u and S belong does not contain the totally symmetric representation, because the Hamiltonian operator H^{SO} is totally symmetric. For both wave functions T_u and S, the overall symmetry is the product of the symmetry of the space part and the spin part. The symmetry of the space part is normally easily determined from the knowledge of the CSFs that enter into the wave function and the symmetries of the orbitals used to construct them. The symmetry of the spin part is even easier. The spin functions that enter T_u, given in Equation 3.25, transform as the components of the rotation operator R_u, and the singlet spin function of Equation 3.26 is totally symmetric. Hence, it is an easy matter to find the vanishing coordinates of **SOC** and, in particular, to determine whether the length of this vector is zero.

For attempts to understand why the computed **SOC** vectors have the length and orientation they have, it is often useful to divide their overall values into contributions from individual atoms and, even beyond that, into contributions from individual orbital pairs. An example of a situation where such an analysis is needed is provided by studies of the heavy atom effect. How much of the overall change in **SOC** that occurs when a heavy atom is introduced into a molecule is actually due to its other effects that have little to do with its atomic number, such as inductive and resonance effects? Why does the introduction of the heavy atom into various positions in the same molecule have vastly different effects on the resulting **SOC**? These and similar questions are difficult to answer if the overall **SOC** vector cannot be subdivided into atomic contributions. According to Equation 3.3, the one-electron part of H^{SO} is a sum over atoms, and the separation into atomic contribution is automatic, but some choices must be made to separate the two-electron part similarly. A separation proposed by Havlas et al.[38] is based on the observation that in the basis of Weinhold's natural hybrid orbitals[39] the only nonnegligible two-electron spin–orbit coupling integrals are those in which two of the four orbital indices are identical and equal to an inner core orbital on a single atom. Physically, these integrals describe the interaction of hybrid pairs screened by inner-shell electrons on that atom, and they are counted as a part of its atomic contribution.

Why do certain atoms make large contributions whereas others of the same atomic number make small ones? In order to address these issues, two groups of authors, Zimmerman and Kutateladze[40] and Michl and collaborators[38,41] independently introduced a decomposition of the total computed spin–orbit coupling into contributions provided by local orbital pairs, choosing the local orbitals to be the natural hybrid orbitals introduced by Weinhold.[39] There are only minor differences between the two treatments. Zimmerman neglects the small three- and four-center contributions in the transformation of the two-electron spin-orbit coupling integrals into the Weinhold basis. He also does not use H^{SS} to calculate zero-field splitting and the magnetic axes and instead uses an arbitrary set of molecular axes for the calculation. This makes it impossible to make statements about the spin-orbit coupling behavior of the three individual sublevels, and only the behavior of the overall magnitude of spin-orbit coupling $|\mathbf{SOC}|$ can be analyzed.

3.5 SINGLET-TRIPLET SPIN-ORBIT COUPLING IN BIRADICALS

Many photochemical reactions proceed through the triplet state. This is always true when the initial excitation of the substrate occurs by triplet sensitization and is often true even when the initial excitation is by direct photon absorption, because in many molecules intersystem crossing from the initially excited singlet manifold of states into the triplet manifold is fast enough to compete with other processes. More often than not, photochemical paths from triplet-excited reactants lead through biradicaloid geometries, that is, those at which a simple description of the low-energy electronic states of the molecules features two nearly nonbonding orbitals occupied by a total of only two electrons. Frequently, the lowest triplet energy surface contains a minimum at such a geometry, and a molecule occupying such a minimum is usually called a triplet biradical. We prefer to use the term *perfect biradical* for instances in which the two orbitals are exactly degenerate and their occupancies equal and the term *biradicaloid* otherwise, and we prefer to use the term *biradical* only in instances that are quite close to perfect biradical. There is no unanimity on this nomenclature among organic photochemists.

After a period of time in a triplet biradical potential energy minimum, which is normally sufficient to establish a vibrational and conformational equilibrium, the molecule undergoes intersystem crossing to the lowest singlet state, which is usually not far in energy, and then relaxes and thermally equilibrates on the S_0 surface to yield the final product or products. In solution reactions, vibrational thermal equilibrium is established quickly, and the molecular geometry tends to limit its excursions to the low-energy region of the potential-energy surface. In triplet biradicals, this region is frequently quite extensive and may contain several shallow minima representing conformational isomers. The nature of the final product is dictated by the geometry at which the intersystem crossing takes place and by the shape of the S_0 surface in that vicinity. The location of the points at which intersystem crossing is most likely is dictated by the shape of the T_1 surface and by the rate of intersystem coupling in each low-energy region. The return to S_0 does not necessarily occur at the minimum in T_1 if the spin flip can occur more readily at other geometries.

In biradicals in which the two radical centers are localized far apart, hyperfine coupling with nuclear magnetic moments plays an important role in determining the inter-system crossing rate, but in others, spin-orbit coupling is the determining factor. It is thus fair to say that an understanding of the structural dependence of the magnitude of spin–orbit coupling in biradicals plays an essential role in attempts to understand organic triplet photochemistry. This was recognized early on, and an important set of qualitative rules was formulated as early as 1972 by Salem and Rowland,[42] based on the 2-in-2 ("3 $\times$ 3") model of bitopic biradical electronic structure (no general rules have been formulated so far for biradicals of higher topicity, in which more than two orbitals are accessible to the unpaired electrons in low-energy states). Recently, the Salem–Rowland rules have been reformulated more rigorously by Michl.[41]

In the corrected form, the rules state that for a large value of $|\mathbf{SOC}|$ between T_1 and S_0, the most localized orthogonal orbitals A and B singly occupied in T_1 should be as follows:

1. They either interact covalently through a nonzero resonance integral or are sufficiently different in energy for one of them to have electron occupancy near two in the S_0 state, and
2. the biradical contains one or more atoms at which one p orbital contributes strongly to A and another to B. The higher the atomic numbers of these atoms, the better, provided that
3. these p orbitals enter into A and B in such a manner that the contributions on all such atoms add rather than cancel.

Condition 1 is a more rigorous statement of the Salem and Rowland requirement of "ionic character" in S_0 and reflects the fact that only the amplitude of the $[A(1)A(2) + B(1)B(2)]$ configuration matters if the biradical is close to perfect. It follows that in a perfect biradical, the spin-orbit coupling of the T_1 and S_0 states vanishes. In systems with A and B at similar energies, condition 1 is generally met at the expense of increasing the energy of T_1 above its minimum, because the required orbital interaction is destabilizing. This is nicely exemplified by twisted triplet ethylene, which has an energy minimum and negligible spin-orbit coupling at orthogonal twist and strong spin-orbit coupling at intermediate twist angles. In biradicals whose radical centers are not located on the same atom, most of the interaction between them is carried by through-bond coupling, that is, by delocalization of the orbitals A and B and not by direct through-space interaction. The results can be rationalized by classical resonance structures.[41] Even in twisted ethylene (a 1,2 biradical), the dominant interaction is not provided by the two-center term involving the singly occupied p orbitals on the two neighboring methylene carbons but by the oppositely signed sum of the terms involving one of these p orbitals at a time along with the small portion of the other singly occupied orbital that is delocalized by hyperconjugation onto the adjacent methylene group.

If condition 1 is to be met by making the energies of A and B different enough for the hole-pair structure $B(1)B(2)$ to dominate S_0, it is essential to make the energy difference large enough to go beyond the critical threshold value of the heterosymmetric perturbation, $2[K'_{AB}(K'_{AB} - K_{AB})]^{1/2}$, where K_{AB} is the exchange integral between the localized orbitals A and B and K'_{AB} is the exchange integral between the real delocalized orbitals $2^{-1/2}[A \pm B]$. The threshold value is smallest when A and B are located at the same atom and increases with their increasing separation. Thus, in a carbene, already a small difference in the content of s character in the two nonbonding orbitals, produced by bending, is sufficient, whereas in a twisted ethylene, one of the carbon atoms needs to be replaced by an atom at least as electronegative as a positively charged nitrogen. In 1,3 biradicals the required electronegativity difference is even larger, and, indeed, a moderate degree of polarization does not enhance the rate of intersystem crossing in 1,3-diaryl-1,3-cyclopentadienyls.[43]

Condition 2 is reminiscent of the classic El Sayed rules.[44] It is helpful in trying to understand the origin of the heavy atom effect: In order for it to operate optimally, one of the orbitals A and B should delocalize into one p orbital on the heavy atom and the other into another p orbital on the heavy atom.

The effects of condition 3 can often be worked out from molecular symmetry. When this is not possible, it is necessary to apply the simple recipes available[41] for deducing the direction of the atomic contribution to **SOC** from inspection of the orbitals A and B expressed in the basis of natural hybrid orbitals. These procedures are essential for understanding the effects of multiple heavy atoms and for evaluating the potential for the presence of an inverse heavy atom effect.

3.6 NONCONJUGATED π SYSTEMS

3.6.1 Ethylene and Alkenes

Most of the studies of these species included an examination of strongly twisted and otherwise severely distorted geometries, at which the double bond has considerable biradicaloid character, and could also be placed in Section 3.9. Three methods, the full Breit–Pauli H^{SO}, one-electron mean field, and the approximate one-electron operator with an effective nuclear charge, and a variety of basis sets, from a minimum of one up to a polarized quadruple zeta, were compared in evaluating the spin-orbit coupling between the T_1 and S_0 states of ethylene as a function of twist (**1a**) and syn (**1b**) and anti (**1c**) pyramidalization distortions.[45] It was found that at least a double zeta basis set must be used for qualitatively correct results. The quality of the wave function is also crucial. Not surprisingly, single-reference CI wave functions are faulty, especially for the twist mode, which produces biradicaloid geometries, and even CISDTQ is not sufficient. In contrast, the multireference methods MCSCF and MRCI perform correctly. Syn pyramidalization enhances spin-orbit coupling to values above 6 cm^{-1}, whereas pure twisting distortion yields about 2 cm^{-1} at most. The mean-field approximation was found to work as well as the full Breit–Pauli H^{SO}, while the one-electron approximation with effective charges is questionable. The anti pyramidalization mode was found to be an important

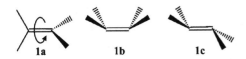

1a   1b   1c

factor for "locked" alkenes, where the constraint is caused by a four-membered ring. Filatov et al.[46] studied a set of 1,2-substituted cyclobutenes (**2**), where substituents varied from strong π acceptors (BH_2) to strong π donors (NH_2). Unlike π acceptors, π donors cause strong anti pyramidalization of the π bond carbons in the triplet state, resulting in small spin-orbit coupling (~0.5 cm⁻¹).

Woeller at al.[47] studied the radiationless decay mechanisms of cyclic alkenes (**3**) with ring sizes four to six, norbornene, and their phenyl derivatives. The potential-energy surfaces of the S_0 and T_1 states have been explored along double-bond twisting and anti pyramidalization reaction pathways to explain the experimentally observed inverse proportionality between ring size and triplet-state lifetime. The spin-orbit coupling constants, calculated at the CASSCF(2,2)/DZP level with the one-electron part of H^{SO} using effective nuclear charge (and checked against CIS-DTQ(12–12) at several points) were used to calculate the T_1

X=H,CH₃,BH₂,NH₂,F

X   X
2

3a   **3b**   **3c**   **3d**   R=H,Ph

R   R   R   R

$\rightarrow S_0$ transition probabilities based on the Fermi golden rule. Only a combined reaction coordinate of anti pyramidalization and twisting at the double bond provides a low-energy pathway that reproduces the experimentally observed transition probabilities, whereas the traditional model invoking only a pure twist around the double bond fails.

The intersystem crossing rate of triplet norbornene ((**3d**) R=H), a state with a very short nonradiative lifetime of 250 ns, was examined by Harvey et al.[48] The potential energy of S_0 and T_1 states along the anti pyramidalization coordinate was evaluated at the DFT level, and spin-orbit coupling matrix elements were calculated at the CASSCF(2,2) level, using a one-electron operator with effective nuclear charges. Both the Fermi golden rule and nonadiabatic RRKM reproduced the experimental triplet lifetime with reasonable accuracy, compared with a previous[49] Landau–Zener treatment. The reason for the failure of the Landau–Zener model was the incorrect assumption that the decay takes place only at the crossing seam between the T_1 and S_0 surfaces. In fact, much of the nonradiative transition occurs in the tunneling regime, well bellow the crossing point, and the surfaces need not even cross for hopping to occur.

3.6.2 Nitrosoalkanes

The predissociation lifetimes of several vibrational states of the S_1 state of two nitrosoalkanes, CH_3NO and $(CH_3)_3CNO$, were evaluated by computing[50] nonadiabatic and spin-orbit couplings. The intersystem crossing rate, evaluated from the Fermi golden rule, included a calculation of the spin-orbit coupling element between the S_1 and T_1 states and its derivatives related to selected normal vibrational modes. The matrix elements of H^{SO} were evaluated using the one-electron Hamiltonian with fitted effective nuclear charges (CASSCF(12,8)/6-31G**). The authors concluded that more accurate calculations of decay dynamics would require the inclusion of more than a few states in the calculation.

3.6.3 Azidoalkanes

Arenas et al.[51] proposed two reaction paths for nitrene production by thermal decomposition of aliphatic azides ($R-N_3$, R = H, Me, Et), using a scaled one-electron approximation calculated from state-averaged CASSCF wave functions. One is a thermally activated singlet path and the other invokes intersystem crossing on the S_0/T_1 crossing seam. A large calculated spin–orbit coupling of 43 cm^{-1} at the minimum of crossing seam, similar gradients for the S_0 and T_1 states, and small suggested velocities (the crossing point is close to the singlet transition state) led the authors to propose that both paths are possible. Similarities in energy and structure of the transition state and the crossing point make both reactions equally probable.

3.7 LINEAR CONJUGATED π SYSTEMS

In the study of the vibronic spectrum of a doublet HCCS radical, Perić et al.[52] calculated the spin-orbit coupling constant at the equilibrium geometry of the radical by using the two-component relativistic no-pair Hamiltonian derived by Samzow et al.[14] In the calculation, truncated (8,8)MRDCI wave functions were used with orbitals optimized for the triplet state of the corresponding cation. The spin-orbit coupling constant of 261 cm^{-1} agreed well with the experimental data.

The full one- and two-electron and the mean-field levels of theory of spin-orbit coupling were compared in Tatchen and Marian's calculation[53] of fine-structure splitting in the $^2\Pi$ electronic ground state of the linear isoelectronic HC_6H^+, NC_5H^+, and NC_4N^+ ions, using the no-pair spin-orbit Hamiltonian of Samzow et al.,[14] TZP basis, and truncated CISD wave functions. Compared with the results of full spin-orbit treatment, the molecular and atomic HF mean-field approximations produced only slight differences, whereas the neglect of all multicenter two-electron integrals gave the right trends but significantly overestimated splittings. When all one- and two-electron multicenter terms were neglected, the results were considerably improved because of a compensation of one- and two-electron terms, and deviated by less than 5% from those obtained with the full Hamiltonian.

3.7.1 Glyoxal

Nakajima and Kato[54] used the CASSCF(8,6)/DZP approxima- tion to examine intersystem crossing from the S_1 state of glyoxal (Structure 3.4) induced by collisions with argon atoms. They evaluated the S_1 and T_1 interaction potentials and the spin-orbit coupling matrix elements between these two states at each geometry by using the full Breit–Pauli H^{SO}. A semi- classical dynamics calculation was used to estimate transition cross-sections and rate constants. The magnitude of the spin–orbit interaction is very small (below 1 cm^{-1}) and depends heavily on the orientation of the path of the Ar atom relative to the glyoxal molecule. An out-of-plane approach of the Ar atom is particularly effective. According to the dynam- ics calculations, the intermolecular spin-orbit interactions increase the transition cross-section more than 100 times over the intramolecular value at room temperature.

3.8 CYCLIC CONJUGATED π SYSTEMS

3.8.1 Benzene

Vahtras et al.[21] used the CASSCF(6,6)/DZP and atomic mean- field approximations to calculate the zero-field splitting parameters of benzene, including both spin-spin coupling to the first order and spin-orbit coupling to the second order of

4

perturbation theory. The relative importance of these two contributions is strongly system dependent, but in the lowest triplet state of benzene the zero-field splitting is determined entirely by spin-spin coupling. The calculated value of the D parameter, 0.158 cm^{-1}, agrees well with the experimental value[55] of 0.159 cm^{-1}.

3.8.2 Pyridine

The observed phosphorescence and high-resolution singlet-to-triplet absorption spectra of pyridine (**5**) were assigned by Cai and Reimers,[56] who performed a vibronic coupling calculation involving six active modes and three near-degenerate electronic states: 1^3A_1, 2^3A_1, and 1^3B_1. The lowest surface has a double minimum of A′ symmetry and is strongly vibronically coupled to the B_1 triplet state. The spin–orbit coupling matrix elements were evaluated at the CASSCF(8,6)/cc-pVDZ level by using a scaled one-electron spin-orbit coupling operator. The strongest coupling is predicted to occur between the $T_2(A_1)$ and $T_3(B_1)$ triplet states and the $S_0(A_1)$ singlet ground state (32.6 and 30.3 cm^{-1}). These values explain the observed rapid nonradiative decay and low phosphorescence quantum yield. The next largest coupling (24.5 cm^{-1}) is predicted between the T_1 and 2^1B_1 states, but this high-lying singlet state is both very weakly allowed and quite distant in energy (E = 9.0 eV) and so is unlikely to contribute to the observed singlet-to-triplet absorption. It was estimated that the T_1 state borrows 3% of the intensity of the $S_2(1B_2)$ and 4% of the $2B_2$ singlet states. The intensity borrowed by the T_3 state is 6% from the

5

$1B_2$, 13% from the $2A_1$, and 4% from each of the $3A_1$ and $2B_2$ singlet states.

3.8.3 Thiophene

Kleinschmidt et al.[33] calculated the spin–orbit coupling matrix elements of thiophene (**6**) within a one-center mean-field scheme based on a DFT/MRCI method. Test calculations performed for several diatomic molecules and 4H-pyran-4-thione (**7**) showed that this method is comparable with MRDCI based on HF orbitals, using the same mean-field scheme. The application to thiophene included calculations of both singlet-triplet and triplet-triplet SOC matrix elements. As expected, the matrix elements between all $\pi \rightarrow \pi^*$ states are small (< 0.1 cm^{-1}). The largest matrix element of 94.7 cm^{-1} corresponds to the interaction of the singlet ground state with an $n\pi^*$ triplet state of A_2 symmetry (with some Rydberg character).

3.8.4   4H-Pyran-4-thione

Fine structure splitting and transition moments for radiative and nonradiative spin-orbit coupling induced processes were calculated by Tatchen et al.[57] for several low-lying singlet and triplet states of 4H-pyran-4-thione (**7**). The spin-independent properties were calculated by the DFT/MRCI method of Grimme and Waletzke,[58] and the spin-dependent part was included at the level of quasi-degenerate perturbation theory

6

7

based on the mean-field Hamiltonian. The calculated properties compared very well with the available experimental data. For the T_1 state, phosphorescence and nonradiative decay via intersystem crossing to the S_0 state compete (rate ~10^4 s^{-1}). The $(T_1)z \rightarrow S_0$ radiative transition gains its intensity from two sources: direct SO coupling of S_0 and T_1 states and strong spin-allowed $S_2 \rightarrow S_0$ transition. The zero-field T_1 splitting of D = -18 cm^{-1} was computed from only the spin-orbit part, and the spin–spin dipolar part was neglected.

3.8.5 Styrene

Bearpark et al.[59] studied styrene (**8**) cis-trans photoisomerization in the CASSCF(8,8)/4-31G approximation with the one-electron approximation for H^{SO} and effective nuclear charges and concluded that it occurs on the S_1 surface. Experimentally, two mechanisms had been suggested,[60,61] with a possible role for the triplet state. The intersystem crossing to the T_2 state is expected to be inefficient because of a small calculated spin–orbit coupling (<1 cm^{-1}).

3.8.6 Phenyl Cation

The minimum energy crossing point on $^1A_1$ and $^3B_1$ surfaces of the phenyl cation (**9**; X=H) was located by Harvey et al.[62] This point with C_{2v} symmetry lies only 0.12 kcal/mol (CCSD[T]) above the triplet state minimum. At this point, spin-orbit coupling, evaluated at the CASSCF(6,7) level by using a one-electron operator with effective nuclear charges,

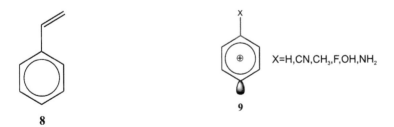

X=H,CN,CH$_3$,F,OH,NH$_2$

8

9

is quite large (6.3 or 6.6 cm^{-1}, depending on which state the orbitals were optimized for, S or T). The study was extended[63] to para-substituted phenyl cations (Structure 3.9; X=H, CN, CH_3, F, OH, NH_2). Except for the OH and NH_2 derivatives, the cations were found to be ground-state singlets. The OH substituted derivative has almost isoenergetic S and T states. The minimum energy crossing points between the two surfaces lie very little above the higher of the two minima in all cases. The spin-orbit coupling strengths (CASSCF (8,7), (6,7)) for the $C_6H_4^+$, 6-31G* basis set, one-electron approximation) are significant: 5.8, 5.4, 5.5, 6.9, 3.3, and 4.7 cm^{-1}, respectively.

3.8.7 Antiaromatics

Because of the biradicaloid nature of these structures, they could also be listed in Section 3.9 under biradicals. Shiota et al.[64] described how molecular distortions enhance the strength of the spin-orbit coupling between the S_0 and T_1 states of the cyclopropenyl anion (**10a**), cyclobutadiene (**10b**), and the cyclopentadienyl cation (**10c**). On the basis of CASSCF/TZ wave functions and the full Breit–Pauli H^{SO}, they found that molecular distortions along certain C-H out-of-plane bending modes significantly enhance spin-orbit coupling in C_4H_4 and $C_5H_5^+$. The out-of-plane motions destroy planarity and lead to rehybridization. In the nonplanar $C_3H_3^-$ anion, the carbon ring distortion that connects the triplet and singlet equilibrium structures is the main factor that dominates the transition between the two states. The SOC values are up to 1 cm^{-1}.

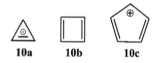

10a   **10b**   **10c**

3.9 BIRADICALS

3.9.1 1,1-Biradicals (Carbenes and Analogues)

The parent carbene and its heavier analogues (**11**; XH_2, X = C, Si, Ge, Sn, Pb) have often been used for testing different methods and algorithms. Havlas et al.[38] reported calculations for CH_2 and SiH_2 as a function of the valence angle by using the full Breit–Pauli H^{SO} and H^{SS} operators. The results for spin-orbit coupling converged quite rapidly with basis set size, and already a DZ basis set was reasonably adequate. At the equilibrium geometry of the triplet, using CASSCF(6,6)/D95**, they found an S_0/T_1 spin-orbit coupling of 12.4 cm^{-1} for carbene, of which 11.1 cm^{-1} originates in the pair of nonbonding Weinhold hybrid orbitals on the carbon atom. For silylene, the numbers were 56.4 and 43.7 cm^{-1}, respectively. The values depended on the valence angle in the way expected from the 3×3 model of biradical structure. Good results for dipolar spin-spin coupling required a large basis set and a high level of electron correlation; CASSCF(6,6) was inadequate, whereas CISD provided good results. The zero-field splitting parameter E in T_1 was dominated by the spin-spin interaction both in carbene and in silylene, as was the D parameter in carbene. In silylene, D was dominated by spin-orbit coupling. The prediction of an unusually large D value, 1 to 3 cm^{-1} depending on the valence angle, for silylenes forced by bulky substituents to have a large valence angle and a triplet ground state, has recently been confirmed experimentally (1.63 cm^{-1}, assuming $E = 0$).[65]

11

Fedorov and Gordon[9] calculated the T_1/S_0 spin-orbit coupling for X up to Pb and compared the values based on the full Breit–Pauli H^{SO} operator with their partial two-electron contribution (P2E) method. The approximation works for heavier elements (where the importance of the two-electron term ultimately vanishes), but for methylene it causes an error of 15%. Their best values for the spin-orbit coupling constant for XH_2 (X = C, Si, Ge) amount to 12.3, 57.1, and 327.3 cm^{-1} (CASSCF(6,6)/TZ). The selection of the SOCI wave function instead of the CAS one, which includes more dynamical correlation, changes the values by less than 3%. A previous study by Matsunaga at al.[66] used a CASSCF(6,6) wave function in conjunction with only the one-electron part of the Breit–Pauli H^{SO}, adopting the effective nuclear charges as adjustable empirical parameters. To account for other relativistic effects (contraction of core orbitals at heavy atoms) they used theSBK relativistic core potential. The calculated SOC values were somewhat smaller than those obtained by Fedorov and Gordon.[9] The need for at least a double zeta basis set is illustrated by the result of Yu et al.,[67] who calculated the spin-orbit coupling between S_0 and T_1 states of bromomethylene at the CASSCF level with a minimum basis set (STO-3G). The result, 140 cm^{-1}, was off by a factor of 2.3 from the value of 322 cm^{-1} determined by fitting to measured spectra. The CASSCF(8,6)/cc-pVDZ calculation of Havlas and Michl[68] yielded a value of about 230 cm^{-1}, still too small. These authors performed calculations for CHX (X = H, F, Cl, Br) at a range of valence angles using full H^{SO} and H^{SS} operators. In addition to making semiquantitative predictions of the zero-field splitting parameters, they were interested in the origin of the heavy atom effect. They analyzed their spin-orbit coupling results in terms of atomic contributions to the spin–orbit coupling vector and found that the vectors provided by the carbon and halogen atoms were roughly parallel. They rationalized their directions in terms of contributions from pairs of Weinhold hybrid orbitals. Guided by these considerations, they then looked for a case in which the carbon and halogen vectors would be approximately opposed, because this would result in an inverse heavy atom effect if their magnitudes were comparable. This was indeed found[11] in the case of bromomethylcarbene, for which weaker spin-orbit coupling is calculated than for methylcarbene, at all values of the dihedral angle.

Christiansen et al.[28] tested the coupled-cluster response theory on spin–orbit coupling constants of substituted silylenes (HSiX, X = F, Cl, Br) in the atomic mean-field approximation. Comparison with the full Breit–Pauli H^{SO} showed that the approximation is quite accurate. The calculated values for $|\langle 1^3 A' |{}^{eff}H^{SO}| 1^1 A'\rangle|$ of 67.5, 91.5, and 261.3 cm^{-1} for X = F, Cl, and Br, respectively, differ from the full Breit–Pauli matrix elements by less than 1 cm^{-1}. Even better agreement was obtained for the matrix element between the lowest triplet and the first excited singlet state, $|\langle 1^3 A' |{}^{eff}H^{SO}| 1^1 A'\rangle|$, which in the case of silylene vanishes by symmetry (F = 0.05, Cl = 1.43, Br = 7.68 cm^{-1}).

3.9.2 Nitrenes

Johnson et al.[69] analyzed spin–orbit coupling in phenyl nitrenes (**12**) using the full Breit–Pauli H^{SO} and CASSCF(8,8) wave functions. Within the C_{2v} point group the T_1 and S_0 states do not couple, and perturbations due to meta substitutions do not increase the spin-orbit coupling value above 0.1 cm^{-1}. Symmetry allows T_1 to couple with the S_1 and S_2 states, and the computed values are of the order of 15 cm^{-1} for the former and 40 cm^{-1} for the latter in various para and meta substi-

12

tuted phenyl nitrenes. The SOC between T_1 and S_1 increases moderately with the decreasing Hammett F value of the substituent. The trend for SOC between T_1 and S_2 is the opposite, but the differences are very small.

3.9.3   1,3-Biradicals

To explain the influence of polar substituents on the intersystem crossing rates in triplet 1,3-biradicals, Ichinose et al.[70] calculated the spin–orbit coupling constants between the S_0 and T_1 states for three model systems: trimethylene (**13a**), 1-amino-3-cyanotrimethylene (**13b**), and (aminomethyl)borane (**13c**). They applied the Salem–Rowland rules in their original form[42] and expected that the substitutions and exchange of end carbon atoms by a boron-nitrogen pair would increase the zwitterionic character of the singlet state and, therefore, the magnitude of spin-orbit coupling. The CASSCF(2,2)/3-21G method with the full Breit–Pauli H^{SO}, yielded enhancement by a factor of only 1.4 for the substituted compound and 2.1 for the boron-nitrogen analogue. The authors concluded that the effects of polar substituents are relatively small. The calculated spin-orbit coupling constants were in the range of 1 to 5 cm^{-1}, depending on the geometry and substitution.

Zimmerman and Kutateladze[71] calculated spin-orbit coupling in linear 1,n-diyl biradicals (n = 3 to 8) at the CASSCF(4,4)/3-21G level with the Breit–Pauli spin-orbit Hamiltonian and analyzed it using natural bond orbitals. For diyls with an even number of carbons, symmetry forces zero SOC, whereas for those with an odd number of carbons the main terms have like signs and add, providing a nonvanishing net SOC. The maximum SOC is found at 90° orientation

a, $R_1=R_2=H, X=Y=C$
b, $R_1=NH_2, R_2=CN, X=Y=C$
c, $R_1=R_2=H, X=B, Y=N$

13

between the p components, because each MO extends through the entire molecules. The SOC decreases from 1.7 cm^{-1} for 1,3-propanediyl to 0.02 cm^{-1} for 1,7-heptanediyl. The earlier calculations by Furlani and King[72] of 1,3-propanediyl at the CASSCF(2,2) level provided qualitatively equivalent results, with SOC almost two times higher.

3.9.4 Norbornadiene Cyclization

Norbornadiene undergoes a photochemical valence isomerization to quadricyclane with unit quantum yield when photosensitized with acetophenone.[73] The proposed reaction scheme (**14**) includes two triplet state intermediates: norbornadiene triplet (**14b**) and the 3,5-nortricyclanediyl biradical (**14c**). Based on the full Breit–Pauli H^{SO} with a rather modest quality wave function (CASSCF(2,2)/3-21G) the spin–orbit coupling matrix element between T$_1$ and S$_0$ states was calculated for different geometries of triplet norbornadiene. The values between 1 and 2 cm^{-1} correspond to that of twisted triplet ethylene, and the equilibrium structure resembles it. The short measured[74] norbornadiene triplet lifetime of 6.2 ns (the lifetime of **14c** is unknown) is consistent with the calculated spin-orbit coupling.[73] Helms and Caldwell[74] concluded that the reaction proceeds through the antipyramidalized exo and endogeometry because of its lowest S/T gap.

3.9.5 Paternò-Büchi Biradicals

These 1,4-biradicals (**15**) are formed in [2+2] cycloaddition of electronically excited carbonyl compounds to alkenes. The conformational dependence of their spin-orbit coupling and

14a   14b   14c   14d

singlet-triplet gap, modeled by the addition of formaldehyde to ethylene, was studied by Kutateladze.[75] He used two models to evaluate the spin-orbit coupling: a one-electron Hamiltonian with effective nuclear charges, with a state-averaged CASSCF wave function, and the full Breit–Pauli H^{SO} operator with fully optimized CASSCF triplet orbitals and frozen-core triplet CASSCF singlet orbitals. The results revealed two distinct areas of elevated spin-orbit coupling values. The first corresponds to the region of cisoid conformations of the C-C-O-C fragment with the singly occupied atomic orbitals closer to each other, and the second corresponds to a partially eclipsed conformation without direct overlap of the singly occupied orbitals. An NBO analysis showed that the enhanced SOC is mediated by the oxygen's lone pair. The S/T separation, the other factor affecting the intersystem crossing rate, was evaluated by the CASMP2 method. The eclipsed (torsion angle ~0°) and partially eclipsed (120°) conformations have a small S/T gap and increased value of SOC (~2 cm⁻¹). The computational findings thus support the model by Griesbeck and Fiege[76] for the stereochemistry of this reaction.

Bertrand et al.[77] studied the 1,4-biradicals (**16**) expected in the model triplet [2+2] photocycloaddition between acrolein and ethylene. According to their computations, the molecules first form the C–C bond on the triplet surface; the system then undergoes an intersystem crossing to the singlet surface and either reverts into the reactants or completes the cycloaddition reaction. To rationalize the intersystem crossing step, they evaluated the spin-orbit coupling between singlet and triplet

15

16

states for various geometries at the CAS(2,2) level, calculating both the one-electron and the two-electron contributions. The coupling elements are relatively small (up to 0.1 cm^{-1}).

Zimmerman and Kutateladze[78] have drawn several qualitative conclusions from a study of spin-orbit coupling in Type B cyclohexenone (**17**) photorearrangement, evaluated at critical points along the presumed reaction path (full Breit–Pauli H^{SO}, CASSCF/STO-3G, NHO analysis). The most important finding is that the S_0/T_1 spin-orbit coupling is dominated by the geminal orbital pairs located on oxygen. The calculated value for the enone reactant is 0.01 cm^{-1} at the CASSCF(6,6) level. It increases to 80 cm^{-1} at the CASSCF(8,8) level, after the proper oxygen orbitals are included and hardly changes any further when the active space is enlarged to (10,10). The same conclusion concerning the importance of the oxygen lone-pair orbitals can be drawn from the NHO analysis. A more general discussion of spin-orbit coupling based on NHO was published in a subsequent paper by the same authors.[71] Of course, at the biradicaloid geometries the S_0/T_1 gap is much smaller, making it hard to estimate where intersystem crossing will be the fastest. The authors seem to have no doubt that it will be at the reactant and product geometries. In the two biradical intermediates, where the two unpaired (π) electrons are distant, spin-orbit coupling is two orders of magnitude smaller than in the starting material and the bicyclic product. The final product formation was proposed to result from T_1 to S_0 intersystem crossing in the region past the second biradical, where the S_0/T_1 spin-orbit coupling increases as the product geometry is approached.

17

3.9.6 Delocalized Biradicals

Although negative-ion photoelectron spectroscopy[79] showed that the ground state of tetramethyleneethane (**18**) in gas phase is a singlet, ~2 kcal/mol below the lowest triplet state, matrix isolation EPR experiments[80,81] showed that under these conditions the ground state is a triplet, or possibly the two states are degenerate, and yielded a zero E parameter that implies D_{2d} symmetry (orthogonal twist). Various calculations (CASSCF,[82] CISD,[83–85] REKS-DFT,[86] CASMP2[86]) placed the singlet state 1 to 2 kcal/mol below the triplet, but an CISD calculation[84] with a larger basis set predicted the triplet state 0.1 kcal/mol below the singlet. An attempt to determine the twist angle in the triplet by comparison of observed[81] and calculated[85] zero-field splitting parameters failed. The D parameter calculated from H^{SS} alone (0.026 cm^{-1}) agrees well with the experimental value but is insensitive to the torsion angle, and the predicted E parameter is considerably overestimated as a result of the limited amount of electron correlation included in the calculation. The spin–orbit correction to the zero-field splitting parameters, computed in the CASSCF(6,6) approximation, does not change the situation.

The $\langle S_0 \,|H^{SO}|\, T_1 \rangle$ matrix element is small, less than 0.08 cm^{-1} using the full Breit–Pauli H^{SO}[85] or 0.05 cm^{-1} using the one-electron operator with effective nuclear charge.[86]

Zero-field splitting in both the T_1 and the T_2 states of *m*-xylylene (**19**), one of the few very large molecules for which such information is available experimentally for a higher triplet state, was examined by Havlas and Michl[87] using both H^{SS}

18

19

and H^{SO} at the CASSCF(6,6)/cc-pVDZ level. The spin-orbit contributions were negligible and the agreement with experiment was excellent. The authors noted that the nonzero spin-orbit coupling elements between the T_x sublevel of triplets with an even number of π electrons and the singlets with the same number of π electrons, which do not contain one-center terms for symmetry reasons and are small, as expected from El Sayed's rule,[44] do not exhibit the usual proportionality between the one-electron and two-electron parts and the total spin-orbit coupling effect. The cancellation is more pronounced than usual, and the use of the standard effective charge approximation would lead to a considerable error.

Kondo et al.[88] suggested a possible mechanism for the spin transitions in bis(phenylmethylenyl)[2.2]paracyclophanes (**20**). Spin-orbit coupling was calculated for models of two diphenylmethylenes, which mimicked different configurations of substituted [2.2]paracyclophanes, using a one-electron H^{SO} operator with effective nuclear charges and CASSCF(4,4) wave functions. The system of two weakly coupled carbenes has close-lying singlet (S), triplet (T), and quintet (Q) states. Spin-orbit coupling between the S_1 and T_2 states, the T_2 and Q_1 states, the Q_2 and T_2 states, and the T_2 and S_0 states was found to be strong (9.7, 6.7, 3.9, and 3.3 cm^{-1}, respectively), demonstrating that the conversion between the low-spin (singlet) and the high-spin (quintet) states can occur via the first excited triplet (T_2).

20

3.10 SUMMARY

The introductory part of this chapter, which provides a survey of Hamiltonian operators and wave function and energy evaluation methods that are currently in use for the calculation of spin-orbit coupling, is fairly general. In particular, we attempt to incorporate the spin-dipole spin-dipole interaction operator as an equal partner, whose appreciation is necessary for spin-orbit coupling calculations on organic molecules.

The remainder of the chapter is strongly biased toward the specific needs of organic photochemists. After a fairly detailed discussion of the spin-orbit coupling of triplets with singlets, it devotes additional attention to the particular case of S/T spin-orbit coupling in triplet biradicals, which are of special interest to photochemists but few others. The selection of examples of recent spin-orbit calculations that follows is also heavily slanted toward organic photochemistry.

Theory can now provide much valuable guidance and interpretive assistance to the mechanistic photochemist, and the evaluation of spin-orbit coupling matrix elements has become relatively routine. For the fairly large molecules of common interest, the level of calculation cannot be very high. In molecules composed of light atoms, the use of effective charges is, however, probably best avoided, and a case is pointed out in which its results are incorrect. It seems that the mean-field approximation is a superior way to simplify the computational effort. The use of at least a double zeta basis set with a method of wave function computation that includes electron correlation, such as CASSCF, appears to be imperative even for calculations that are meant to provide only semiquantitative results. The once-prevalent degenerate perturbation theory is now obsolete for quantitative work but will presumably remain in use for qualitative interpretations.

In addition to studies in which the numerical values of spin-orbit coupling energies are the primary objective, there is room for additional work on the relation of molecular structure and spin-orbit coupling. The operation of the heavy atom effect would seem to deserve considerable additional attention, as would the construction of new simple models for interpretive work.

ACKNOWLEDGMENTS

Our work on spin-orbit coupling has been supported by the U.S. National Science Foundation (CHE-0140478), by project LN 00A032, and by the research project Z4 055 905 from the Ministry of Education of the Czech Republic.

REFERENCES

1. Langhoff, S.R. and Kern, C.W., in *Applications of Electronic Structure Theory,* Shaefer, H.F. III, Ed., Plenum Press, New York, 1977, pp. 381–437.

2. Richards, W.G., Trivedi, H.P, and Cooper, D.L., *Spin-Orbit Coupling in Molecules,* Clarendon Press, Oxford, 1981.

3. Ermler, W.C., Ross, R.B., and Christiansen, P.A., *Adv. Quantum Chem.,* 19, 139–182, 1988.

4. Yarkony, D.R., *Int. Rev. Phys. Chem.,* 11, 195–242, 1992.

5. Hess, B.A., Marian, C.M., and Peyerimhoff, S.D., in *Modern Electronic Structure Theory,* Yarkony, D.R., Ed., World Scientific, Singapore, 1995, pp. 152–278.

6. Ågren, H., Vahtras, O., and Minaev, B., *Adv. Quantum Chem.,* 27, 71–162, 1996.

7. Marian, C.M., in *Reviews in Computational Chemistry,* Vol. 17, Lipkowitz, K.B. and Boyd, D.B., Eds., Wiley-VCH, New York, 2001, pp. 99–204.

8. McClure, D.S., *J. Chem. Phys.,* 17, 905–909, 1949.

9. Fedorov, D.G. and Gordon, M.S., *J. Chem. Phys.,* 112, 5611–5623, 2000.

10. Nicklass, A., Peterson, K.A., Berning, A., Werner, H.-J., and Knowles, P.J., *J. Chem. Phys.,* 112, 5624–5632, 2000.

11. Havlas, Z. and Michl, J., *J. Am. Chem. Soc.,* 124, 5606–5607, 2002.

12. Sjøvoll, M., Gropen, O., and Olsen, J., *Theor. Chem. Acc.*, 97, 301–312, 1997.

13. Hess, B.A., *Phys. Rev. A*, 32, 756–763, 1985.

14. Samzow, R. and Hess, B.A., *J. Chem. Phys., Lett.*, 184, 491–496, 1991.

15. Marian, C.M. and Wahlgren, U., *Chem. Phys. Lett.*, 251, 357–364, 1996.

16. Hess, B.A., Marian, C.M., Wahlgren, U., and Gropen, O., *Chem. Phys. Lett.*, 251, 365–371, 1996.

17. Tatchen, J. and Marian, C.M., *Chem. Phys. Lett.*, 313, 351–357, 1999.

18. Koseki, S., Gordon, M.S., Schmidt, M.W., and Matsunaga, N., *J. Phys. Chem.*, 99, 12764–12772, 1995.

19. Koseki, S., Schimdt, M.W., and Gordon, M.S., *J. Phys. Chem. A*, 102, 10430–10435, 1998.

20. Davidson, E.R., in *Methods and Techniques in Computational Chemistry: METECC-94*, Vol. B. *Medium Size Systems*, Clementi, E., Ed., STEF, Cagliari, 1993, pp. 209–274.

21. Vahtras, O., Loboda, O., Minaev, B., Ågren, H., and Ruud, K., *Chem. Phys.*, 279, 133–142, 2002.

22. Havriliak, S.J. and Yarkony, D.R., *J. Chem. Phys.*, 83, 1168–1172, 1985.

23. Yarkony, D.R., *J. Chem. Phys.*, 84, 2075–2078, 1986.

24. Jensen, J.O. and Yarkony, D.R., *Chem. Phys. Lett.*, 141, 391–396, 1987.

25. Ågren, H., Vahtras, O., and Minaev, B., *Adv. Quantum Chem.*, 27, 71–162, 1996.

26. Vahtras, O., Ågren, H., Jørgensen, P., and Jensen, H.J.Aa., Helgaker, T., and Olsen, J., *Chem. Phys.*, 96, 2118–2126, 1992.

27. Vahtras, O., Ågren, H., Jørgensen, P., Jensen, H.J.Aa., Helgaker, T., and Olsen J., *Chem. Phys,*. 97, 9178–9187, 1992.

28. Christiansen, O., Gauss, J., and Schimmelpfennig, B., *Phys. Chem. Chem. Phys.*, 2, 965–971, 2000.

29. Furlani, T.R. and King, H.F., *J. Chem. Phys.*, 82, 5577–5583, 1985.

30. Fedorov, D.G. and Finley, J.P., *Phys. Rev. A*, 64, 042502, 2001.

31. Teichteil, C., Pelissier, M., and Spiegelmann, F., *Chem. Phys.*, 81, 273–282, 1983.

32. Berning, A., Schweizer, M., Werner, H.-J., Knowles, P.J., and Palmieri, P., *Mol. Phys.*, 98, 1823–1833, 2000.

33. Kleinschmidt, M., Tatchen, J., and Marian, C.M., *J. Comput. Chem.*, 23, 824–833, 2002.

34. Ilias̆, M., Kellö, V., Visscher, L., and Schimmelpfennig, B., *J. Chem. Phys.*, 115, 9667–9674, 2001.

35. Wahlgren, U., Sjøvoll, M., Fagerli, H., Gropen, O., and Schimmelpfennig, B., *Theor. Chem. Acc.*, 97, 324–330, 1997.

36. Sjøvoll, M., Fagerli, H., Gropen, O., Almlöf, J., Schimmelpfennig, B., and Wahlgren, U., *Theor. Chem. Acc.*, 99, 1–7, 1998.

37. Fleig, T., Olsen, J., and Marian, C.M., *J. Chem. Phys.*, 114, 4775–4790, 2001.

38. Havlas, Z., Downing, J.W., and Michl, J., *J. Phys. Chem. A*, 102, 5681–5692, 1998.

39. Reed, A.E., Curtiss, L.A., and Weinhold, F., *Chem. Rev.*, 88, 899–926, 1988.

40. Zimmerman, H.E. and Kutateladze, A.G., *J. Am. Chem. Soc.*, 118, 249–250, 1996.

41. Michl J., *J. Am. Chem. Soc.*, 118, 3568–3579, 1996.

42. Salem, L. and Rowland, C., *Angew Chem. Int. Ed. Engl.*, 11, 92–111, 1972.

43. Kita, F., Nau, W.M., Adam, W., and Wirz J., *J. Am. Chem. Soc.*, 117, 8670–8671, 1995.

44. El-Sayed, M., *J. Chem. Phys.*, 38, 2834–2838, 1963.

45. Danovich, D., Marian, C.M., Neuheuser, T., Peyerimhoff, S., and Shaik, S., *J. Phys. Chem. A*, 102, 5923–5936, 1998.

46. Filatov, M., Shaik, S., Woeller, M., Grimme, S., and Peyerimhoff, S.D., *Chem. Phys. Lett.*, 316, 135–140, 2000.

47. Woeller, M., Grimme, S., Peyerimhoff, S.D., Danovich, D., Filatov, M., and Shaik, S., *J. Phys. Chem. A,* 104, 5366–5373, 2000.

48. Harvey, J.N., Grimme, S., Woeller, M., Peyerimhoff, S.D., Danovich, D., and Shaik, S., *Chem. Phys. Lett.,* 322, 358–362, 2000.

49. Grimme, S., Woeller, M., Peyerimhoff, S.D., Danovich, D., and Shaik, S., *Chem. Phys. Lett.,* 287, 601–607, 1998.

50. Toniolo, A. and Persico, M., *J. Chem. Phys.,* 115, 1817–1827, 2001.

51. Arenas, J.F., Marcos, J.I., Otero, J.C., Tocón, I.L., and Soto, J., *Int. J. Quantum Chem.,* 84, 241–248, 2001.

52. Perić, M., Marian, C.M., and Peyerimhoff, S.D., *J. Chem. Phys.,* 114, 6086–6099, 2001.

53. Tatchen, J. and Marian, C.M., *Chem. Phys. Lett.,* 313, 351–357, 1999.

54. Nakajima, T. and Kato, S., *J. Phys. Chem. A.,* 105, 10657–10663, 2001.

55. Vergrart, P.J. and van der Waals, J.H., *Chem. Phys. Lett.,* 36, 283–289, 1975.

56. Cai, Z.-L. and Reimers, J.R., *J. Phys. Chem. A,* 104, 8389–8408, 2000.

57. Tatchen, J., Waletzke, M., Marian, C.M., and Grimme, S., *Chem. Phys.,* 264, 245–254, 2001.

58. Grimme, S. and Waletzke, M., *J. Chem. Phys.,* 111, 5645–5655, 1999.

59. Bearpark, M.J., Olivucci, M., Wisley, S., Bernardi, F., and Robb, M.A., *J. Am. Chem. Soc.,* 117, 6944–6953, 1995.

60. Lewis, F.D. and Bassani, D.M., *J. Am. Chem. Soc.,* 115, 7523–7524, 1993.

61. Lewis, F.D., Bassani, D.M., Caldwell, R.A., and Unett, D.J., *J. Am. Chem. Soc.,* 116, 10477–10485, 1994.

62. Harvey, J.N., Aschi, M., Schwarz, H., and Koch, W., *Theor. Chem. Acc.,* 99, 95–99, 1998.

63. Aschi, M. and Harvey, J.N., *J. Chem. Soc. Perkin Trans.,* 2, 1059–1062, 1999.

64. Shiota, Y., Kondo, M., and Yoshizawa, K., *J. Chem. Phys.*, 115, 9243–9254, 2001.

65. Sekiguchi, A., Tanaka, T., Ichinohe, M., Akiyama, K., and Tero-Kubota, S., *J. Am. Chem. Soc.*, 125, 4962–4963, 2003.

66. Matsunaga, N., Koseki, S., and Gordon, M.S., *J. Chem. Phys.*, 104, 7988–7996, 1996.

67. Yu, H.-G., Gonzalez-Lezana, T., Marr, A.J., Muckerman, J.T., and Sears, T.J., *J. Chem. Phys.*, 115, 5433–5444, 2001.

68. Havlas, Z. and Michl, J., *Collect. Czech. Chem. Commun.*, 63, 1485–1497, 1998.

69. Johnson, W.T.G., Sullivan, M.B., and Cramer, C.J., *Int. J. Quantum Chem.*, 85, 492–508, 2001.

70. Ichinose, N., Mizuno, K., Otsuji, Y., Caldwell, R.A., and Helms, A.M., *J. Org. Chem.*, 63, 3176–3184, 1998.

71. Zimmerman, H.E. and Kutateladze, A.G. *J. Am. Chem. Soc.*, 118:249-250, 1996.

72. Furlani, T.R. and King, H.F., *J.* Chem. Phys. 82:5577-5583, 1985.

73. Turro, N.J., Cherry, W.R., Mirbach, M.F., and Mirbach, M.J.,. *J. Am. Chem. Soc.*, 99, 7388–7390, 1977.

74. Helms, A.M. and Caldwell, R.A., *J. Am. Chem. Soc.*, 117, 358–361, 1995.

75. Kutateladze, A.G., *J. Am. Chem. Soc.*, 123, 9279–9282, 2001.

76. Griesbeck, A.G. and Fiege, M., in *Molecular and Supramolecular Photochemistry*, Vol. 6, Ramamurthy, V. and Schanze, K.S., Eds, Marcel Dekker, New York, 2000, pp., 33–100.

77. Bertrand, C., Bouquant, J., Pete, J.P., and Humbel, S., *J. Mol. Struct. (THEOCHEM)*, 538, 165–177, 2001.

78. Zimmerman, H.E. and Kutateladze, A.G., *J. Org. Chem.*, 60, 6008–6009, 1995.

79. Clifford, E.P., Wenthold, P.G., Lineberger, W.C., Ellison, G.B., Wang, C.X., Grabowski, J.J., Vila, F., and Jordan, K.D., *J. Chem. Soc. Perkin Trans.*, 2, 1015–1022, 1998.

80. Dowd, P., *J. Am. Chem. Soc.*, 92, 1066–1068, 1970.

81. Dowd, P., Chang, W., and Paik, Y.H., *J. Am. Chem. Soc.,* 108, 7416–7417, 1986.

82. Nachtigall, P. and Jordan, K.D., *J. Am. Chem. Soc.,* 114, 4743–4747, 1992.

83. Du, P. and Borden, W.T., *J. Am. Chem. Soc.,* 109, 930–931, 1987.

84. Nachtigall, P. and Jordan, K.D., *J. Am. Chem. Soc.,* 115, 270–271, 1993.

85. Havlas, Z. and Michl, J., *J. Mol. Struct. (TEOCHEM),* 398, 281–291, 1997.

86. Filatov, M. and Shaik, S., *J. Phys. Chem. A,* 103, 8885–8889, 1999.

87. Havlas, Z. and Michl, J., *J. Chem. Soc. Perkin Trans.,* 2, 2299–2303, 1999.

88. Kondo, M., Shiota, Y., and Yoshizawa, K., *J. Phys. Chem. A,* 106, 7915–7920, 2002.

4

Photochemistry from First Principles and Direct Dynamics

A. TONIOLO,
BENJAMIN G. LEVINE,
ALEXIS L. THOMPSON,
JASON QUENNEVILLE,
M. BEN-NUN, JANE M. OWENS,
SETH OLSEN, LESLIE MANOHAR,
AND TODD J. MARTINEZ

Department of Chemistry and
The Beckman Institute,
University of Illinois at
Urbana-Champaign,
600 S. Mathews,
Urbana, IL 61801

CONTENTS

4.1. INTRODUCTION

As amply demonstrated by other contributions in this volume, current thinking about photochemical mechanisms is strongly influenced by conical intersections.[1–5] The lowest-lying of these points of exact degeneracy between two or more[6] electronic states can be thought of as photochemical analogs of the transition state in a thermal reaction.[5] Hence, the widespread efforts to develop efficient algorithms for locating these points[7,8] and the efforts to elucidate them for photochemical reactions are not surprising. However, one must not forget that there are some crucial differences between intersections and transition states. Transition states in thermal reactions are generally reached from below by rare trajectories with sufficient energy to scale the barrier. In contrast, intersections in photochemical reactions are usually reached from above by trajectories resembling downhill skiers careening out of control. One can easily imagine that it will therefore be very dangerous to apply thermal reasoning to photochemical reactions, and, in fact, the lowest energy point along an intersection seam could easily have little or no relevance to the photochemistry.

Molecular dynamics always plays an essential role in chemical reactions, but the considerations above lead to the expectation that it is even more important in photochemical reactions than in thermal reactions. This significantly complicates the understanding of photochemical reactions. First, multiple potential-energy surfaces (PESs) and their couplings, as opposed to a few stationary points (which would be sufficient for a transition-state theory estimate of a thermal reaction rate, for example), are required. Second, molecular dynamics must be carried out on these PESs, a problem that can be as challenging as their determination. The two problems have traditionally been treated separately for both thermal and photochemical reactions. However, a major shift in theoretical chemistry over the past decade has been to combine the solution of the electronic structure and molecular dynamics problems.[9-14] The introduction of Car–Parrinello molecular dynamics[14] accelerated this transition, focusing on classical molecular dynamics and a density functional theory (DFT)[15] description of electronic structure. Both the classical nature of the nuclei and the DFT treatment of the electronic structure prevented the application of these methods to photochemical dynamics. Attempts to extend DFT to electronically excited states have a long history,[16] but it is only recently that these have begun to meet with success.[17-22] Although a few preliminary applications of Car–Parrinello dynamics on electronic excited states have been discussed,[23,24] the most experience with excited electronic states is certainly in the context of conventional *ab initio* methods, and most first principles "on-the-fly" treatments of photochemical reaction dynamics have used *ab initio* methods.[25-33]

In this chapter, we will first discuss some of the important features of dynamical methods that purport to treat multiple electronic states, that is, beyond the Born–Oppenheimer approximation which separates electronic and nuclear motion. Then we will discuss the various possibilities for representing the PESs and their couplings, from *ab initio* to semiempirical and hybrid quantum mechanical and molecular mechanical methods. In both of these discussions, the emphasis will be on approaches that are suited to simultaneous solution of the dynamics and electronic structure problems. Finally, we illustrate with several example applications.

4.2. NONADIABATIC DYNAMICS

The central equation representing the time evolution of the correlated motion of nuclei and electrons in a molecule is the time-dependent Schrödinger equation (TDSE):

$$i\hbar \frac{\partial \psi(\mathbf{q}, \mathbf{Q})}{\partial t} = H\psi(\mathbf{q}, \mathbf{Q}) \tag{4.1}$$

where $\psi(\mathbf{q}, \mathbf{Q})$ is the wave function describing the state of the system composed of N electrons ($\mathbf{q}$ collects all their space and spin coordinates) and M nuclei (with space and spin coordinates $\mathbf{Q}$). Here, H is the nonrelativistic molecular Hamiltonian operator:

$$H = \sum_s^M \frac{P_s^2}{2M_s} + \frac{1}{2m} \sum_i^N p_i^2 - e \sum_s^M \sum_i^N \frac{Z_s}{r_{si}} + \sum_{t<s}^M \frac{Z_s Z_t}{r_{st}} + e^2 \sum_{i<j}^N \frac{1}{r_{ij}} \tag{4.2}$$

and the indices t and s refer to nuclei, whereas i and j refer to electrons. M_s and Z_s represent nuclear masses and charges respectively, and m and e are the analogous electronic quantities. P and p represent nuclear and electronic linear momentum operators, whereas distances are denoted by r, appropriately subscripted.

Even for the simplest molecules, Equation 4.1 is very difficult to solve. Therefore, some approximations are needed to use it for practical calculations. In the following, we present an overview of some of the methods commonly exploited for the study of molecular dynamics, especially in the context of photochemistry. We give an overview of semiclassical and quantum mechanical approaches to dynamics, making use of some well-known concepts and methods such as factorization of the wave function and diabatic states, which will be briefly introduced.

4.2.1   Born–Oppenheimer Approximation

We start by defining a set of coordinates from Cartesian displacements about some reference geometry:

$$Q_r = \sum_s^M L_{rs} \sqrt{M_s \Delta x_s} \ . \tag{4.3}$$

Three of the Q_r coordinates can be identified as translations and three more as rotations. The other M-6 coordinates represent true internal coordinates. Choosing the matrix $\mathbf{L}$ of Equation 4.3 to be unitary, the Hamiltonian operator can be rewritten in the translating-rotating molecular frame as

$$H = \sum_s^{M-6} \frac{1}{2} P_s^2 + \frac{1}{2m} \sum_i^N p_i^2 + V\left(\mathbf{q}, \mathbf{Q}\right) \ . \tag{4.4}$$

The potential-energy term couples nuclear and electronic coordinates and cannot be neglected. Therefore, electronic and nuclear motion should be coupled. However, because nuclei are much heavier than electrons, the Born–Oppenheimer approximation[34] is generally valid. In the Born–Oppenheimer approximation, the molecular wave function is written as a sum of products of electronic wave functions $\Psi_k(\mathbf{q};\mathbf{Q})$, explicitly dependent on electronic coordinates and parametrically dependent on nuclear coordinates, and nuclear wave functions $\chi_u(\mathbf{Q})$:

$$\psi(\mathbf{q},\mathbf{Q}) = \sum_{Ku} \psi_K(\mathbf{q};\mathbf{Q})\chi_u(\mathbf{Q}). \tag{4.5}$$

Then the Hamiltonian matrix elements become

$$\langle \psi_K \chi_u | H | \psi_L \chi_v \rangle = \int \chi_u^*(\mathbf{Q}) \Bigg[\Big\langle \psi_K \Big| \frac{1}{2m} \sum_i p_i^2 + V(\mathbf{q},\mathbf{Q}) \Big| \psi_L \Big\rangle +$$

$$-\frac{\hbar^2}{2} \sum_s^{M-6} \Bigg(\langle \psi_K | \psi_L \rangle \frac{\partial^2}{\partial Q_s^2} + 2\Big\langle \psi_K \Big| \frac{\partial}{\partial Q_s} \Big| \psi_L \Big\rangle \frac{\partial}{\partial Q_s} + \tag{4.6}$$

$$\Big\langle \psi_K \Big| \frac{\partial^2}{\partial Q_s^2} \Big| \psi_L \Big\rangle \Bigg) \Bigg] \chi_v(\mathbf{Q})d\mathbf{Q}.$$

If these Born–Oppenheimer product wave functions are to approximate Hamiltonian eigenvectors, we have to minimize all off-diagonal matrix elements ($K \neq L$ and $u \neq v$). To this end, the electronic wave functions are chosen to be eigenvectors of a part of the Hamiltonian operator called the electronic Hamiltonian H_{el} (adiabatic states):

$$H_{el} = \frac{1}{2m} \sum_i p_i^2 + V(\mathbf{q},\mathbf{Q})$$

$$\tag{4.7}$$

$$H_{el}|\psi_K\rangle = U_K(\mathbf{Q})|\psi_K\rangle$$

Electronic wave functions are then defined for each fixed nuclear configuration and depend parametrically on nuclear coordinates. Now the Hamiltonian matrix elements become

$$\langle \psi_K \chi_u | H | \psi_L \chi_v \rangle = \int \chi_u^*(\mathbf{Q}) \Bigg[\delta_{KL} \Bigg(U_K(Q) - \frac{\hbar^2}{2} \sum_s^{M-6} \frac{\partial^2}{\partial Q_s^2} \Bigg)$$

$$\tag{4.8}$$

$$-\frac{\hbar^2}{2} \sum_s^{M-6} \Bigg(2h_{KL}^{(s)} \frac{\partial}{\partial Q_s} + t_{KL}^{(s)} \Bigg) \Bigg] \chi_v(Q)d\mathbf{Q}.$$

where $h_{KL}^{(s)} = \left\langle \psi_K \left| \dfrac{\partial}{\partial Q_s} \right| \psi_L \right\rangle$ is the nonadiabatic coupling vector

and $t_{KL}^{(s)} = \left\langle \psi_K \left| \dfrac{\partial^2}{\partial Q_s^2} \right| \psi_L \right\rangle$ is a small diagonal term that is usu-

ally neglected. Now we require the nuclear wave functions to be eigenfunctions of the vibrational Hamiltonian:

$$H_{vib} = U_K(\mathbf{Q}) - \frac{\hbar^2}{2} \sum_s^{M-6} \frac{\partial^2}{\partial Q_s^2}$$

$$H_{vib} \left| \chi_{Ku} \right\rangle = E_{Ku} \left| \chi_{Ku} \right\rangle.$$

(4.9)

In the vibrational Hamiltonian, the electronic energy $U_K(\mathbf{Q})$ plays the role of the PES where nuclei move; therefore, the nuclear wave functions have to be labeled with the index of the electronic state to which it belongs. Note that the Born–Oppenheimer wave function does not diagonalize the molecular Hamiltonian, because there are nonvanishing off-diagonal terms represented by the nonadiabatic coupling. As a consequence, the Born–Oppenheimer states cannot be pure states, in which the molecule resides indeterminately, but they are each coupled through the nonadiabatic coupling terms and evolve in time.

4.2.2 Conical Intersections and Diabatic States

The Born–Oppenheimer approximation is valid when the non-adiabatic coupling terms are very small, as is generally the case for most organic molecules near equilibrium geometries in their ground electronic state. However, it fails when the energy gap between two electronic PESs vanishes, as can be shown by the Hellmann–Feynman formula for nonadiabatic coupling:

$$h_{KL}^{(s)} = \frac{\left\langle \psi_K \left| \dfrac{\partial H_{el}}{\partial Q_s} \right| \psi_L \right\rangle}{U_L - U_K} \qquad (4.10)$$

which shows that the nonadiabatic coupling becomes large when the energy gap becomes small unless the numerator also vanishes. The possible existence of nuclear geometries in which two different electronic levels have the same energy is subject to the noncrossing rule: It is easy to show[35] that if N is the number of internal coordinates, two PESs can coincide in a space of at most dimension N-2 when the states have the same symmetry, that is, belong to the same irreducible representation of the symmetry spatial group. Therefore, in the case of a single internal coordinate, such as in a diatomic molecule, two states of the same symmetry can get close but never coincide. This is known as an avoided crossing. However, when the number of internal coordinates is greater than one, as is the case for all polyatomic molecules, two eigenvalues of the electronic Hamiltonian can be degenerate in a space of dimension N-2, often called a "seam." In the two remaining dimensions, known as the "branching plane," the two surfaces must separate.[36] The points of degeneracy along the seam are often called conical intersections,[1-4] although, strictly speaking, the adjective "conical" should be reserved for cases that have a conical shape in the immediate vicinity of the point of degeneracy.

It is important to note that the "N-2" rule provides only an upper bound to the dimensionality of the nuclear subspace that maintains electronic degeneracy. There is no good reason to suppose that there are only two nuclear displacement coordinates that break the electronic degeneracy around an intersection, and several analytic examples where the branching plane is greater than two dimensional have been provided recently.[37,38] The lowest energy point in a seam is called a minimal energy conical intersection (MECI), and there exist a number of algorithms to locate such points efficiently.[7,8] When the branching plane is two dimensional, it can be shown that the nonadiabatic coupling (h) and difference gradient (g)

vectors are the nuclear displacements that break the degeneracy in first order:

$$\vec{h}_{KL} = \left\langle \psi_K \left| \frac{\partial}{\partial \vec{Q}} \right| \psi_L \right\rangle \tag{4.11}$$

$$\vec{g}_{KL} = \frac{\partial}{\partial \vec{Q}} \left(\left\langle \psi_K \left| H_{el} \right| \psi_K \right\rangle - \left\langle \psi_L \left| H_{el} \right| \psi_L \right\rangle \right). \tag{4.12}$$

Around a seam, the PESs can undergo rapid variations, and when the intersections are conical they will have cusps. The adiabatic electronic wave functions will change suddenly with nuclear displacements in the branching plane, and the nonadiabatic coupling can diverge, as indicated by the vanishing energy gap in the denominator of Equation 4.10. These features are all quite inconvenient, and thus it is common to introduce diabatic states that are electronic states that only weakly depend on nuclear configuration. This leads to small nonadiabatic coupling matrix elements, but the price to be paid is that they no longer diagonalize the electronic Hamiltonian. Thus, the transition from adiabatic to diabatic electronic states shifts the coupling between electronic states from the kinetic energy part of the Hamiltonian to the potential-energy part. The most rigorous definition of diabatic states η_K is the condition that the nonadiabatic coupling matrix elements vanish:

$$h_{KL}^{(D,s)} = \left\langle \eta_K \left| \frac{\partial}{\partial Q_s} \right| \eta_L \right\rangle = 0. \tag{4.13}$$

As a consequence, in the basis of electronic diabatic states, the molecular Hamiltonian can be written as

$$\left\langle \psi_K \chi_{Kv} \left| H \right| \psi_L \chi_{Lv} \right\rangle = \left\langle \chi_{Ku} \left| H_{KL} \right| \chi_{Lv} \right\rangle \tag{4.14}$$

where $H_{KL} = \left\langle \psi_K \left| H_{el} \right| \psi_L \right\rangle$. The consequence of the definition of Equation 4.13, applied to a complete set of diabatic states, is that they are completely independent of the nuclear coordinates. As a consequence, their nature remains unchanged when the nuclei move:

$$\frac{\partial}{\partial Q_s}\left|\eta_K\right\rangle = \sum_L \left|\eta_L\right\rangle\left\langle\eta_L\left|\frac{\partial}{\partial Q_s}\right|\eta_K\right\rangle = 0. \qquad (4.15)$$

However, this strict (and useless) definition can be applied only in the case of a complete set of diabatic states. When the study is restricted to a finite number of states of interest, this rule does not apply and the diabatic states must depend on the nuclear coordinates. They can be obtained from the corresponding adiabatic states through a unitary transformation:

$$\left|\psi_K\right\rangle = \sum_j \left|\eta_j\right\rangle C_{jK}. \qquad (4.16)$$

Now, recalling the definition of Equation 4.13 we can rewrite the nonadiabatic coupling element between adiabatic states K and L along the nuclear coordinate s:

$$h_{KL}^{(s)} = \left\langle\psi_K\left|\frac{\partial}{\partial Q_s}\right|\psi_L\right\rangle = \sum_{ij} C_{Ki}^* C_{jL}\left\langle\eta_K\left|\frac{\partial}{\partial Q_s}\right|\eta_L\right\rangle +$$

$$\sum_{ij} C_{Ki}^* \frac{\partial C_{jL}}{\partial Q_s}\delta_{KL} = \sum_i C_{Ki}^* \frac{\partial C_{iL}}{\partial Q_s}. \qquad (4.17)$$

In matrix formalism:

$$\frac{\partial C}{\partial Q_s} = \mathbf{C}\mathbf{g}^{(s)}. \qquad (4.18)$$

This problem can be fulfilled only in the case of one internal coordinate, but a general solution does not exist, essentially because the number of conditions exceeds the number of variables.[39] The single internal coordinate for which a diabatic transformation is exact can be the path traced by a classical trajectory; the idea of relating the adiabatic to diabatic transformation to a given path in the coordinate space can be traced back to the work of Baer,[40] Gadéa and Pélissier,[41] Petsalakis et al.,[42] and more recently Granucci and coworkers[43] with their local diabatization. Here, the unitary matrix transforming the adiabatic into the diabatic basis set is obtained directly along the classical nuclear path. In this approach, the diabatic states are chosen so as to eliminate the dynamic coupling in a given subspace of electronic states of interest, under the assumption that the coupling with the external space is negligible. The locally diabatic basis is obtained at each time step: at the beginning of the step t_0, the matrix $\mathbf{C}$ is chosen to be the identity matrix, so $\left| \eta_1 \left(t_0 \right) \right\rangle = \left| \psi_1 \left(t_0 \right) \right\rangle$. The $\mathbf{C}$ matrix at time Δt is related to the overlap between adiabatic functions at the beginning and at the end of the time step:

$$S_{KL} = \left\langle \psi_K \left(t_0 \right) \middle| \psi_L \left(t_0 + \Delta t \right) \right\rangle = \sum_{J}^{N} \left\langle \eta_K \left(t_0 \right) \middle| \eta_J \left(t_0 + \Delta t \right) \right\rangle C_{ij} \left(t_0 + \Delta t \right).$$

$$(4.19)$$

For each time step, it is possible to obtain the transformation matrix that diabatizes the electronic functions in the given interval. In this way, the transformation matrix from the adiabatic states at time t_0 to diabatic states at time t is simply the product of the intermediate unitary matrices.

4.2.3 Quantum Dynamics and Path Integrals

Let us define the temporal evolution operator $U(t_1, t_0)$ as the operator that, applied to the molecular wave function at time t_0, produces the wave function at time t_1:

$$\left| \Psi \left(t_1 \right) \right\rangle = U \left(t_1, t_0 \right) \left| \Psi \left(t_0 \right) \right\rangle. \qquad (4.20)$$

In order to preserve the normalization of the wave function, the temporal evolution operator has to be unitary. From the TDSE, for an infinitesimal time step Δt:

$$U\left(t_0 + \varepsilon, t_0\right) = e^{-iH\varepsilon/\hbar} \tag{4.21}$$

We introduce the propagator $K\left(\mathbf{x}_0 + \delta\mathbf{x}, t_0 + \varepsilon; \mathbf{x}_0, t_0\right)$ that gives the probability for the molecular system to move electronic and nuclear coordinates by the vector $\delta\mathbf{x}$ in the time ε:

$$K\left(\mathbf{x}_0 + \delta\mathbf{x}, t_0 + \varepsilon; \mathbf{x}_0, t_0\right) = \sum_J \left\langle \mathbf{x}_0 + \delta\mathbf{x} | \Psi_J \right\rangle e^{-iH\varepsilon/\hbar} \left\langle \Psi_J | \mathbf{x}_0 \right\rangle =$$
$$\left\langle \mathbf{x}_0 + \delta\mathbf{x} | e^{-iH\varepsilon/\hbar} | \mathbf{x}_0 \right\rangle = \left\langle \mathbf{x}_0 + \delta\mathbf{x}, t_0 + \varepsilon; \mathbf{x}_0, t_0 \right\rangle \tag{4.22}$$

Being interested in molecular systems, we will distinguish between the two sets of nuclear and electronic coordinates, $\mathbf{Q}$ and $\mathbf{q}$. Moreover, to focus on transitions between electronic states, we define the "reduced propagator," where we integrate out the electronic coordinates:

$$K_{\beta\alpha}\left(\mathbf{x}_0 + \delta\mathbf{x}, t_0 + \varepsilon; \mathbf{x}_0, t_0\right) = \int d\left(q_0 + \delta q\right) \int dq_0 \Psi_\beta^* \left(q_0 + \delta q, \mathbf{Q}_0 + \delta\mathbf{Q}\right) \cdot$$
$$K\left(\mathbf{x}_0 + \delta\mathbf{x}, t_0 + \varepsilon; \mathbf{x}_0, t_0\right) \Psi_\alpha \left(q_0, \mathbf{Q}_0\right) =$$
$$\left\langle \mathbf{Q}_0 + \delta\mathbf{Q}, \beta | e^{-iH\varepsilon/\hbar} | \mathbf{Q}_0, \alpha \right\rangle . \tag{4.23}$$

The ket $\left| \mathbf{Q}_0, \alpha \right\rangle$ is a position eigenstate for the nuclei and a Hamiltonian eigenstate for the electrons. In the case of finite time, we can always divide it into N small time steps ε and integrate over the intermediate nuclear positions:

$$K_{\beta\alpha}\left(\mathbf{Q}_N, t_N; \mathbf{Q}_0, t_0\right) = \int d\mathbf{Q}_{N-1} \int d\mathbf{Q}_{N-2} \cdots d\mathbf{Q}_1 \left\langle \mathbf{Q}_N \beta | e^{-iH\varepsilon/\hbar} | \mathbf{Q}_{N-1} \right\rangle \cdot$$
$$\left\langle \mathbf{Q}_{N-1} | e^{-iH\varepsilon/\hbar} | \mathbf{Q}_{N-2} \right\rangle \cdots \cdot \left\langle \mathbf{Q}_1 | e^{-iH\varepsilon/\hbar} | \mathbf{Q}_0 \alpha \right\rangle . \tag{4.24}$$

Feynman and Hibbs[44] related this quantity with the classical Lagrangian L_{cl} along all possible paths connecting the point $\mathbf{Q}_0$ to $\mathbf{Q}_N$:

$$K_{\beta\alpha}\left(\mathbf{Q}_N, t_N; \mathbf{Q}_0, t_0\right) = \int\limits_{\mathbf{Q}_{0(t_0)}}^{\mathbf{Q}_N(t_N)} \mathrm{D}\left[\mathbf{Q}(t)\right]\exp\left[\frac{i}{\hbar}\int\limits_{t_0}^{t_N} L_{cl}\left(\mathbf{Q}, \dot{\mathbf{Q}}\right)dt\right] \quad (4.25)$$

where the infinite-dimensional integral operator

$\int\limits_{\mathbf{Q}_{0(t_0)}}^{\mathbf{Q}_N(t_N)} \mathrm{D}\left[\mathbf{Q}(t)\right]$ accounts for all the possible paths starting from

$\mathbf{Q}_0$ and ending in $\mathbf{Q}_N$.[45] Miller and George[46] provided a more useful representation for the reduced propagator:

$$K_{\beta\alpha}\left(\mathbf{Q}_N, t_N; \mathbf{Q}_0, t_0\right) = \int\limits_{\mathbf{Q}_{0(t_0)}}^{\mathbf{Q}_N(t_N)} \mathrm{D}\left[\mathbf{Q}(t)\right] e^{iS_0\left[\mathbf{Q}(t)\right]/\hbar} \, T_{\beta\alpha}\left[\mathbf{Q}(t)\right] \quad (4.26)$$

where $S_0\left[\mathbf{Q}(t)\right] = \int\limits_{t_0}^{t_N} dt\mathrm{D}\sum_k \frac{1}{2}M_k\dot{Q}_k^2(t)$ is the action functional

for nuclear motion and

$$T_{\beta\alpha}\left[\mathbf{Q}(t)\right] = \left\langle \Psi_\beta\left(t_N\right)\middle| e^{-\frac{i}{\hbar}\int\limits_{t_0}^{t_N} H_{el}(\mathbf{Q}(t))dt} \quad - \middle| \Psi_\alpha\left(t_0\right) \right\rangle$$

is the amplitude for the transition $\alpha \to \beta$ as the atoms travel along the path $\mathbf{Q}(t)$. Equation 4.26 is a fundamental and important relation, giving the following prescription for constructing the amplitude for propagation from nuclear positions $\mathbf{Q}_0$ and electronic state $\left|\Psi_\alpha\right\rangle$ to nuclear positions $\mathbf{Q}_N$ and electronic state $\left|\Psi_\beta\right\rangle$. One first solves the time-dependent electronic problem for a generic nuclear path $\mathbf{Q}(t)$ and then, after supplying the phase factors that are the action associated with the nuclear kinetic energy, integrates over all nuclear paths $\mathbf{Q}(t)$ with boundary conditions $(\mathbf{Q}_0, \mathbf{Q}_N)$. This has two important interpretations. First, it places in an exact framework the various approximate time-dependent models often used in photochemical problems. Equation 4.26 shows that the exact solution requires integration over all possible nuclear paths. Second, it has a "dynamic Born–Oppenheimer" interpretation analogous to the Born–Oppenheimer representation of stationary states. In the latter case, one first fixes the nuclear positions and solves for electronic eigenvalues; here, one fixes the trajectory of the nuclei and solves for the electronic transition amplitudes. In the former case the electronic eigenvalues are functions of the nuclear positions, whereas in Equation 4.26 the electronic transition amplitudes are functions of the nuclear path.

4.2.4 Semiclassical Approximation

From Equation 4.25, it is clear that, because $\hbar$ is a very small number, the exponential rapidly oscillates so that contributions from near paths cancel each other. Therefore, only few paths effectively contribute to the path integral in Equation 4.25. The important exception occurs when the phase reaches a minimum — at this stationary phase point the contributions from similar paths no longer cancel. Consequently, the most important contributions to the integral come from paths close to the one for which

$$\delta \int_{t_0}^{t_N} L_{cl}(\mathbf{Q}, \dot{\mathbf{Q}}) dt = 0. \tag{4.27}$$

This condition defines the classical path from $\mathbf{Q}_0$ to $\mathbf{Q}_N$. More specifically, a stationary phase approximation to Equation 4.26 can be considered by equating to zero the first-order variation of the phase about the classical path $\mathbf{Q}(t)$:

$$\delta\left(S_0\left[Q(t)\right] + \hbar \Im \ln T_{\beta\alpha}\left[Q(t)\right]\right) = 0 \tag{4.28}$$

From this condition, Pechukas obtained the force acting on the j-th nuclear coordinate[47]:

$$F_j(t) = M_j \ddot{Q}_j(t) = -\Re\left\langle \psi_\beta(t_N, t) \middle| \Delta_j H_{el}(t) \middle| \psi_\alpha(t_0, t) \right\rangle, \tag{4.29}$$

where $\Im$ and $\Re$ denote imaginary and real parts respectively.

The effective potential corresponding to the force in Equation 4.29 is then

$$V_{eff}(t) = \Re \left\langle \psi_\beta(t_N,t) \left| H_{el}(t) \right| \psi_\alpha(t_0,t) \right\rangle. \qquad (4.30)$$

Therefore, the paths of the nuclear coordinates that constitute the most important contribution to the path integral expression in Equation 4.26 are those of stationary phase $\mathbf{Q}(t)$, which satisfy Equation 4.29. Here, $\left| \psi_\alpha(t_0,t) \right\rangle$ is the mixed-state wave function at time t that started out at time t_0 in the pure state $\left| \Psi_\alpha \right\rangle$, and $\left| \psi_\beta(t_N,t) \right\rangle$ started out at time t_N in the pure state $\left| \Psi_\beta \right\rangle$; the time-dependent electronic Hamiltonian $H_{el}(\mathbf{Q}(t))$ evaluated along the stationary phase path has been used to propagate these boundary states forward and backward in time to t. The nuclear wave function has to be reduced to the nuclear position at time t. This can be obtained by replacing the nuclear density with a product of delta functions centered at the positions of the nuclei. As a consequence, the expectation value of the position operator for the nuclei becomes simply

$$\int dQ \chi_u^*(Q,t) Q_I(t) \chi_u Q_I(t). \qquad (4.31)$$

The Pechukas equation of motion (Equation 4.29) must be solved iteratively due to nonlocality of the forces; that is, the force at time t can be determined only by knowing the full trajectory $\mathbf{Q}(t)$. These methods have been explored both as numerical methods[48,49] and as a means of understanding the character of surface hopping approximations.[50]

4.2.5 Classical Trajectories

A set of purely classical trajectories can be seen as the most drastic approximation to Equation 4.29. In fact, it is often found that classical-trajectories provide an accurate description of chemical reaction dynamics, which is not surprising because nuclear masses are large. The simplest way to apply classical trajectories in the context of photochemistry is to start with a certain number of trajectories on the ground electronic state. Each of them is then vertically excited into the state of interest and propagated according to the classical Hamilton's equations, and confined on a single electronic PES. The electronic motion is simply integrated out, and the nuclear motion results from the gradient of the PES:

$$F_j(t) = -\frac{\partial}{\partial Q_j}\left\langle \psi_\beta(t)\middle| H_{el}(q, Q.t)\middle| \psi_\beta(t)\right\rangle \qquad (4.32)$$

This approach can give important qualitative information about the excited state dynamics, at least until a region of strong nonadiabatic coupling is reached when classical mechanics becomes meaningless. Classical trajectories still provide a useful model to determine the portion of the PES accessed by the molecules and roughly the time required to observe certain events. In particular, this approach has been exploited in the first applications of *ab initio* molecular dynamics. This is the only alternative when the *ab initio* calculation is too expensive to allow the calculation of the nonadiabatic coupling or the integration of the electronic equation of motion.

4.2.6 Mean-Field Approach

Once the nuclear wave function has been collapsed into a product of delta functions that identify the nuclear positions, the time-dependent electronic wave function can be expanded in the adiabatic basis:

$$\left| \psi(t)\right\rangle = \sum_K A_K(t) e^{-\frac{i}{\hbar}\gamma_K(t)}\left| \psi_K(t)\right\rangle \qquad (4.33)$$

with

$$\gamma_K(t) = \int_0^t E_K(Q(t))dt'. \qquad (4.34)$$

Substituting this expansion in the TDSE, we can derive the time derivative of $A_K(t)$:

$$\dot{A}_K(t) = -\sum_{L \neq K} A_L(t) e^{-\frac{i}{\hbar}(\gamma_L - \gamma_K)} \sum_j \dot{Q}_j h_{KL}^{(j)}.$$ (4.35)

The transition probability depends on the dot product between the velocity $\dot{Q}$ and the nonadiabatic coupling h_{KL} vectors. It is easy to obtain the corresponding derivative with respect to the j-th nuclear coordinate:

$$\frac{\partial A_K(t)}{\partial Q_j} = -\sum_{L \neq K} A_L(t) e^{-\frac{i}{\hbar}(\gamma_L - \gamma_K)} h_{KL}^{(j)}.$$ (4.36)

Now we can compute the force acting on the j-th nuclear coordinate in the local time approximation of Equation 4.29:

$$F_j(t) = -\Re \frac{\partial}{\partial Q_j} \langle \psi(t) | H_{el} | \psi(t) \rangle = -\Re \frac{\partial}{\partial Q_j} \left[\sum_K |A_K|^2 E_K \right] =$$

$$-\sum_K \sum_{L \neq K} \Re \left(A_k^* A_L e^{-\frac{i}{\hbar}(\gamma_L - \gamma_K)} h_{KL}^{(j)} \right) (E_L - E_K) - \sum_K |A_K|^2 \frac{\partial E_K}{\partial Q_j} =$$

$$-\sum_K \sum_{L \neq K} \Re \left(A_k^* A_L e^{-\frac{i}{\hbar}(\gamma_L - \gamma_K)} h_{KL}^{(j)} \right) \left\langle \psi_K \left| \frac{\partial H_{el}}{\partial Q_j} \right| \psi_L \right\rangle - \sum_K |A_K|^2 \frac{\partial E_K}{\partial Q_j}.$$

(4.37)

In mean-field or Ehrenfest methods,[51-56] the forces result from the contribution of two terms: the first is related to the nonadiabatic coupling, the second is an average of the gradients of the potentials of the populated electronic states. Therefore, the forces acting on the nuclei depend directly on the population of the electronic states. The electronic problem and the nuclear dynamics have to be solved simultaneously. The time step must be sufficiently small to account for the time variation of the electronic wave functions. In this case, the solution of the TDSE can be propagated as[57]

$$A(t) = \exp\left(-\frac{i}{\hbar} H_{el} t \right) A(0) = \left[\sum_K X_K e^{-\frac{i}{\hbar} E_K} X_K^t \right] A(0).$$ (4.38)

This equation, written in matrix form, shows that the time-evolution operator can be expressed in terms of the electronic Hamiltonian eigenvectors X_K.

Unlike purely classical trajectories, the Ehrenfest approach allows for nonadiabatic transitions between electronic states. As in the quantum wavepacket methods, the system is thought to exist simultaneously in different electronic states, with different weights. Contrary to quantum treatments, however, mean field methods force the nuclei to occupy the same position for all the states: The nuclear trajectory is determined by the average potential, whereas, in regions where the nonadiabatic coupling is negligible, the quantum wavepackets travel independently on each PES. As a result, although the mean-field methods behave correctly in surface-crossing situations, they may give rise to artifacts when electronic states are well separated in energy. To some extent, this can be remedied by on-the-fly binning techniques, which can restore the correct state-specific character to the dynamics.[52,54,55]

4.2.7 Surface Hopping Swarm Dynamics

The main practical problem in the implementation of the mixed quantum-classical dynamics method described in Section 4.2.4 is the nonlocal nature of the force in the equation of motion for the stationary-phase trajectories (Equation 4.29). Surface hopping methods provide an approximate, intuitive, stochastic alternative approach that uses the average dynamics of swarm of trajectories over the coupled surfaces to approximate the behavior of the nonlocal stationary-phase trajectory. The surface hopping method of Tully and Preston[58] and Tully[59] describes nonadiabatic dynamics even for systems with many particles. Commonly, the nuclei are treated classically, but it is important to consider a large number of trajectories in order to sample the quantum probability distribution in the phase space and, if necessary, a statistical distribution over states. In each of the many independent trajectories, the system evolves from the initial configuration for the time necessary for the description of the event of interest. The integration of a trajectory is done numerically with small time steps. The nuclei move according to the classical equations of motion; in particular, the force acting on nucleus j at each time step is the same shown in Equation 4.32.

The time evolution of the electronic wave function can be obtained in the adiabatic or in the diabatic basis set. At each time step, one evaluates the transition probabilities between electronic states and decides whether to hop to another surface. When hopping occurs, nuclear velocities have to be adjusted to keep the total energy constant. After hopping, the forces are calculated from the potential of the newly populated electronic state. To decide whether or not to hop, a Monte Carlo technique is used: Once the transition probability is obtained, a random number in the range (0,1) is generated and compared with the transition probability. If the number is less than the probability, a hop occurs; otherwise, the nuclear motion continues on the same surface as before. At the end of the simulation, one can analyze populations, distribution of nuclear geometries, reaction times, and other observables as an average over all the trajectories.

Even if a hop is made, the total energy of the system has to be conserved. Let us call ΔE the potential-energy difference between final and initial surface. In order to preserve the total energy, the kinetic energy must be varied by $-\Delta E$. Usually, only the component of the nuclear momenta along the direction of the nonadiabatic coupling vector is changed.[60] After the hop, the component of the nuclear velocity perpendicular to the nonadiabatic coupling has to be unchanged, whereas the parallel one has to counterbalance the energy difference between the two surfaces. Because it is possible to hop to a state that is higher in energy than the initial one, ΔE could be greater than zero. In this case, the nuclear velocity has to decrease. But if there is not enough "parallel" kinetic energy to compensate for the hop, the hop is commonly rejected. These classically forbidden transitions are believed to be the origin of the discrepancy often observed in surface hopping methods between the fraction of trajectories on each state and the corresponding average quantum probability determined by propagation of the quantum amplitudes.[61,62]

As we have seen, the surface hopping method can respond with a hop when a region of strong nonadiabatic coupling is encountered. However, because quantum transitions are inherently nonadiabatic processes, the adiabatic approximation underlying the classical equation of motion does not hold in those regions of the surfaces where the transitions take place. Mixed methods that combine surface hopping and mean-field approaches (the latter applied only in the regions of strong coupling) have been recently proposed.[63–65]

4.2.8 Full Multiple Spawning

The full multiple spawning (FMS) method[66,67] has been developed as a genuine quantum mechanical method based on semiclassical considerations. The FMS method can be seen as an extension of semiclassical methods that brings back quantum character to the nuclear motion. Indeed, the nuclear wave function is not reduced to a product of delta functions centered on the nuclear positions but retains a minimum uncertainty relationship. The nuclear wave function is expressed as a sum of Born–Oppenheimer states:

$$\Psi(q,Q,t) = \sum_K \psi_K(q;Q)\chi_K(Q,t). \tag{4.39}$$

Thus, each electronic state K has a corresponding time-dependent nuclear wave function $\chi_K(Q, t)$. As introduced by Heller, the nuclear wavepacket can be expressed as a sum of multidimensional traveling frozen Gaussian basis functions with time-dependent coefficients:

$$\chi_K(R;t) = \sum_{j=1}^{N_K(t)} C_j^K(t)\chi_j^K\left(R;\bar{R}_j^K(t),\bar{P}_j^K(t),\bar{\gamma}_j^K(t),\alpha_j^K\right) \tag{4.40}$$

where the index j labels nuclear basis functions on electronic state K, $N_K(t)$ is the number of nuclear basis functions on electronic state K at time t (this number is allowed to change during the propagation), and we have explicitly denoted the time-dependent parameters of the individual basis functions.

Individual, multidimensional, nuclear basis functions are expressed as a product of one-dimensional Gaussian basis functions

$$\chi_j^K(R;t) = \left(R;\bar{P}_j^K(t),\bar{\gamma}_j^K(t),\alpha_j^K\right) =$$

$$e^{\bar{\gamma}_j^K(t)t}\prod_{p=1}^{3N}\left(\frac{2\alpha_{pj}^K}{\pi}\right)^{1/4}\exp\left[-\alpha_{pj}^K\left(R_{pj}-\bar{R}_{pj}^K(t)\right)^2 + i\bar{P}_{pj}^K(t)\left(R_{pj}-\bar{R}_{pj}^K(t)\right)\right]$$

$$(4.41)$$

where the index p enumerates the 3N coordinates of the molecule, typically chosen to be Cartesian coordinates. The frozen Gaussian basis functions are parameterized with a time-independent width (α_{pj}^K) and time-dependent position, momentum, and nuclear phase ($\left[\bar{R}_{pj}^K(t),\bar{P}_{pj}^K(t),\bar{\gamma}_j^K,\text{ respectively}\right]$).

The centroids of the nuclear basis functions ($\left(\bar{R}_{pj}^K(t),\bar{P}_{pj}^K(t)\right)$) have been chosen to move classically over the surface of the corresponding K-th electronic state. This choice for the motion of the nuclear basis functions is arbitrary. Although other choices are possible, the classical movement of the basis functions appears to be the most computationally convenient.

When the Equations 4.39 through 4.41 are inserted in the TDSE, the following set of differential equations for the time-dependent coefficients $C_j^I(t)$ results:

$$\frac{dC^I(t)}{dt} = -i\left(S_{II}^{-1}\right)\left\{\left[H_{II} - i\dot{S}_{II}\right]C^I + \sum_{J\neq I}H_{IJ}C^J\right\}. \quad (4.42)$$

For compactness and clarity, Equation 4.42 is written in matrix notation. because the nuclear basis set in FMS is time dependent and nonorthogonal, the nuclear overlap matrix **S** and its time derivative, as well as the Hamiltonian matrix **H**, have to be evaluated.[69]

If the two indices K and $N_K(t)$ in the FMS method run over infinite terms, the method is exact. However, if both indices are allowed to be at most equal to one, the method can be considered purely classical. In the usual case, the number of electronic states included in the calculation is limited to the number of states of interest plus, if necessary, the ones higher in energy that may interact with the lower energy subspace. The number of nuclear basis functions varies during the simulation. Generally, at the beginning of the simulation, one places all the nuclear basis functions on a given surface where their positions evolve classically. Once a region of strong nonadiabatic coupling is reached, new nuclear basis functions are "spawned" on the coupled surface. Then, the TDSE (Equation 4.42) determines how much population will be transferred to the newly generated basis function.

In other words, the FMS, method in principle, can converge to an exact description of the nuclear dynamics if an infinite number of nuclear basis functions is included that span the entire phase space of an infinite number of electronic states. The approximation consists of allowing only relatively few of these basis functions to have a nonvanishing coefficient $C_j^K(t)$. The algorithm attempts to select the nuclear basis functions that are the most important for the description of the wavepackets.

4.3 POTENTIAL-ENERGY SURFACES AND THEIR COUPLINGS

4.3.1 *Ab Initio* Methods

If the PESs and their couplings are to be evaluated as needed during the photodynamics simulation, it becomes imperative that the method for generating these PESs be computationally tractable. However, one must be very careful not to sacrifice accuracy, at least in the qualitative sense, in the process of finding an efficient scheme for generating the PESs. Perhaps the best choice for photochemical dynamics from the standpoint of accuracy is *ab initio* quantum chemistry. PESs for multiple electronic states can be difficult to represent correctly with anything less than direct solution of the electronic Schrödinger equation. Nevertheless, the computational price paid is high, and currently the largest molecules we have studied in this way are stilbene and azobenzene. Certainly such simulations of photochemistry in solution or protein environments are out of the question at the present time using *ab initio* electronic structure theory.

The two important requirements in a correct treatment of excited electronic states are the multireference character and the need to avoid variational bias to a particular state. Excited states are often found to have significant multireference character, but regardless of any debate that might exist on this point, a conical intersection cannot be described correctly without a multireference wave function. Complete active space methods are perhaps the best choice here when a suitable active space can be identified.[70] Avoiding bias to a particular state can be accomplished by using state-averaging techniques. In these methods, the orbitals are determined to minimize a weighted average of electronic state energies. The resulting orbitals are often called a "best-compromise" set since they are not optimal for any single electronic state. Once the orbitals have been determined, they must be allowed to relax separately for each electronic state. One way to do this is by including single excitations in a configuration interaction (CI) wave function. In this approach, single excitations are taken from the same set of reference configurations that was used to determine the orbitals in the state-averaged multi-configuration SCF (MCSCF). Variants of this technique, such as the first-order CI method of Schaefer and Bender[71] and the POL-CI method of Hay et al.,[72] have been successfully used in past treatments of excited states.

4.3.2 Time-Dependent DFT

Several extensions to DFT that allow for the treatment of excited states have recently been developed and implemented. The implementation of time-dependent DFT (TDDFT) in Gaussian98[73] has led to an explosion in the number of calculations using this particular variant of excited-state DFT. The performance of TDDFT for vertical excitation energies to valence excited states is often quite good,[20,21,74–77] with an apparent accuracy similar to some of the best *ab initio* methods. When the excited state of interest has significant Rydberg character, there are often problems related to the improper asymptotic properties of most functionals.[20,78] An efficient approximate formulation of TDDFT has been presented[79] that makes the computational cost almost identical to the CIS method.

These considerations make TDDFT a very attractive method in the context of first-principles photodynamics calculations. However, there are problems that need to be resolved. Most current implementations of TDDFT use the adiabatic approximation, which states that the density functional is independent of excitation energy. The severity of this approximation is not known. Furthermore, the TDDFT equations are often linearized for efficient solution. Again, the importance of the higher terms is not well characterized. This linearization leads to a formal similarity of TDDFT to CIS and has made many workers suspect that TDDFT may not properly represent doubly excited states.[20] Althought it has been shown that such an assessment may be overly pessimistic,[74,80] some amount of caution is nevertheless recommended.

Because almost all assessments of TDDFT have focused on vertical excitation energies, we were interested to see how the method would perform if applied to the photodynamics of ethylene. As discussed in Section 4.4.1 in this chapter, we have extensive experience with this molecule, and important aspects of the excited-state surface are well established. In particular, the important coordinates in the excited-state dynamics are pyramidalization and torsion. In Figure 4.1a we show a two-dimensional cut of the S_0 and S_1 PESs focusing on these variables, using a SA-2-CAS(2/2)*PT2 wave function in the 6-31G* basis set. There is little doubt that this level of electronic structure theory will give qualitatively correct PESs for the valence states of this molecule, as we have demonstrated using larger basis sets and different correlation methods.[26,28,81] In Figure 4.1b through Figure 4.1d, we use contour plots to emphasize the shape of S_1 for CASPT2, CIS, and TDDFT. The CIS and TDDFT calculations are both spin-unrestricted, the 6-31G** basis set is used, and the functional in TDDFT is the hybrid B3LYP functional,[82] which has shown

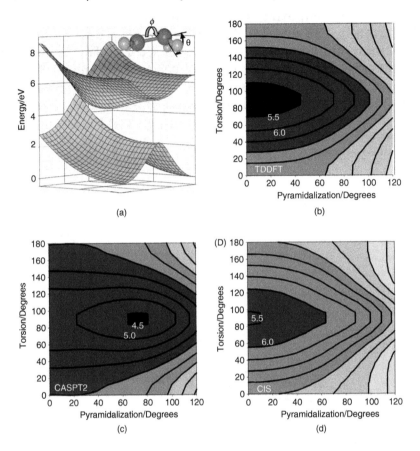

(a)

(b)

(c)

(d)

Figure 4.1 Important features of ground- and excited-state PESs for ethylene photodynamics and demonstration of the inadequacy of TDDFT and CIS methods for this problem. (a) S_0 and S_1 PESs for ethylene in the pyramidalization and torsion coordinates (defined in the inset) that dominate the photodynamics. This surface was calculated using multireference perturbation theory — CAS(2/2)*PT2. The global minimum on S_1 occurs at twisted and pyramidalized geometries. (b–d) A quantitative comparison of the S_1 PES obtained with CAS(2/2)*PT2, TDDFT/B3LYP, and CIS, respectively. All calculations use the 6-31G* basis set. The TDDFT and CIS calculations are performed in a spin-unrestricted formalism. Contour values are given in eV, and in all cases the energies are referenced to the S_0 equilibrium geometry at the corresponding level of theory. Only the multireference calculation captures the S_1 minimum correctly.

promise in computing vertical excitation energies.[76,77,83] Importantly, CIS and TDDFT both fail to capture the pyramidalized minimum on S_1 (see also Figure 4.2). This is not at all surprising in CIS; the origin of the twisted and pyramidalized minimum on S_1 is a charge transfer state, which is formally a double excitation from the ground state. Hence, this state cannot be well described, and instead the minimum observed on S_1 is the expected minimum of the $\pi\pi^*$ state (purely twisted). The question will be how to fix this problem

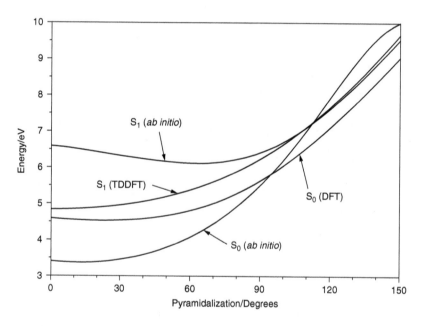

Figure 4.2 Potential-energy curves for the S_0 and S_1 electronic states of 90° twisted ethylene as a function of CH_2 monopyramidalization. TDDFT and *ab initio* [SA-2-CAS(2/2)] results are compared. The 6-31G basis set was used with both methods. The zero of energy is chosen as in Figure 4.1. All coordinates not involved in pyramidalization were fixed at the values obtained by optimizing the geometry on S_1 at the respective level of theory subject to the constraint that it be twisted by 90° with no pyramidalization. The S_0/S_1 energy gap is unaffected by pyramidalization in the TDDFT method, in contrast to CAS(2/2), where pyramidalization is a dominant coordinate in tuning the energy gap to reach a conical intersection.

in future variants of TDDFT: If the problem really is that doubly excited states are poorly or not at all represented, this will need to be addressed before TDDFT photodynamics is generally applicable. A number of workers are extending TDDFT beyond the adiabatic approximation and linear response[84,85] and addressing deficiencies in the functionals.[86] This will be an exciting area for first-principles photochemistry as a workable method emerges.

4.3.3 Semiempirical Methods

The extreme computational expense of *ab initio* methods appropriate for excited states poses a serious limitation to photodynamical simulations. In particular, molecules with more than 30 atoms are at the edge of what is possible with current computational resources. Timescales exceeding a picosecond are also difficult to compute. This can be a serious constraint when calculating product branching ratios because confident assignment of quantum mechanical flux to a particular product channel requires considerable evolution on the ground state. Thus, one should seek alternatives that retain the advantages of *ab initio* multiple spawning as far as possible, while decreasing the computational cost associated with the electronic structure. Semiempirical methods are quite promising in this regard. Numerous workers have demonstrated the utility of these methods for ground-state problems, and there have been a few applications to excited states. However, one needs to be extremely cautious when embarking on this road. In particular, it should always be kept in mind that dynamics can only be as good as the underlying PESs and couplings. There have been few assessments of the accuracy of semiempirical methods for excited-state PESs, and there is little reason to believe that existing parameterizations should be adequate in the photochemical context. In particular, previous parameterizations have emphasized ground-state properties and sometimes vertical excitation energies, not global features of excited-state PESs.

Very few semiempirical methods and parameterizations were explicitly intended to treat excited electronic states.

Most of these were designed to reproduce vertical excitation energies as estimated from the absorption maxima in electronic absorption spectra. Examples include the CNDO/S[87] and INDO/S[88] methods — all of which rely on a single reference determinant when configuration interaction is included. The MNDOC method[89–91] has been used with an MRSDCI wave function[92–95] but is built on molecular orbitals derived from a single determinant SCF calculation, Hartree–Fock. An interesting multireference semiempirical model has been proposed by Cullen[96] but was applied only to ground states. As discussed above, the procedure expected to give the best results in the *ab initio* context is a state-averaged MCSCF followed by MRSDCI. There has been much discussion in the literature about the role of semiempirical parameterization in capturing correlation effects, which we do not repeat here. However, though it seems reasonable to believe that the parameterization of integrals could capture dynamic correlation effects, it is hard to see how this parameterization could possibly capture static correlation effects that are strongly geometry dependent. Hence, we adopt an analog of the *ab initio* approaches discussed above in the semiempirical context. The floating occupation molecular orbital (FOMO) method replaces occupation averaging for true state averaging; this is roughly equivalent to an energy-weighted state-averaging procedure. The resulting orbitals are best-compromise orbitals, and thus the procedure must be followed by CI. We perform a CASCI for this purpose.[97,98]

4.3.4   Optimization of Semiempirical Parameters

We base semiempirical methods on CI wave functions by using FOMOs for the investigation of the electronic structure. because we are eventually interested in dynamics simulations, this scheme is adopted in the framework of one of the popular semiempirical NDDO methods:[99] MNDO,[100,101] AM1,[102] or PM3.[103] However, the semiempirical parameters have been optimized to reproduce many properties — for example, equilibrium geometries, heats of formation, and isomerization energies — for many molecules in the ground electronic state computed with SCF wave functions. As a consequence, they may not be adequate when applied to the study of excited states using FOMO-CI wave functions. In addition, because the parameters have been optimized by comparing results from SCF wave functions with experimental results, the semiempirical parameters may already incorporate correlation effects — which are, of course, present in any experiment — within the SCF wave function. Then any correlation added to configuration interaction might be counting dynamic correlation twice — once explicitly through configuration interaction and once implicitly through the parameterization. This is difficult to quantify, but the most important point is a practical one: The results of semiempirical FOMO-CI calculations for excited states using standard parameterizations are usually somewhat disappointing when compared with experimental and *ab initio* results.[97,104,105] In the most favorable cases (usually the lowest states of organic molecules), the qualitative behavior of the PESs of excited states is well reproduced, but in any case the errors in excitation energies and relative energies of different points on the same electronic state prevent the use of standard parameters for routine calculations.

For this reason a different strategy has been adopted: We apply high-level *ab initio* methods at the most important geometries of the molecule under study (ground- and excited-state minima and MECIs) in all the electronic states of interest. Then we reoptimize the semiempirical parameters to reproduce the *ab initio* results and possibly also experimental results, if available. These target results can include not only energies but also geometries; dipole moments; and, in general, all the features we want to reproduce. Such system-specific reparameterization was almost routine in the early implementations of semiempirical methods and was most recently revived under the acronym SRP ("specific reaction parameters") for rate calculations.[106] In certain cases, the same atom may be assigned different parameters according to connectivity in a molecule. For example, the acetone molecule might be better represented if the acylic and the alkylic carbons had distinct sets of parameters. One must be cautious in this respect, because such an asymmetric treatment of the same atoms will have unphysical consequences if it is possible for reactions to exchange such "equal but different" atoms.

In practice, the reoptimization of parameters is carried out by introducing a "target function." This function represents a measure of the difference between *ab initio* or experimental data and the corresponding values obtained with the semiempirical method:

$$f(\mathbf{X}) = \sqrt{\frac{\sum_i \left(\frac{y_{0,i} - y_{s,i}}{y_{0,i}}\right)^2 w_i}{\sum_i w_i}}. \qquad (4.43)$$

Here, $y_{0,i}$ and $y_{s,i}$ are *ab initio* and semiempirical values, respectively, and w_i is a weight. The vector $\mathbf{X}$ contains all the semiempirical parameters to be optimized. The reparameterization problem becomes the search for the minimum of a surface in a space whose dimension is the number of parameters to be optimized. At this point, different algorithms could be chosen to minimize the target function. For example, if one chooses an algorithm based on the gradient, the local minimum closest to the initial point would be obtained. However, simulated annealing and genetic algorithms are methods designed to find the global minimum of the surface.[107] The simplex method can be seen as an intermediate strategy because it is driven to the closest minimum, but it can overcome barriers of limited height to access lower regions of the surface.

It should be emphasized that a smaller final value of the target function does not necessarily mean that the electronic structure of the molecule has been well represented. In fact, our goal is to obtain a good representation of the PESs for our molecule not only in the proximity of the points included in the optimization, but also more globally — at least in all the regions of the PESs important to the photodynamics. Sometimes the absolute minimum of the target function corresponds to a set of semiempirical parameters that are very different from the original ones. Because the original set was well characterized and usually known to be reasonable across a wide range of molecules and geometries, one might worry that sets that stray dramatically from the original set could exhibit regions of unphysical behavior far from the regions of the PES fit during the reoptimization. Penalty functions can be introduced to prevent the parameters from straying too far from their original values, but this introduces ambiguities that are difficult to resolve. How far is too far? The answer to this question is almost certainly different for each parameter and probably from case to case. In our experience, the use of global optimization techniques is successful only with the addition of such penalty functions. On the contrary, the methods that find local minima close to the initial guess (which in our case is one of the standard parameterizations) ensure that the results have been improved without undue alteration of the parameters. In the end, the choice of the method depends on the specific problem and on the target results, but in any case, once the parameters have been optimized in some way, they must be checked carefully. Potential energy curves and surfaces have to be computed in between the reference points, but also outside the region they span, and then compared with additional *ab initio* calculations. If the resulting curves do not exhibit any unphysical behavior and the agreement with the new *ab initio* data (which were not included in the reoptimization) is satisfactory, the new set of parameters can be accepted.

The results of such reparameterization are illustrated here for an analog of retinal protonated Schiff base (RPSB). In Figure 4.3, we show the important points that were targeted in the optimization: the vertical excitation energy, relative energies of two conical intersections that are competing decay pathways, relaxation energy while maintaining planarity, and relative energies of minima on the excited state. The solid lines indicate the *ab initio* results, obtained with SA-2-CAS(10/10) wave functions in a 6-31G* basis set. The results obtained using the standard PM3 parameter set and a FOMO-CI wave function are shown as dotted lines. The important point is that the intersection corresponding to twisting around the $C_{13}=C_{14}$ bond lies above the Franck–Condon point. Thus,

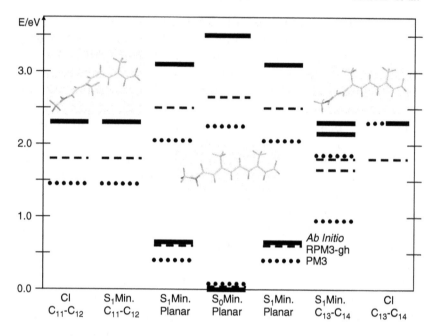

Figure 4.3 Comparison of energies for important points on the S_0 and S_1 PESs for an analog of RPSB. The S_1Min-Planar structure is not a true minimum but rather the lowest energy geometry obtained on S_1 while constraining the molecule to remain planar. The S_1Min-C_{11}-C_{12} and S_1Min-C_{13}-C_{14} structures are minima found after twisting around the C_{11}-C_{12} or C_{13}-C_{14} bonds, respectively. The CI-C_{11}-C_{12} and CI-$C_{13}C_{14}$ structures are minimal energy conical intersections found after twisting around the indicated bonds. *Ab initio* results obtained with SA-2-CAS(10/10) wave functions in the 6-31G* basis set are indicated by solid lines. Semiempirical FOMO-CI results obtained with the PM3 parameter set are indicated by dotted lines. The results obtained after reparameterization of the semiempirical FOMO-CI method (RPM3-gh as described in the text) are indicated by dashed lines.

the molecule can get trapped in an S_1 minimum corresponding to a geometry twisted about the $C_{13}=C_{14}$ bond. This is physically unreasonable because it implies fluorescence from the chromophore, which is not observed. Furthermore, it contradicts the *ab initio* results. After reparameterization, we obtain the dashed lines, which now exhibit the desired degeneracy between the intersections corresponding to twisting around the $C_{11}=C_{12}$ and $C_{13}=C_{14}$ bonds. Notice that we do not exactly achieve the target optimization criteria, because we search only for the nearest local minimum. Thus, the vertical excitation energy is still lower in the reparameterized semiempirical method than it is in the *ab initio* case.

Reparameterizing to relative energies of intersections and excited-state minima can be quite useful, but one might then also ask to what extent the shape of the conical intersections is preserved. In Figure 4.4, we show the topography around the $C_{11}=C_{12}$ intersection in RPSB. Figure 4.4a shows the results obtained with the *ab initio* wave function — a clear peaked form is observed. However, the form obtained from the original PM3 parameter set and FOMO-CI wave functions is less peaked. After reparameterizing, the plot in Figure 4.4c is obtained, which has a decidedly sloped character. This is quite different from the shape observed in the *ab initio* case. We did not bias the target function in any way to get the same shape of the PESs around the conical intersection, so one may not be surprised at the lack of agreement. However, we then proceeded to add the g and h vectors to the target function. In other words, we computed the overlap between the *ab initio* branching plane and the semiempirical branching plane and used this as a component of the target function in reparameterization. This is not a direct fit of the topography of the PESs around the conical intersection. Interestingly, this improved the agreement of the intersection topography — compare Figure 4.4a and Figure 4.4d. This indicates that the character of the nuclear displacements involved in the branching plane may be an important characteristic to get correct.

In this reparameterization strategy, we abandon the hope that a single set of parameters could be sufficient to describe all atoms of the same type in all possible molecules. On the contrary, we develop different parameters for each system we study. Several molecules have been reoptimized according to this strategy: ClOOCl,[105] dimethylnitrosamine,[108] ethylene, benzene, retinal protonmated Schiff base, and the chromophore of the green fluorescent protein. This approach is no

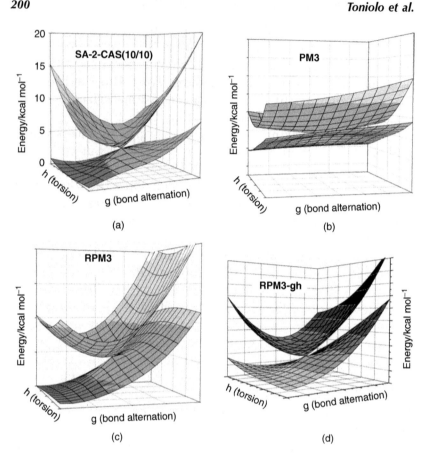

(a)

(b)

(c)

(d)

Figure 4.4 The ground states and first excited states, computed with different methods, of the RSPB analog as a function of displacement from the twisted $C_{11}=C_{12}$ conical intersection along the gradient difference vector, g, and the nonadiabatic coupling vector, h. (a) The conical intersection geometry, vectors, and PESs were calculated by *ab initio* electronic structure methods. The remaining panels used semiempirical FOMO-CI methods. The standard PM3 parameters were used in (b). A set of parameters that was reoptimized from PM3 to replicate *ab initio* RSPB analog energy levels was used in (c). The set of parameters used in (d) were reoptimized from the set used in (c) by additionally requiring agreement with the *ab initio* g and h vectors for the $C_{11}=C_{12}$ and $C_{13}=C_{14}$ conical intersections. See text for details of the reoptimizations.

longer general, in the sense that a parameter set is in principle good only for one molecule and possibly only for one set of reactions involving that molecule. However, we have indications that the parameters are, at a certain level, transferable. In the case of the azobenzene molecule, Inglese and coworkers replaced the carbon atom parameters with ones we had obtained by reparameterizing the FOMO-CI method (starting with the PM3 parameter set) to reproduce important points in benzene photochemistry. They noticed a clear improvement over the results obtained with the standard PM3 parameters. Consequently, they optimized the nitrogen parameters only, leaving the carbon ones fixed.

Semiempirical calculations for ethylene with PM3 or AM1 parameters obtain incorrect energy orderings for various points on the excited-state surface. In particular, these calculations do not find the low-energy pyramidalized conical intersection found in *ab initio* calculations.[81] In an effort to reparameterize ethylene, we have included various terms into the target function. Among the types of constraints used were energy gaps at different geometries, energy differences between geometries on the excited state, overlap with *ab initio* g and h vectors, and root-mean-square deviations between optimized semiempirical and *ab initio* geometries. The *ab initio* energies used in these target functions are calculated at the CAS–SDCI level, and the *ab initio* geometries are optimized using CAS, as previously reported.[81] We have tried various combinations of these terms and obtain very similar results as long as one of the constraints is a comparision between the *ab initio* conical intersection geometry and the calculated semiempirical one. Although the different parameter sets have very similar behavior in terms of energies and PES cuts, the actual values of the parameters can differ significantly between the different optimizations.

A more challenging test of transferability would eschew any subsequent reparameterization in the new molecule. For example, one could obtain new parameters for carbon atoms intended to reproduce important features in ethylene photochemistry. Then, one might apply these parameters without change to butadiene. Ideally, the new parameters would improve the description of butadiene photochemistry and be close to an optimal set that might be obtained by direct reparameterization of butadiene. In fact, we have indications that this is the case for butadiene and ethylene. However, here we focus on a slightly more complicated example — reparameterization for the stilbene molecule. We had carried out a reparameterization of ethylene and benzene separately for other purposes. Hence, a natural test of parameter transferability was to use these sets in the context of stilbene.

The transferability of the reoptimized parameter sets is tested by comparing stilbene calculations by using these new parameter sets with the *ab initio* results. The *ab initio* results are that the excited-state energy decreases after surmounting a small barrier, from the essentially planar excited-state *trans* minimum (S_1TransMin) to the rigidly twisted D_{2d} geometry (S_1Min90). As one of the ethylenic carbons undergoes pyramidalization, there is a second minimum on the excited state, S_1GlobalMin, which is lower in energy than is the nearly planar minimum. Very close to this pyramidalized minimum, there is a low-energy conical intersection that is twisted and pyramidalized (PyrCI). The relative energies of these points are indicated in Figure 4.5, along with the results obtained with various parameter sets, as we now discuss. Using either the AM1 or the PM3 parameters, the twisted D_{2d} geometry was incorrectly found to be a true minimum, and all the pyramidalized geometries are significantly higher in energy. When the benzene-optimized parameters are used on the phenyl carbons and AM1 parameters are used for the remainder of the atoms, the energy decreases significantly from the *trans* excited-state minimum to the D_{2d} geometry, but all the pyramidalized geometries remain as high in energy as they are with AM1 or PM3 parameters. In contrast, when the ethylene-optimized parameters are used on the ethylenic carbons and AM1 parameters are used on the remaining atoms, the energy increases from the *trans* minimum to the D_{2d} geometry. However, the energy decreases as one of the ethylenic carbons becomes pyramidalized and a low-energy pyramidalized conical intersection is found. Finally, when the benzene-optimized parameters are used on the phenyl carbons and the ethylene-optimized parameters are used on the ethylenic car-

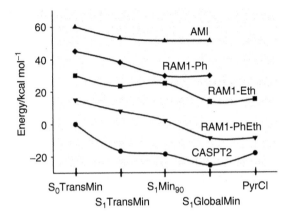

Figure 4.5 Demonstration of aspects of transferability of semiempirical reparameterization. Stationary points for stilbene are indicated on the x axis, and the curves above show the relative energies on S_1 of these points by using various methods. Curves are offset from each other vertically for ease of comparison. Points are connected by curves only to guide the eye. The lowest curve shows SA-3-CAS(2/2)-PT2 results, which are considered the benchmark for judging the remaining curves. The uppermost curve shows relative energies obtained using FOMO(2/2)-CI with the standard AM1 parameter set. The global minimum is erroneously purely twisted, no pyramidalized conical intersection can be found, and the planar and twisted minima on S_1 are isoenergetic, that is, there is no driving force for the molecule to twist around the carbon double bond. Using the reparameterized benzene parameters for phenyl carbons leads to the curve labeled RAM1-Ph. This leads to a stabilization of the twisted geometry. Using reparameterized ethylene parameters for ethylenic carbons leads to the RAM1-Eth curve, where the twisted geometry is erroneously unfavored, but the global minimum on S_1 is correctly pyramidalized and a pyramidalized conical intersection is found. Finally, combining the two sets of parameters gives the RAM1-PhEth curve, which is in qualitative agreement with the CASPT2 curve.

bons, the character of the *ab initio* surface is well reproduced. The energy decreases between the planar minimum and the twisted minimum (likely due to the benzene-optimized parameters), and a low-energy pyramidalized conical intersection is found (likely due to the ethylene-optimized parameters). A comparison of the potential-energy curves from the D_{2d} geometry with the pyramidalized conical intersection can be seen in Figure 4.6.

4.3.5 Hybrid Quantum Mechanical and Molecular Mechanical Approaches

In cases involving very large molecules, such as proteins or solution-phase photochemistry, even reparameterized semiempirical methods are computationally prohibitive. Thus, one needs to begin considering further simplification of the elec-

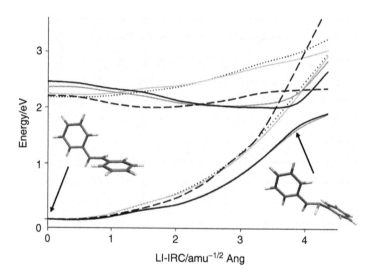

Figure 4.6 Comparison of *ab initio* [SA-3-CAS(2/2)-PT2/6-31G, black dashed lines] and semiempirical FOMO-CI potential-energy curves for stilbene. The geometries are obtained by linear interpolation from the purely twisted D_{2d} saddle point on S_1 to the global minimum on S_1 to the PyrCI geometry. Using the original AM1 parameters (black dotted lines), the D_{2d} geometry is erroneously predicted to be a minimum, and there is no low-lying pyramidalized conical intersection. Using benzene-optimized parameters for the phenyl carbons (gray solid lines) does not remedy the problem. Using ethylene-optimized parameters for the ethylenic carbons (dark gray solid lines) leads to the correct behavior, and this persists when the ethylene-optimized and benzene-optimized parameters are combined (black solid lines). See text for details.

tronic structure problem. The mixed quantum mechanical and molecular mechanical (QM/MM) methods are very promising in this regard. First implemented by Warshel and Levitt,[109] the basic idea is to describe a relatively small region of the system quantum mechanically (this could use either *ab initio* or semiempirical methods or even some combination), whereas the remainder is described with a classical force field. For example, in a photoactive protein, the QM region could be limited to the chromophore. In this case we first consider the chromophore, saturated with hydrogen atoms or methyl groups, and optimize the parameters for this limited part of the molecule as already described. In our development version of the MOPAC semiempirical code,[110] the QM part of the system can be linked to the rest (MM) through "connection atoms".[111] In this case, we repeat the same calculations performed for optimization of the chromophore for a larger QM fragment of the molecule, using only the previously optimized parameters for the chromophore and standard parameters for the remainder. Now we consider these results as a reference and reoptimize the parameters of the connection atoms only with the QM/MM technique for the same system. In this way, one can hope to obtain satisfactory parameters for the entire system.

4.4 EXAMPLES

4.4.1 Ethylene

One of the first molecules we studied with the ab initio multiple spawning (AIMS) method was ethylene. The photochemistry of ethylene can be considered paradigmatic for *cis-trans* isomerization in unsaturated hydrocarbons. The rough outlines of the photochemistry of ethylene are easily summarized. Upon absorption of a photon, an electron is promoted from a bonding π molecular orbital (MO) to an antibonding π^* MO. The ground electronic state of ethylene is planar and stable with respect to twisting, but the $\pi\pi^*$ state favors a twisted D_{2d} geometry. Therefore, electronic excitation results in geometric relaxation toward a twisted geometry. However, there is also a doubly excited state of charge transfer character that is lower in energy than is the optically accessed state in the twisted region. The charge transfer character of this state favors pyramidalization after twisting. In practice, the twisting and pyramidalization occur practically simultaneously. Resonance Raman experiments support this description of the initial dynamics,[112] as does comparison of our simulation results and ultrafast pump-probe experiments.[113,114]

Using a GVB-OA-CAS(2/2)*S wave function and a double-ζ quality basis set, we simulated ethylene photochemistry following $\pi \rightarrow \pi^*$ excitation. The AIMS simulations treat the excitation as instantaneous and centered at the absorption maximum. Hence, the initial-state nuclear basis functions are sampled from the ground-state Wigner distribution in the harmonic approximation. Ten basis functions are used to describe the initial state. Overall, approximately 100 basis functions are spawned during the dynamics, and we follow the dynamics up to 0.5 ps (picoseconds) (using a time-step of 0.25 fs [femtoseconds]).

Detailed discussion of the resulting photodynamics can be found in our previous work.[26,27,30,32,81] For now, we focus on one aspect of the dynamics. A number of different kinds of conical intersections lie below the Franck–Condon point in ethylene. The one that dominates the quenching to the ground state has a twisted and monopyramidalized geometry, which we have called PyrCI. This geometry is depicted in the inset of the left panel in Figure 4.7. A related intersection geometry previously found by Ohmine[115] is the "H-migration" intersection, H-m-CI. The molecular geometry is depicted in the inset of the right panel of Figure 4.7. Looking at the PyrCI geometry in detail reveals that one of the C–H bonds is extended, that is, there is some H-migration character already present. This suggests that the H-m-CI is actually closely related to the PyrCI geometry, and in fact they belong to the same seam.[116] In Figure 4.7, we show the topography of these two intersections. The PES around the PyrCI geometry is more clearly "peaked" than it is around the H-m-CI geometry. This type of topographical consideration has been suggested to be important to the efficiency of quenching around a conical intersection; the more strongly peaked an intersection is, the more

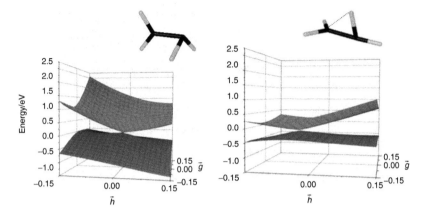

Figure 4.7 Topography around two of the important types of intersections in ethylene, displacing geometries along the g and h vectors that form the "branching plane." (Left panel) The monopyramidalized intersection denoted PyrCI. This is a true minimal energy intersection. (Right panel) An intersection, denoted H-m-CI, where it is evident that one of the hydrogen atoms is migrating to form ethylidene. The topographies around the intersections are quite different, with Pyr-CI being more strongly peaked and H-m-CI being better characterized as sloped. This is expected to have consequences for the relative quenching efficiency of these intersections, as discussed in the text and shown in Figure 4.8.

likely it is to "funnel" population down to the lower state.[117-119] Thus, it is interesting to inquire into the relationship between intersection topography and quenching efficiency in this case.

In Figure 4.8, we show the geometries at which quenching occurs. Every time a nonadiabatic event occurs and a basis function is spawned, we determine the distance between the "migrating" hydrogen atom and the opposite carbon atom and the pyramidalization angle of the most pyramidalized carbon atom. The basis function is followed until the end of the simulation, and its final population determines the radius of the circle that is plotted. The resulting plot shows a tendency toward "clumping" in the upper right and lower left parts of the graph. These areas correspond roughly to the Pyr-CI and H-m-CI geometries. However, the distribution is quite broad, indicating that the minimal energy conical intersection is of

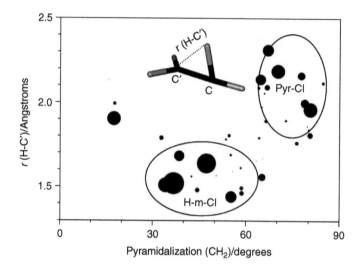

Figure 4.8 The distance between the migrating hydrogen atom (H) and the carbon atom (C) to which it is migrating plotted against the pyramidalization angle of the carbon atom (C) to which H was originally bonded for each of the nonadiabatic events in the C_2H_4 simulations. The size of each circle represents the weight of the basis function by population. There are two groupings of nonadiabatic events possible representing separate families of conical intersections. These correspond to the H-m-CI and Pyr-CI geometries shown in Figure 4.7 and are indicated with ovals.

limited importance, as alluded to in the introduction. In this case, one sees roughly the same amount of population transfer occurring around the two intersection geometries. This shows that, although topography may be important to the resulting quenching efficiency, one also needs to know the details of the dynamics before a clear assessment can be made. In this case, both of the intersections are peaked, so perhaps one should not expect a dramatic difference in the final quenching efficiency.

It is also interesting to inquire about kinematic effects. If the quenching is dominated by monopyramidalization and H-migration, one might expect significant isotope effects. In Figure 4.9, we again show the quenching efficiency as in Figure 4.8 but now in terms of the pyramidalization angle around the left and right carbon atoms. Furthermore, we now

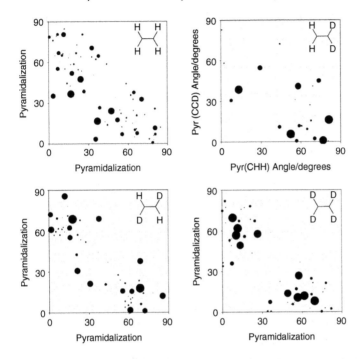

Figure 4.9 The pyramidalization angle of one of the methylene groups plotted against the pyramidalization of the other for each S_1/S_0 nonadiabatic event of the four isotopomers studied. As in Figure 4.8, the size of each circle represents the weight of the ground state basis function at the end of the simulation. Notice that in 1,1-$C_2H_2D_2$ pyramidalization of the CH_2 pyramidalization leads to greater population quenching than does CD_2 pyramidalization.

show four panels, corresponding to different isotopomers. First, we compare C_2H_4 and C_2D_4. The distribution of quenching events is broader for C_2H_4 than for C_2D_4 (the circles are more spread out), and each quenching event transfers more population for C_2D_4 (the circles are larger). This is what might be expected: Because C_2H_4 is lighter, it will access more of the phase space and hence the broader distribution. Comparing CH_2CD_2 with C_2H_4, we see that there is a pronounced tendency for the CH_2 side to pyramidalize. This is again as should be expected from kinematic considerations: The lighter CH_2 can pyramidalize faster, and hence this side of the molecule will tend to pyramidalize first. Finally, the CHDCHD case is intermediate.

4.4.2 Butadiene

As the smallest conjugated organic molecule, 1,3-butadiene has received considerable attention. The *trans* isomer, in particular, offers a great challenge, because the two lowest lying valence excited states are nearly degenerate. There has been considerable controversy concerning the correct ordering of these states. Currently, most high level *ab initio* calculations[120,121] put the bright 1^1B_u ($\pi \rightarrow \pi^*$) state just below the dark 2^1A_g ($\pi^2 \rightarrow \pi^{*2}$) state. With a few exceptions,[122] this is also the ordering deduced from experiments.[123–127] The electronic absorption spectra of butadiene is quite broad in the region of the 1^1B_u excitation,[128,129] and there is no detectable fluorescence. Both of these factors suggest the existence of a fast and efficient radiationless process to relax from the 1^1B_u state. Dinur et al.[130] suggest a nonplanar excited-state minimum and significant vibronic coupling that results in this quick decay from the bright state. Resonance Raman experiments on isoprene by Trulson and Mathies[131] suggest that this relaxation is so fast that any significant geometric distortion (i.e., isomerization) must occur on the dark state. Vibronic coupling models constructed by Ostojic and Domcke[132] and Krawczyk et al.,[133] which accurately reproduced absorption and resonance Raman spectra, demonstrate a fast decay from the 1^1B_u state as well. Recently, the timescale of this process has been investigated with pump-probe spectroscopy, finding that the 1^1B_u state decays in less than 50 fs.[134,135] Zerbetto and Zgierski[136] used quantum chemical and model calculations to study the relaxation to the ground state. They suggest that out-of-plane distortion of butadiene and hexatriene dramatically increases the rate of internal conversion to the ground state. These short polyenes have nonplanar excited state minima, while octatetraene, whereas does have significant fluorescence quantum yield, has a planar excited-state minimum.

Much experimental and theoretical work has also focused on determining reaction mechanisms for the photoinduced isomerization processes of butadiene. In particular, the stereochemical products of the electrocyclic reactions of butadiene have been the topic of detailed examination. The Woodward–Hoffman rules,[137] and other related principles,[138–142] predict the stereochemistry of the products of thermal and photoinduced electrocyclic reactions on the basis of orbital phase. Although these rules have been extremely successful in predicting the products of thermal reactions, they are less effective for predicting photoproducts, especially for reactions involving cyclobutenes.[143] This could be due to fast conversion from the bright excited state to a state of different electronic character, where the Woodward–Hoffman rules do not apply. Dauben and Ritscher[144] found that the photoinduced cyclization of ethylidenecyclooctene results in products with stereochemistry opposite that predicted by the Woodward–Hoffman rules.[144] They suggest the importance of an allyl anion-methyl cation form of butadiene, which is twisted 90° between the allyl and the methyl units. Theoretical work by Bruckmann and Salem[145] proposed the existence of two oppositely charged zwitterionic forms of butadiene on the excited state: the one described by Dauben and Ritscher, and an allyl cation-methyl anion form. They suggest that the relative energy of these two forms depends on the substitution of the terminal carbons, and that violation of the Woodward–Hoffman rules results when the allyl anion form is favored. More recent work by Garavelli et al.[146] demonstrated the strong effect substitution of C_2 and C_3 can have on the PES of butadiene as well.

A great deal of work has been done to study the reaction paths leading to all the various photoproducts of butadiene. Early experiments demonstrated that the most common photoisomerization processes in butadienes are cyclobutene formation and double bond isomerization.[147,148] *Ab initio* calculations by Aoyagi and coworkers[149] and Aoyagi and Osamura[150] showed that twisting around the double bond is likely an important motion in both the singly and doubly excited states of butadiene, and that the PES of the doubly excited state is likely strongly effected by substituent and solvent effects. This work also suggests the importance of a methyl anion–allyl cation form in the photochemistry of buta-diene. On the basis of their low temperature, experimental investigation of deuterated 1,3-dienes, Squillacote and Semple, proposed a PES that accounts for the branching between double bond isomerization and cyclobutene formation and suggests a nonsynchronous disrotatory ring closure path-way.[151] Olivucci et al.[152] and Celani et al.[153] have carried out a detailed study of the reaction paths and conical intersections associated with these photoisomerization processes.[152,153] Their work on 2-cyanobutadiene and 2,3-dimethylbutadiene suggests that substituent effects on the ground-state PES and conservation of dynamical momentum play an important role in photoproduct distribution.[154] Their recent molecular mechanics–valence bond (MMVB) study of the unsubstituted molecule demonstrated that dynamical paths exist from elec-tronically excited s-cis butadiene to several photoproducts (s-trans butadiene, bicyclobutane, Z,Z-butadiene) that lie far from previously discussed minimum energy paths.[155] Simi-larly, *ab initio* calculations reported by Sakai[156] pointed out the importance of charge transfer between C_2 and C_3 as buta-diene quenches to the ground state for photocyclization pro-cesses and suggest that conservation of dynamical momentum is the primary factor favoring cyclobutene formation over bicyclobutane formation in unsubstituted butadiene.[157] How-ever, dynamics calculations were not performed, making it difficult to determine this with certainty.

We have modeled the short-time dynamics of *trans*-buta-diene by using *ab initio* multiple spawning. The results described come from 10 initial conditions, each starting with a single nuclear basis function placed on the brighter of the two excited states. The position and momentum centers of these initial basis functions were chosen from the ground-state Wigner distribution an S_0 equilibrium geometry and vibrational frequencies determined using DFT with the B3LYP functional. The classical equations of motion were integrated using a Velocity-Verlet integrator with a timestep of 0.5 fs, and all simulations were run for 200 fs. The adiabatic PESs and nonadiabatic coupling matrix elements used are calculated on-the-fly using the CASSCF method with analytical gradients as implemented in the MOLPRO molecular electronic structure package. An active space with four electrons in three orbitals was chosen because of its balanced treatment of the two low-lying valence excited states (the bright 1^1B_u state is 0.5 eV below the dark 2^1A_g state). These calculations used state averaging over three states to eliminate variational bias toward the ground state, and the 6-31G electronic basis set was used.

These simulations provide an opportunity to demonstrate the iterative process that we invariably use in application of AIMS. Because of the extreme computational cost, it is rarely practical to use the best-available electronic structure method to generate PESs and couplings. Instead, we start with an investigation of the electronic states in the Franck–Condon region. One can generally apply a number of different methods that are of known reliability in this region, such as CASPT2 and equation-of-motion coupled cluster (EOM-CCSD). Additionally, there is ample information about bright states in the Franck–Condon region through the electronic absorption spectra. In this case, much previous work on the electronic states in the Franck–Condon region has already been reported in the literature. Thus, as discussed above, we know that S_1 and S_2 (we ignore Rydberg states in our labeling here) should be nearly degenerate and that S_1 should be the bright state. A few preliminary calculations show that this result is obtained with an SA-3-CAS(4/4)-PT2 calculation in a 6-31G* basis set. This is already a considerably smaller calculation than those reported by Serrano-Andres et al.[120] but still not ideal for AIMS because of the difficulty of computing gradients with current CASPT2 implementations.

It is natural to try the SA-3-CAS(4/4) method in a 6-31G basis set, but then the S_1/S_2 energy gap at the Franck–Condon point increases to over 2 eV. At this point, the active space can be modified to probe the sensitivity of the target result (in this case the ordering and energy gap for S_1 and S_2). Using SA-3-CAS(4/3) in the 6-31G basis set does give the desired result, matching the S_1/S_2 energy gap from CASPT2 quite closely. However, it does not match the CASPT2 S_0/S_1 energy gap. Indeed, one generally expects (and finds) the vertical excitation energies to be too large in CASSCF methods. Such is the case here, but because our initial inquiry concerns the interaction of S_1 and S_2, we should first be concerned about the S_1/S_2 energy gap. Proceeding with the dynamics after photoexcitation, the bond alternation coordinate is important (switching the single and double bond character of the carbon backbone). We will discuss this in more detail, but for the moment the important point is that the next step is to revalidate the SA-3-CAS(4/3)/6-31G method we have chosen for the electronic structure theory. Having identified the coordinate that is most important at the earliest times, the natural test is to compare the SA-3-CAS(4/3)/6-31G and SA-3-CAS(4/4)-PT2/6-31G* methods along this coordinate. The result of this is shown in Figure 4.10, where we demonstrate that the chosen electronic structure method reproduces the more expensive CASPT2 results almost quantitatively. This bootstrapping process is continued throughout an AIMS study of photochemistry; initial benchmarking in the Franck–Condon region provides a practical electronic structure theory method that generates dynamical results that are then validated using higher levels of electronic structure theory.

The dominant motion in the short time dynamics of *trans*-butadiene in the excited state is bond alternation. Figure 4.11 shows the C–C bond lengths averaged over the 10 simulations and clearly demonstrates the shortening of the central C–C bond accompanied by lengthening of the two outer C–C bonds. This bond alternation motion clearly favors the 2^1A_g state (see Figure 4.10) resulting in the nuclear wavepacket encountering an intersection almost immediately after excitation. However, because the molecule is still nearly symmetric during these early moments of the dynamics, and S_1 and S_2 belong to different irreducible representations in the point group of the molecule at the Franck–Condon point, we expect nearly diabatic population transfer from S_1 to S_2.

The next motion that "lights up" after bond alternation is torsion. Significant twisting around the terminal C–C bonds is observed, as can be seen in Figure 4.12. Here, we define the twist angle of the terminal CH_2 group as the angle

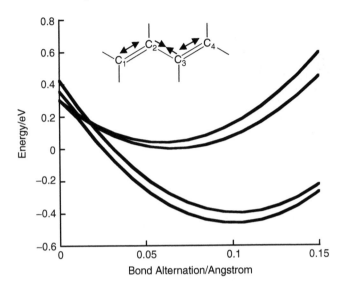

Figure 4.10 Potential-energy curves for S_1 and S_2 along the bond alternation coordinate from the Franck–Condon point. The arrows in the inset define the bond alternation coordinate, which involves equal compression and extension of the indicated bonds. All other coordinates are held constant at their values at the ground-state equilibrium geometry. The results obtained with an SA-3-CAS(4/3) wave function in a 6-31G basis set (black lines) are compared with results using SA-3-CAS(4/4)-PT2 with a 6-31G* basis set (gray lines). The zero of energy is arbitrary, and the curves are shifted such that the S_1 energies of the CAS and CASPT2 calculations coincide at the Franck–Condon point. The S_1/S_2 energy gap with an SA-3-CAS(4/4) wave function and the 6-31G basis set is 2.28eV, compared with <0.15eV obtained with the more reliable SA-3-CAS(4/4)-PT2/6-31G* method. These results suggest the SA-3-CAS(4/3)/6-31G wave function as a good choice for AIMD studies, as discussed in the text.

between the plane containing both central and one-terminal carbon atoms and the line connecting the two terminal hydrogen atoms. Initially, the terminal CH_2 groups twist in concert. This motion, along with the continued bond alternation, leads to quenching through conical intersections such as a disrotatory twisted intersection (not shown). Eventually, the symmetry of the terminal group torsion is broken and the molecule

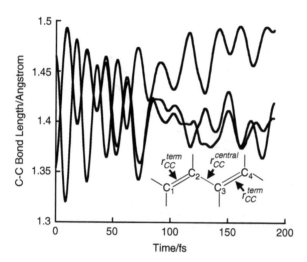

Figure 4.11 Carbon–carbon bond length as a function of time after photoexcitation of butadiene, averaged over all runs. The black line shows the C_2–C_3 bond length, $r_{CC}^{central}$, and the gray lines represent the C_1–C_2 and C_3–C_4 bond lengths, r_{CC}^{term}. Motion in the C–C–C–C bond alternation mode is very important in the first 50 fs after excitation.

tends toward the S_2 minimum, which has a planar allylic group and a 90° twisted terminal methylene group. Because we have the electronic wave function at our disposal, it is natural to characterize the electronic distribution that accompanies the observed dynamics. As shown in Figure 4.13, considerable charge transfer occurs, beginning at approximately 50 fs. The more twisted of the two ethylenic groups carries the negative charge, consistent with the allyl cation–methyl anion type of resonance structure that might have been anticipated on the basis of our ethylene results. This echoes earlier suggestions of Bruckmann and Salem.[145] The gap at the S_2 minimum that we have characterized is very small (0.05 eV), suggesting that a conical intersection may lie very close by. The time it takes to quench back to S_1 varies greatly among the 10 simulations because it depends on how directly the trajectories hit the seam of conical intersections.

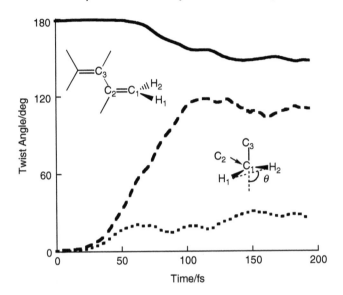

Figure 4.12 C1–C2–C3–C4 dihedral angle (solid line) and terminal C–C bond twist angles (dashed and dotted lines) as a function of time after photoexcitation of butadiene. Results are averaged over 10 initial conditions. The dashed and dotted lines are the average of the twist angle for the more- and less-twisted terminal bonds, respectively. The definition of the terminal C–C bond twist angle is illustrated in the inset and described in the text. Asymmetric twisting around a single terminal C–C bond begins about 40 fs after excitation. Eventually, the more-twisted bond tends toward 90° whereas the less-twisted bond remains nearly planar.

Once the molecules are back on S_1, they quickly move toward various S_1-S_0 conical intersections, sometimes quenching in as little as 25 fs. The S_1 minimum is twisted and pyramidalized around one terminal carbon, but the PES is quite flat in this region, and in our simulations, pyramidalization around both of the carbons in the twisted ethylenic unit is observed. Most of the quenching in our simulation occurs at geometries near intersections with significant H-migration character, but quenching at other geometries is also observed. The population dynamics is shown in Figure 4.14. Almost 70% of the population is transferred to S_2 immediately

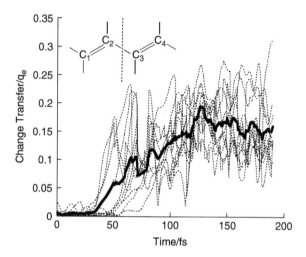

Figure 4.13 Difference in charge (in units of electronic charge) between the two ethylenic units (defined as left and right of the dashed line in the inset) as a function of time after photoexcitation of butadiene. Dashed lines represent the values from individual runs, and the bold line represents the average over all runs. Significant charge transfer begins about 40 fs after excitation.

after photoexcitation. This is really diabatic passage through the S_1/S_2 intersection seam and occurs because the two states are strongly coupled in the Franck–Condon region. The quenching from S_2 back to S_1 takes about 75 fs and occurs simultaneously with population transfer from S_1 to S_0. Because of our short simulation time, we cannot say much about the fate of ground-state molecules after quenching, although efforts to elucidate this are in progress. The major message from the AIMS simulations is that charge transfer is important in butadiene photochemistry, just as it is for ethylene. Furthermore, we see very fast transfer from S_2 to S_1, in accord with Kasha's rule.

4.4.3 Stilbene

In the context of photoinduced *cis-trans* isomerization, stilbene has been one of the most intensely studied molecules,

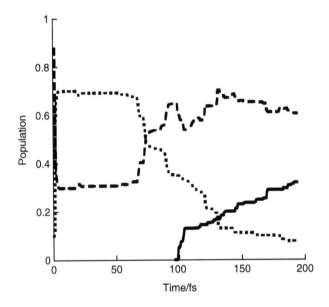

Figure 4.14 Population on S_0 (solid line), S_1 (dashed line), and S_2 (dotted line) as a function of time after photoexcitation of butadiene. Results are averaged over 10 runs. Immediately upon excitation to the bright excited state, there is nearly diabatic population transfer to S_2. Quenching from S_2 to S_1 begins about 70 fs after photoexcitation and quenching to the ground state begins roughly 30 fs later.

in large part because it absorbs light in an experimentally convenient wavelength region. Much of the discussion of stilbene photochemistry has centered on a one-dimensional reaction coordinate that is primarily torsional.[158–161] Although there have been only a handful of suggestions that the reaction coordinate was multidimensional in the case of *trans*-stilbene,[162,163] a number of workers have suggested that the photochemistry of the *cis* isomer cannot be one dimensional.[164–172] In particular, steric repulsion forces the ground-state geometry of the *cis* isomer to be nonplanar, and both photoisomerization and photocyclization pathways are accessed.[169,173–176] Thus, a few theoretical studies of photodynamics in the *cis* isomer have allowed for degrees of freedom beyond simple torsion.[177–182] However, these have emphasized phenyl rotation, and the pyramidalization coordinate that we found to be important in ethylene has gone mostly unnoticed.

Recently, we have carried out an *ab initio* study of the ground- and excited-state PESs of stilbene that showed that pyramidalization is likely to be an important coordinate in stilbene photoisomerization.[183] Indeed, the similarities between the PESs of stilbene and ethylene are quite striking, especially in the region near the lowest-energy S_0/S_1 conical intersection. It is therefore interesting to determine the extent to which these similarities carry over to the photodynamics. We discuss the results of an excited-state molecular dynamics simulation consisting of a single trajectory, starting from the Franck–Condon point of *cis*-stilbene. Potential-energy surfaces and gradients were calculated with SA-2-CAS(2/2) wave functions in the 6-31G basis set. The simulation began on the S_1 PES at the S_0 *cis*-stilbene equilibrium geometry with zero kinetic energy, which models photoexcitation in the gas phase at 0K. Snapshots at points along this trajectory are shown in Figure 4.15. The initial motion involves both bond alternation and ethylenic torsion. At this level of theory, we have found a local minimum on S_1 corresponding to a *cis*-oid structure. Thus, there is a small barrier to torsion from the Franck–Condon point, which we have estimated to occur at a 50° ethylenic torsional angle and to be less than 2 kcal/mol. The evolution of the dihedral angles is shown in Figure 4.16, where the molecule has passed the S_1 *cis-trans* isomerization barrier after less than 40 fs. This is consistent with the short ($\approx$300 fs) experimental lifetimes measured in the gas phase for the *cis* isomer.[184–186] Ten femtoseconds later, the ethylene moiety becomes perpendicular, and the molecule enters the region of the global S_1 minimum — the "phantom state."[159–161,187,188]

The details of the ethylenic twisting are quite interesting. Indeed, it is somewhat misleading to focus on the dihedral angles shown in Figure 4.16. As shown in Figure 4.17, the phenyl rings are nearly stationary during the first 150 fs. All the changes (~35) in the phenyl rotation angles over this time period are due to geometrical changes localized in the ethylenic part of the molecule. This is in excellent agreement with

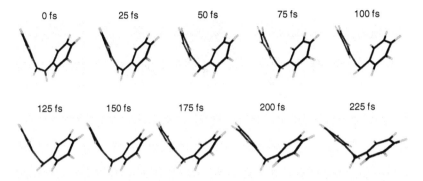

Figure 4.15 Snapshots of the dynamics of stilbene after $\pi \rightarrow \pi^*$ excitation.

the resonance Raman studies of Myers and Mathies.[168] They fit their spectra using reduced dimensionality harmonic models and concluded that the short-time dynamics of *cis*-stilbene would be dominated by ethylenic torsion with the phenyl rings acting as spectators. Indeed, they predicted that the ethylenic H–C–C–H dihedral angle would change by 25° in 20 fs, in very good agreement with the behavior seen in Figure 4.16. These results show that even a single excited-state trajectory can provide useful information about photochemical dynamics, although many more trajectories and the inclusion of population decay through spawning are needed for detailed comparisons with experiment.

4.5 CONCLUSIONS

The field of first-principles photochemical simulation is clearly still in its infancy. We have given an overview of the issues involved in the dynamics and electronic structure aspects of the problem and presented a number of applications to gas-phase photochemistry. At the same time, we have tried to give the reader a sense of what is to come: The implementation of QM/MM techniques and reparameterized semiempirical methods will make "nearly" first-principles simulations of photodynamics possible in solution and protein environ-

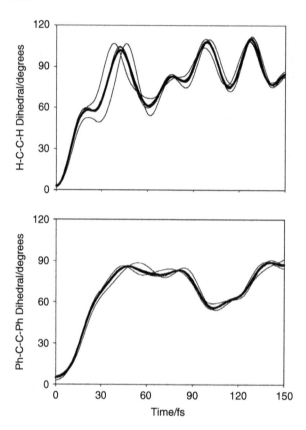

Figure 4.16 Evolution of dihedral angles in *cis*-stilbene after photoexcitaiton. Light lines represent the two possible dihedral angles in each case, and dark lines are averages of these. Dihedral angles are representative of a twisted molecule within 50 fs.

ments. This will make it possible to begin using computational techniques to *design* photochemical "devices." It is anticipated that the first efforts along these lines will probably consist of predicting the outcome of alterations to a well-defined photochemical structure, for example, point mutations in a photoactive protein.

We intended to make a few specific points in our choice of applications. The role of intersection topography is alluded to throughout this paper, but a specific example is provided in the case of ethylene, where the different types of intersec-

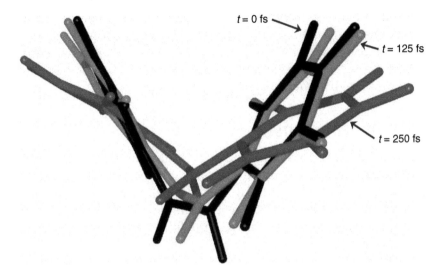

Figure 4.17 Superimposed snapshots of photoexcited *cis*-stilbene at three points along an *ab initio* molecular dynamics simulation started at the Frank–Condon geometry (black). In 125 fs (light gray), the Ph-C-C-Ph dihedral (70°) indicates a twisted geometry, as shown in Figure 4.16. However, the phenyl rings have barely moved. This is because the initial dynamics is dominated by large amplitude motion of the ethylenic core, and the dihedral angles are somewhat misleading. After 250 fs (dark gray), the phenyl rings have become perpendicular.

tions are of different importance because of differences in their "sloped" or "peaked" character. In the case of butadiene, we provided a detailed example of the iterative procedure that we use to ensure that the electronic structure treatment being used is adequate, without requiring that all calculations be performed with the absolute best-possible quantum chemical methods. The stilbene example highlights the importance of detailed investigation of the molecular dynamics. The molecule begins twisting immediately, but kinematic effects keep the phenyl rings (which form the bulk of the molecule) essentially stationary during much of the reaction. One can easily imagine how these considerations will be important for photoisomerization in solution.

Even though significant advances can be expected first-principles photochemistry is currently practical and capable of providing new insights into photochemical mechanisms. The extension of these approaches into condensed phases and complex environments is already underway.

ACKNOWLEDGMENTS

It is a pleasure to acknowledge partial support of this work from the National Science Foundation and the Department of Energy. TJM is grateful to the Beckman, Packard, and Dreyfus Foundations for a Beckman Young Investigator Award, Packard Fellowship, and Teacher-Scholar Award, respectively.

REFERENCES

1. Teller, E., *J. Phys. Chem.*, 41, 109–116, 1937.

2. Teller, E., *Isr. J. Chem.*, 7, 227–235, 1969.

3. Yarkony, D.R., *J. Phys. Chem.*, 105, 6277–6293, 2001.

4. Yarkony, D.R., *Rev. Mod. Phys.*, 68, 985–1013, 1996.

5. Robb, M.A., Bernardi, F., and Olivucci, M., *Pure Appl. Chem.* 67, 783–789, 1995.

6. Matsika, S. and Yarkony, D.R., *J. Chem. Phys.*, 117, 6907–6910, 2002.

7. Bearpark, M.J., Robb, M.A., and Schlegel, H.B., *Chem. Phys. Lett.*, 223, 269–274, 1994.

8. Manaa, M.R. and Yarkony, D.R., *J. Chem. Phys.*, 99, 5251–5256, 1993.

9. Leforestier, C., *J. Chem. Phys.*, 68, 4406–4410, 1978.

10. Maluendes, S.A. and Dupuis, M., *Int. J. Quantum Chem.*, 42, 1327–1338, 1992.

11. Baldridge, K.K., Gordon, M.S., Steckler, R., and Truhlar, D.G., *J. Phys. Chem.*, 93, 5107–5119, 1989.

12. Hartke, B. and Carter, E.A., *Chem. Phys. Lett.*, 189, 358–362, 1992.

13. Hartke, B. and Carter, E.A., *J. Chem. Phys.*, 97, 6569–6578, 1992.

14. Car, R. and Parrinello, M., *Phys. Rev. Lett.*, 55, 2471–2474, 1985.

15. Parr, R.G. and Yang, W., *Density-Functional Theory of Atoms and Molecules*, Vol. 16, 1st ed., Oxford University Press, Oxford, 1989.

16. Theophilou, A., *J. Phys., C*, 12, 5419–5430, 1979.

17. Grimme, S., *Chem. Phys. Lett.*, 259, 128–137, 1996.

18. Goerling, A., *Phys. Rev. A: At. Mol. Opt. Phys.*, 54, 3912–3915, 1996.

19. Hsu, C.-P., Hirata, S., and Head-Gordon, M., *J. Phys. Chem.*, 105A, 451–458, 2001.

20. Tozer, D.J. and Handy, N.C., *Phys. Chem. Chem. Phys.*, 2, 2117–2121, 2000.

21. Grabo, T., Petersilka, M., and Gross, E.K.U., *J. Mol. Struct. (THEOCHEM)*, 501, 353–367, 2000.

22. Hessler, P., Maitra, N.T., and Burke, K., *J. Chem. Phys.*, 117, 72–81, 2002.

23. Frank, I., Hutter, J., Marx, D., and Parrinello, M., *J. Chem. Phys.*, 108, 4060–4069, 1998.

24. Marx, D. and Doltsinis, N.L., *Phys. Rev. Lett.*, 88, 166402, 2002.

25. Vreven, T., Bernardi, R., Garavelli M., Olivucci, M., Robb, M.A., and Schlegel, H.B., *J. Am. Chem. Soc.*, 119, 12687–12688, 1997.

26. Ben-Nun, M., Quenneville, J., and Martinez, T.J., *J. Phys. Chem. A*, 104, 5161–5175, 2000.

27. Ben-Nun, M. and Martínez, T.J., *Chem. Phys. Lett.*, 298, 57–65, 1998.

28. Quenneville, J., Ben-Nun, M., and Martínez, T.J., *J. Photochem. Photobiol.*, 144A, 229–235, 2001.

29. Martínez, T.J., *Chem. Phys. Lett.*, 272, 139–147, 1997.

30. Ben-Nun, M. and Martínez, T.J., *J. Phys. Chem.*, 103, 10517, 1999.

31. Ben-Nun, M. and Martínez, T.J., *J. Am. Chem. Soc.*, 122, 6299, 2000.

32. Ben-Nun, M. and Martinez, T.J., *ACP*, 121, 439–512, 2002.

33. Thompson, K. and Martínez, T.J., *J. Chem. Phys.*, 110, 1376, 1998.

34. Born, M. and Oppenheimer, R., *Ann. Physik*, 84, 457–484, 1927.

35. Landau, L.D. and Lifschitz, E.M., *Quantum Mechanics*, Pergamon, New York, 1965.

36. Longuet-Higgins, H.C., *Proc. R. Soc. Lond. A*, 344, 147–156, 1975.

37. Matsika, S. and Yarkony, D.R., *J. Chem. Phys.*, 115, 2038–2050, 2001.

38. Matsika, S. and Yarkony, D.R., *J. Chem. Phys.*, 115, 5066–5075, 2001.

39. Mead, C.A. and Truhlar, D.A., *J. Chem. Phys.*, 77, 6090–6098, 1982.

40. Baer, M., *Chem. Phys.*, 15, 49, 1976.

41. Gadéa, F.X. and Pélissier, M., *J. Chem. Phys.*, 93, 545–551, 1990.

42. Petsalakis, I.D., Toedorakopoulos, G., and Nicolaides, C.A., *J. Chem. Phys.*, 97, 7623–7628, 1992.

43. Granucci, G., Persico,M., and Toniolo, A., *J. Chem. Phys.*, 114, 10608–10615, 2001.

44. Feynman, R.P. and Hibbs, A.R., *Quantum Mechanics and Path Integrals*, McGraw-Hill, New York, 1964.

45. Makri, N., *Comp. Phys. Commun.*, 63, 389–414, 1991.

46. Miller, W.H. and George, T.F., *J. Chem. Phys.*, 56, 5637, 1972.

47. Pechukas, P., *Phys. Rev.*, 181, 174, 1969.

48. Webster, F.J., Rossky, P.J., and Friesner, R.A., *Comp. Phys. Commun.*, 63, 494, 1991.

49. Webster, R., Wang, E.T., Rossky, P.J., and Friesner, R.A., *J. Phys. Chem.*, 104, 4835, 1994.

50. Coker, D.F. and Xiao, L., *J. Chem. Phys.*, 102, 496, 1995.

51. Desouter-Lecomte, M., Leclerc, J.C., and Lorquet, J.C., *Chem. Phys.*, 9, 147–156, 1975.

52. Meyer, H.-D. and Miller, W.H., *J. Chem. Phys.*, 70, 3214–3223, 1979.

53. Meyer, H.-D., *Chem. Phys.*, 82, 199–205, 1983.

54. Stock, G., *J. Chem. Phys.*, 103, 1561–1573, 1995.

55. Sun, X. and Miller, W.H., *J. Chem. Phys.*, 106, 6346–6353, 1997.

56. Klein, S., Bearpark, M.J., Smith, B.R., Robb, M.A., Olivucci, M., and Bernardi, F., *Chem. Phys. Lett.*, 292, 259–266, 1998.

57. Park, T.J. and Light, J., *J. Chem. Phys.*, 85, 5870–5876, 1986.

58. Tully, J.C. and Preston, R.K., *J. Chem. Phys.*, 55, 562–572, 1971.

59. Tully, J.C., *Int. J. Quant. Chem.*, 25S, 299–309, 1991.

60. Herman, M.F., *J. Chem. Phys.*, 81, 754–763, 1984.

61. Müller, U. and Stock, G., *J. Chem. Phys.*, 107, 6230–6245, 1997.

62. Fang, J.Y. and Hammes-Schiffer, S., *J. Phys. Chem. A*, 103, 9399–9407, 1999.

63. Prezhdo, O.V. and Rossky, P.J., *J. Chem. Phys.*, 107, 825–834, 1997.

64. Volubuev, Y.L., Hack, M.D., Topaler, M.S., and Truhlar, D.G., *J. Chem. Phys.*, 112, 9716–9726, 2000.

65. Nakamura, M.D. and Truhlar, D.G., *J. Chem. Phys.*, 114, 9305–9314, 2001.

66. Ben-Nun, M., Quenneville, J., and Martínez, T.J., *J. Phys. Chem.*, 104A, 5161–5175, 2000.

67. Ben-Nun, M. and Martinez, T.J., *Adv. Chem. Phys.*, 121, 439–512, 2002.

68. Heller, E.J., *J. Chem. Phys.*, 75, 2923–2931, 1981.

69. Martínez, T.J., Ben-Nun, M., and Levine, R.D., *J. Phys. Chem.*, 100, 7884–7895, 1996.

70. Roos, B.O., The complete active space self-consistent field method and its applications in electronic structure calculations, in *Advances in Chemical Physics: Ab Initio Methods in Quantum Chemistry II*, Lawley, K.P., Ed., John Wiley and Sons, New York, 1987, pp. 399–445.

71. Schaefer, H.F. and Bender, C.F., *J. Chem. Phys.*, 55, 1720, 1971.

72. Hay, P.J., Dunning, T.H., Jr., and Goddard, W.A. III. *J. Chem. Phys.*, 62, 3912–3924, 1975.

73. Frisch, M.J., Trucks, G.W., Schlegel, H.B., Scuseria, G.E., Robb, M.A., Cheesman, J.R., Zakrzewski, V.G., Montgomery, J.A., Jr., Stratmann, R.E., Burant, J.C., Dapprich, S., Millam, J.M., Daniels, A.D., Kudin, K.N., Strain, M.C., Farkas, O., Tomasi, J., Barone, B., Cossi, M., Cammi, R., Mennucci, B., Pomelli, C., Adamo, C., Clifford, S., Ochterski, J., Petersson, G.A., Ayala, P.Y., Cui, Q., Morokuma, K., Malick, D.K., Rabuck, A.D., Raghavachari, K., Foresman, J.B., Cioslowski, J., Oritiz, J.V., Stefanov, B.B., Liu, G., Liashenko, A., Piskorz, P., Komaromi, I., Gomperts, R., Martin, R.L., Fox, D.J., Keith, T., Al-Laham, M.A., Peng, C.Y., Nanayakkara, A., Gonzalez, C., Challacombe, M., Gill, P.M.W., Johnson, B., Chen, W., Wong, M.W., Andres, J.S., Gonzalez, C., Head-Gordon, M., Replogle, E.S., and Pople, J.A., *Gaussian 98, Revision A.1*, Gaussian, Inc., Pittsburgh, 1998.

74. Adamo, C. and Barone, V., *Chem. Phys. Lett.*, 314, 152–157, 1999.

75. Hirata, S., Lee, T.J., and Head-Gordon, M., *J. Chem. Phys.*, 111, 8904–8912, 1999.

76. Stratmann, R.E., Scuseria, G.E., and Frisch, M.J., *J. Chem. Phys.*, 109, 8218–8224, 1998.

77. Bauernschmitt, R. and Ahlrichs, R., *Chem. Phys. Lett.*, 256, 454–464, 1996.

78. Casida, M.E. and Salahub, D.R., *J. Chem. Phys.*, 113, 8918–8935, 2000.

79. Hirata, S. and Head-Gordon, M., *Chem. Phys. Lett.*, 314, 291–299, 1999.

80. Hirata, S. and Head-Gordon, M., *Chem. Phys. Lett.*, 302, 375–382, 1999.

81. Ben-Nun, M. and Martínez, T.J., *Chem. Phys.*, 259, 237, 2000.

82. Becke, A.D., *J. Chem. Phys.*, 98, 5648–5652, 1993.

83. Wiberg, K.B., Stratmann, R.E., and Frisch, M.J., *Chem. Phys. Lett.*, 297, 60–64, 1998.

84. Dobson, J.F., Buner, M.J., and Gross, E.K.U., *Phys. Rev. Lett.*, 79, 1905–1908, 1997.

85. Maitra, N.T., Burke, K., and Woodward, C., *Phys. Rev. Lett.*, 89, 023002, 2002.

86. Casida, M.E., Gutierrez, F., Guan, J., Gadea, F.X., Salahub, D.R., and Daudey, J.P., *J. Chem. Phys.*, 113, 7062–7071, 2000.

87. Del Bene, J. and Jaffe, H.H., *J. Chem. Phys.*, 48, 1807–1813, 1968.

88. Ridley, J. and Zerner, M., *Theor. Chem. Acc.*, 32, 111–134, 1973.

89. Thiel, W., *J. Am. Chem. Soc.*, 103, 1413–1420, 1981.

90. Thiel, W., *J. Am. Chem. Soc.*, 103, 1420–1425, 1981.

91. Schweig, A. and Thiel, W., *J. Am. Chem. Soc.*, 103, 1425–1431, 1981.

92. Reinsch, M., Howeler, U., and Klessinger, M., *Angew. Chem., Int. Ed.*, 99, 250–252, 1987.

93. Reinsch, M., Howeler, U., and Klessinger, M., *J. Mol. Struct. (THEOCHEM)*, 167, 301–306, 1988.

94. Klessinger, M., Potter, T., and Wullen, C.V., *Theor. Chem. Acc.*, 80, 1–17, 1991.

95. Dreyer, J. and Klessinger, M., *J. Chem. Phys.*, 101, 10655–10665, 1994.

96. Cullen, J.M., *Int. J. Quantum Chem.*, 56, 97–113, 1995.

97. Toniolo, A., Ben-Nun, M., and Martinez, T.J., *J. Phys. Chem.*, 106A, 4679–4689, 2002.

98. Granucci, G. and Toniolo, A.,. *Chem. Phys. Lett.*, 325, 79, 2000.

99. Segal, G.A., *Semiempirical Methods of Electronic Structure Calculation*, Plenum Press, New York, 1977.

100. Dewar, M.J.S. and Thiel, W., *J. Am. Chem. Soc.,* 99, 4899–4907, 1977.

101. Dewar, M.J.S. and Thiel, W., *J. Am. Chem. Soc.,* 99, 4907–4917, 1977.

102. Dewar, M.J.S., Zoebisch, E.G., Healy, E.F., and Stewart, J.J.P., *J. Am. Chem. Soc.,* 107, 3902–3909, 1985.

103. Stewart, J.J.P., *J. Comp. Chem.,* 10, 209–220, 1989.

104. Peslherbe, G.H. and Hase, W.L., *J. Chem. Phys.,* 104, 7882–7894, 1996.

105. Toniolo, A., Granucci, G., Inglese, S., and Persico, M., *Phys. Chem. Chem. Phys.,* 3, 4266–4279, 2001.

106. Rossi, I. and Truhlar, D.G., *Chem. Phys. Lett.,* 233, 231–236, 1995.

107. Press, W.H., Teukolsky, S.A., Vetterling, W.T., and Flannery, B.P., *Numerical Recipes,* Cambridge University Press, Cambridge, 1992.

108. Inglese, S., Granucci, G., and Persico, M., in preparation.

109. Warshel, A. and Levitt, M., *J. Mol. Biol.,* 103, 227–249, 1976.

110. Stewart, J.J.P., *MOPAC2000,* Fujitsu Limited, Tokyo, 1999.

111. Antes, I. and Thiel, W., *J. Phys. Chem.,* 103A, 9290–9295, 1999.

112. Sension, R.J. and Hudson, B.S., *J. Chem. Phys.,* 90, 1377–1389, 1989.

113. Farmanara, P., Stert, V., and Radloff, W., *Chem. Phys. Lett.,* 288, 518–522, 1998.

114. Mestdagh, J.M., Visticot, J.P., Elhanine, M., and Soep, B., *J. Chem. Phys.,* 113, 237–248, 2000.

115. Ohmine, I., *J. Chem. Phys.,* 83, 2348, 1985.

116. Wilsey, S. and Houk, K.N., *J. Am. Chem. Soc.,* 124, 11182–11190, 2002.

117. Atchity, G.J., Xantheas, S.S., and Ruedenberg, K., *J. Chem. Phys.,* 95, 1862, 1991.

118. Ben-Nun, M., Molnar, F., Schulten, K., and Martinez, T.J.,. *Proc. Natl. Acad. Sci.,* 99, 1769–1773, 2002.

119. Yarkony, D.R., *J. Chem. Phys.*, 114, 2601–2613, 2001.

120. Serrano-Andres, L., Merchan, M., Nebot-Gil, I., Lindh, R., and Roos, B.O., *J. Chem. Phys.*, 98, 3151–3162, 1993.

121. Graham, R.L. and Freed, K.F., *J. Chem. Phys.*, 96, 1304–1316, 1992.

122. Chadwick, R.R., Gerrity, D.P., Hudson, B.S., *Chem. Phys. Lett.*, 115, 24–28, 1985.

123. Vaida, V., Turner, R.E., Casey, J.L., and Colson, S.D., *Chem. Phys. Lett.*, 54, 25–29, 1978.

124. Doering, J.P. and McDiarmid, R., *J. Chem. Phys.*, 73, 3617–3624, 1980.

125. Rothberg, L.J., Gerrity, D.P., and Vaida, V., *J. Chem. Phys.*, 73, 5508–5513, 1980.

126. Doering, J.P. and McDiarmid, R., *J. Chem. Phys.*, 75, 2477–2478, 1981.

127. McDiarmid, R. and Doering, J.P., *Chem. Phys. Lett.*, 88, 602–606, 1982.

128. Leopold, D.G., Pendley, R.D., Roebber, J.L., Hemley, R.J., and Vaida, V., *J. Chem. Phys.*, 81, 4218–4229, 1984.

129. Vaida, V., *Acc. Chem. Res.*, 19, 114–120, 1986.

130. Dinur, U., Hemley, R.J., and Karplus, M., *J. Phys. Chem.* 87, 924–932, 1983.

131. Trulson, M.O. and Mathies, R.A., *J. Phys. Chem.*, 94, 5741–5747, 1990.

132. Ostojic, B. and Domcke, W., *Chem. Phys.*, 269, 1–10, 2001.

133. Krawczyk, R.P., Malsch, K., Hohlneicher, G., Gillen, R.C., and Domcke, W., *Chem. Phys. Lett.*, 320, 535–541, 2000.

134. Assenmacher, F., Gutmann, M., Hohlneicher, G., Stert, V., and Radloff, W., *J. Chem. Phys.*, 2981–2982, 2001.

135. Fuss, W., Schmid, W.E., and Trushin, S.A., *Chem. Phys. Lett.*, 342, 91–98, 2001.

136. Zerbetto, F. and Zgierski, M.Z., *J. Chem. Phys.*, 93, 1235–1245, 1990.

137. Hoffmann, R. and Woodward, R.B., *J. Am. Chem. Soc.,* 87, 2046–2048, 1965.

138. Zimmerman, H.E., *J. Am. Chem. Soc.,* 88, 1564–1565, 1966.

139. Zimmerman, H.E., *J. Am. Chem. Soc.,* 88, 1566–1567, 1966.

140. Longuet-Higgins, H.C. and Abrahamson, E.E., *J. Am. Chem. Soc.,* 87, 2045–2046, 1965.

141. van der Lugt, W.T.A.M. and Oosterhoff, L.J., *J. Am. Chem. Soc.,* 91, 6042–6049, 1969.

142. Goddard, W.A., *J. Am. Chem. Soc.,* 94, 793–807, 1972.

143. Leigh, W.J., Jr., Postigo, J.A., and Venneri, P.C., *J. Am. Chem. Soc.,* 117, 7826–7827, 1995.

144. Dauben, W.G. and Ritscher, J.S., *J. Am. Chem. Soc.,* 92, 2925–2926, 1970.

145. Bruckmann, P. and Salem, L.,. *J. Am. Chem. Soc.,* 98, 5037–5038, 1976.

146. Garavelli, M., Frabboni, B., Fato, F., Celani, P., Bernardi, F., Robb, M.A., and Olivucci, M., *J. Am. Chem. Soc.,* 121, 1537–1545, 1999.

147. Srinivasan, R., *J. Am. Chem. Soc.,* 85, 4045–4046, 1963.

148. Srinivasan, R., *J. Am. Chem. Soc.,* 90, 4498–4499, 1968.

149. Aoyagi, M., Osamura, Y., and Iwata, S., *J. Chem. Phys.,* 83, 1140–1148, 1985.

150. Aoyagi, M. and Osamura, Y., *J. Am. Chem. Soc.,* 111, 470–474, 1989.

151. Squillacote, M. and Semple, T.C., *J. Am. Chem. Soc.,* 112, 5546–5551, 1990.

152. Olivucci, M., Ragazos, I.N., Bernardi, F., and Robb, M.A., *J. Am. Chem. Soc.,* 115, 3710–3721, 1993.

153. Celani, P., Bernardi, F., Olivucci, M., and Robb, M.A., *J. Chem. Phys.,* 102, 5733–5742, 1995.

154. Olivucci, M., Ragazos, I.N., Bernardi, F., and Robb, M.A., *J. Am. Chem. Soc.,* 116, 2034–2048, 1994.

155. Garavelli, M., Bernardi, F., Olivucci, M., Bearpark, M.J., Klein, S., and Robb, M.A., *J. Phys. Chem.*, 105A, 11496–11504, 2001.

156. Sakai, S., *Chem. Phys. Lett.*, 287, 263–269, 1998.

157. Sakai, S., *Chem. Phys. Lett.*, 319, 687–694, 2000.

158. Saltiel, J., *J. Am. Chem. Soc.*, 89, 1036–1037, 1967.

159. Orlandi, G. and Siebrand, W., *Chem. Phys. Lett.*, 30, 352–354, 1975.

160. Tavan, P. and Schulten, K., *Chem. Phys. Lett.*, 56, 200–204, 1978.

161. Waldeck, D.H., *Chem. Rev.*, 91, 415–436, 1991.

162. Park, N.S. and Waldeck, D.H., *Chem. Phys. Lett.*, 168, 379–384, 1990.

163. Lee, M., Haseltine, J.N., Smith, A.B. III, and Hochstrasser, R.M., *J. Am. Chem. Soc.*, 111, 5044–5051, 1989.

164. Todd, D.C., Jean, J.M., Rosenthal, S.J., Ruggiero, A.J., Yang, D., and Fleming, G.R., *J. Chem. Phys.*, 93, 8658–8668, 1990.

165. Abrash, S., Repinec, S., and Hochstrasser, R.M., *J. Chem. Phys.*, 93, 1041–1053, 1990.

166. Rodier, J.-M., and Myers, A.B., *J. Am. Chem. Soc.*, 115, 10791–10795, 1993.

167. Todd, D.C., Fleming, G.R., and Jean, J.M., *J. Chem. Phys.*, 97, 8915–8925, 1992.

168. Myers, A.B. and Mathies, R.A., *J. Chem. Phys.*, 81, 1552–1558, 1984.

169. Repinec, S.T., Sension, R.J., Szarka, A.Z., and Hochstrasser, R.M., *J. Phys. Chem.*, 95, 10380–10385, 1991.

170. Sension, R.J., Szarka, A.Z., and Hochstrasser, R.M., *J. Chem. Phys.*, 97, 5239–5242, 1992.

171. Sension, R.J., Repinec, S.T., Szarka, A.Z., and Hochstrasser, R.M., *J. Chem. Phys.*, 98, 6291–6315, 1993.

172. Szarka, A.Z., Pugliano, N., Palit, D.K., and Hochstrasser, R.M., *Chem. Phys. Lett.*, 240, 25–30, 1995.

173. Muszkat, K.A. and Fisher, E., *J. Chem. Soc. B*, 662–678, 1967.

174. Petek, H., Yoshihara, K., Fujiwara, Y., Lin, Z., Penn, J.H., and Frederick, J.H., *J. Phys. Chem.*, 94, 7539–7543, 1990.

175. Molina, V., Merchan, M., and Roos, B.O., *Spect. Acta A,* 55, 433–446, 1999.

176. Warshel, A., *J. Chem. Phys.*, 62, 214–221, 1975.

177. Frederick, J.H., Fujiwara, Y., Penn, J.H., Yoshihara, K., and Petek, H., *J. Phys. Chem.*, 95, 2845–2858, 1991.

178. Berweger, C.D., van Gunsteren, W.F., and Muller-Plathe, F., *J. Chem. Phys.*, 108, 8773–8781, 1998.

179. Berweger, C.D., van Gunsteren, W.F., and Muller-Plathe, F., *J. Chem. Phys.*, 111, 8987–8999, 1999.

180. Berweger, C.D., van Gunsteren, W.F., and Muller-Plathe, F., *Angew. Chem. Int. Ed.*, 38, 2609–2611, 1999.

181. Vachev, V.D., Frederick, J.H., Grishanin, B.A., Zadkov, V.N., and Koroteev, N.I., *Chem. Phys. Lett.*, 215, 306–311, 1993.

182. Vachev, V.D., Frederick, J.H., Grishanin, B.A., Zadkov, V.N., and Koroteev, N.I., *J. Phys. Chem.*, 99, 5247–5263, 1995.

183. Quenneville, J. and Martinez, T.J., *J. Phys. Chem.*, 107A, 829–837, 2003.

184. Greene, B.I. and Farrow, R.C., *J. Chem. Phys.*, 76, 3336–3338, 1983.

185. Baumert, T., Frohnmeyer, T., Kiefer, B., Niklaus, P., Strehle, M., Gerber, G., and Zewail, A.H., *Appl. Phys. B,* 72, 105–108, 2001.

186. Pederson, S., Banares, L., and Zewail, A.H., *J. Chem. Phys.*, 97, 8801–8804, 1992.

187. Saltiel, J. and D'Agostino, J.T.,. *J. Am. Chem. Soc.*, 94, 6445–6456, 1972.

188. Saltiel, J., D'Agostino, J.T., Megarity, E.D., Metts, L., Neuberger, L.R., Wrighton, M., and ZaFiriou, O.C., *Org. Photochem.*, 3, 2–113, 1973.

5

The Study of Nitrenes by Theoretical Methods

NINA P. GRITSAN,* MATTHEW S. PLATZ,** AND
WESTON THATCHER BORDEN***

*Institute of Chemical Kinetics and
Combustion and Novosibirsk State University,
Novosibirsk, Russia

**Department of Chemistry, Ohio State
University
Columbus, OH

***Department of Chemistry, Box 351700,
University of Washington,
Seattle, WA

CONTENTS

5.1 INTRODUCTION

Compounds containing neutral, monovalent nitrogen atoms are known as nitrenes. The parent structure, NH, is called imidogen. Because most stable compounds of neutral nitrogen have a valence of three, it is no surprise that nitrenes are typically very reactive and short lived and, hence, are intermediates.

Nitrenes are commonly generated by decomposition of organic and inorganic azides, although other precursors are known. Azides form bonds to many elements; consequently, many types of nitrenes are known or can be imagined. This review will be limited to the application of theoretical methods to only those nitrenes commonly encountered in organic chemistry.[1]

Nitrenes are involved in many useful transformations in classical synthetic organic chemistry.[1] Physical organic chemists seek to understand the role of nitrenes in these reactions and how the structures of nitrenes control their reactivity.[2] Biochemists append azide groups to the natural ligands of biological macromolecules. Photolysis of the complex between ligand-bound azide and a biomolecule often leads to covalent attachment of the ligand to the biomolecule. This technique, invented by Singh et al.[3] and adapted for use with azides by Bayley and Knowles,[4] has come to be known as photoaffinity labeling.[5,6] Materials chemists use nitrene chemistry to attach probes to surfaces.[7] Finally, industrial scientists use azide photochemistry and the nitrenes thus formed in lithography as photoresists.[8]

Nitrenes have attracted the interest of spectroscopists. Spectroscopic studies of nitrenes have been performed in the gas phase, in inert gas matrices, and in solution by using time resolved methodologies.

The energy separations between ground and excited states of nitrenes are relatively small in comparison with the energy separations between these states in closed-shell molecules. This difference dramatically increases the level of complexity and the difficulty in assigning the chemistry and spectroscopy of a transient nitrene to a particular electronic state. In this regard, the proper use of theoretical methods has been invaluable.

This chapter will discuss the use of theory to calculate the minimum-energy geometries of singlet and triplet nitrenes and the energy separations between these states. We will demonstrate how theoretical methods can now routinely be utilized to simulate the vibrational and electronic spectra of nitrenes and, thus, to assign the experimental spectra to one or both of these spin states. Finally, we will review the ways in which theory has explained how the electronic structures of nitrenes controls their reactivity.

There has been recent, dramatic progress in our experimental and theoretical understanding of organic nitrenes.[9–11] This review will attempt to highlight these developments and to illustrate the synergism between theory and experiment, which has led to our current level of understanding of nitrenes.

5.2 ELECTRONIC STRUCTURES AND PROPERTIES OF THE SIMPLEST NITRENE, IMIDOGEN RADICAL NH

Very detailed spectroscopic, kinetic, and thermodynamic information has been reported for the simplest azide and nitrene, HN_3 and NH, respectively.[12-20] The parent nitrene, imidogen, can be produced by photolysis,[13,15,18-21] thermolysis,[22] or multiphoton dissociation[23,24] of hydrazoic acid (HN_3). Photodecomposition of ammonia[12,16] or isocyanic acid[20] yields NH as well.

Photolysis of HN_3 with 266- and 248-nm light generates NH almost exclusively in the lowest singlet state ($^1\Delta$) with a quantum yield near unity.[15,18-20] Formation of NH in different excited states was observed upon photolysis of HN_3 with light of shorter wavelengths.[13,15,20] For instance, NH in $X\,^3\Sigma^-$, $a^1\Delta$, $b^1\Sigma^+$, $A^3\Pi$, and $c^1\Pi$ states was found to be formed by UV photolysis of HN_3 at 193 nm and at 300 K with quantum yields ≤ 0.0019, 0.4, 0.017, 0.00015, and 0.00061, respectively.[20]

NH in its lowest singlet state ($a^1\Delta$) inserts readily into paraffin CH bonds. In this electronic state the nitrene also abstracts hydrogen atoms from hydrocarbons and undergoes quenching to the ground triplet state.[25,26] For example, the ratio of these channels is 0.6:0.1:03 in the case of the reaction of ^{1}NH with ethane:[26]

$$NH\left(^1\Delta\right) + C_2H_6 \quad \rightarrow C_2H_5NH_2^* \qquad \text{(insertion, } \sim 60\%\text{)}$$
$$\rightarrow \cdot C_2H_5 + \cdot NH_2 \qquad \text{(abstraction, } \sim 10\%\text{)}$$
$$\rightarrow NH(^3\Sigma^-) + C_2H_6 \qquad \text{(quenching, } \sim 30\%\text{)}.$$

In the context of this review, the electronic structure and spectroscopy of the simplest nitrene, NH, are very important because they will be useful in the analysis of the more complicated nitrenes. The electronic structure of NH can be

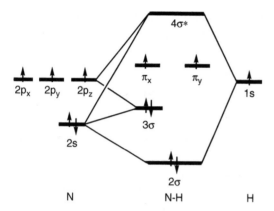

Figure 5.1 Molecular orbitals of nitrene NH. The 1σ-orbital, which is not shown, is the 1s AO on nitrogen.

understood on the basis of elementary molecular orbital (MO) considerations (Figure 5.1).

In NH, two valence MOs, corresponding to the N–H bond (2σ) and lone pair on nitrogen (3σ), are occupied by electron pairs. Two more valence electrons must be distributed between two degenerate, nonbonding molecular orbitals (NBMOs), π_x, and π_y, which consist of the $2p_x$ and $2p_y$ AOs on nitrogen. The three lowest electronic states of NH — $^3\Sigma^-$, $^1\Delta$, and $^1\Sigma^+$ — all arise from the electronic configurations in which the two electrons are distributed between these two NBMOs. The Pauli exclusion prevents electrons with the same spin from simultaneously appearing in the same region of space. Thus, the triplet has the lowest Coulombic repulsion energy of all the low-lying states; hence, it is the ground state of NH (Figure 5.2).

The "closed-shell" component of $^1\Delta$ is a linear combination (with a minus sign) of two configurations in which the two nonbonding electrons occupy the same 2p orbital, whereas in the "open-shell" component one electron occupies each of the 2p AOs. The two components of a $^1\Delta$ state (Figure 5.3) may appear different, but symmetry shows that they are degenerate. If the x and y axes are rotated by 45° and the new $2p_x$ and $2p_y$ orbitals are expressed as the sum and the difference of the old $2p_x$ and $2p_y$ orbitals, the closed-shell singlet in the

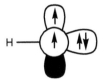

Figure 5.2 Schematic depiction of one of the three spin components of the lowest triplet states of NH.

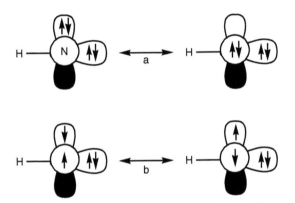

Figure 5.3 Schematic depictions of the (a) closed-shell and (b) open-shell components of the lowest singlet state ($^1\Delta$) of imidogen, NH.

old coordinate system turns into the open-shell singlet and vice versa.

The third of the lowest lying electronic states of NH, $^1\Sigma^+$, would be schematically depicted like the closed-shell singlet in Figure 5.3a. However, the two configurations, which are mixed with a minus sign in the closed-shell component of $^1\Delta$, are combined with a plus sign in the $^1\Sigma^+$ state. The motions of the nonbonding electrons are "anticorrelated" in the $^1\Sigma^+$ state, so they have a higher Coulombic repulsion energy than in $^1\Delta$. This is why $^1\Sigma^+$ is a higher-energy electronic state than $^1\Delta$.

Figure 5.4 presents the state level diagram of NH, constructed on the basis of spectroscopic data.[12,13,15–21] A value of 1.561 eV (36 kcal/mol) for the singlet-triplet splitting (ΔE_{ST}) in NH was obtained very accurately from spectro-

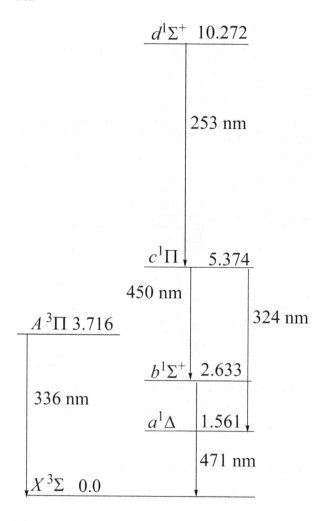

Figure 5.4 Energy level diagram of imidogen, NH, constructed on the basis of the spectroscopic data.[12,13,15–21]

scopic data.[16] A series of bands was observed in the absorption and emission spectra of the HN_3 and NH_3 photolysis products, which were assigned to the transitions between different NH states (Figure 5.4). The experimental absorption spectra of NH in the lowest triplet ($X\ ^3\Sigma^-$) and singlet ($a^1\Delta$) states have similar bands in the near UV region with maxima at 336 and 324 nm, respectively.[12,13,16]

The electronic absorption spectra of NH in the lowest singlet and triplet states were calculated[26a] using the CASSCF(6,5)/CASPT2 procedure with a large atomic natural orbital (ANO-L) basis set at the experimental bond distances.[27] The active space consisted of the three σ-orbitals and two π-orbitals (Figure 5.1) and the six electrons in them. The allowed long-wavelength transitions of singlet and triplet NH (Table 5.1) were predicted at 340 and 308 nm[26a], in good agreement with experiment (Figure 5.4). Both transitions are associated with electron promotion from a 3σ (lone pair) orbital to a non-bonding orbital (Figure 5.1). The next transitions in the observed spectra of triplet and singlet NH are associated with electron promotion from a π-orbital to σ*-orbital and are in the vacuum UV region at 143 and 166 nm, respectively. The calculated value of ΔE_{ST} is equal to 45.5 kcal/mol at the complete active space self-consistent field (CASSCF) level and 38.1 kcal/mol when dynamic electron corrolation is added using second-order perturbation theory (CASPT2). The latter value is within about 2 kcal/mol of the experimental value.

5.3 PHOTOCHEMISTRY OF ALKYL AZIDES AND PROPERTIES OF ALKYLNITRENES

5.3.1 Experimental Studies

The first experimental evidence for the existence of alkylnitrenes was provided by Evectron Spin Resonance (ESR) spectroscopy in 1964.[28] Wasserman et al.[28] detected the ESR spectra of triplet n-propyl, 2-octyl, cyclohexyl, cyclopentyl, α-carbethoxybenzyl and *tert*-butylnitrenes, which were characterized by D values in the range of 1.575 to 1.659 cm^{-1}. Two years earlier the same authors were unable to detect the ESR spectrum of cyclohexylnitrene.[29] In every case the ESR signals were very weak, and sensitized photolysis of alkyl azides was found to be the most appropriate procedure to generate triplet alkylnitrenes.[30] Some triplet alkylnitrenes have subsequently been generated by intramolecular sensitization.[31,32]

The photolysis of alkyl azides at room temperature cleanly forms imines as products.[33] For example, Kyba and

TABLE 5.1 Vertical Excitation Energies of Singlet (R_{NH} = 1.034 Å) and Triplet (R_{NH} = 1.0362 Å) NH Calculated Using the CASSCF(6,5)/CASPT2 Procedure

State	ΔE_{CASSCF} (eV)	ΔE_{CASPT2} (eV)	Ref. Weight	(nm)	Oscillator Strength	Configurations
$^3\Sigma^-$	0.0	0.0	0.95	—	—	—
$1^3\Pi$	3.57	3.65	0.95	340	1.8×10^{-2}	$3\sigma \rightarrow \pi_x,\pi_y$ 99.4%
$2^3\Pi$	7.68	8.65	0.96	143	2.2×10^{-2}	$\pi_x,\pi_y \rightarrow 4\sigma^*$ 97.3%
$^1\Delta$	0.0	0.0	0.96	—	—	—
$^1\Sigma^+$	0.87	1.05	0.96	1178	0	—
$1^1\Pi$	5.04	4.03	0.93	308	9.9×10^{-3}	$3\sigma \rightarrow \pi_x,\pi_y$ 98.6%
$2^1\Pi$	7.18	7.46	0.95	166	4.3×10^{-2}	$\pi_x,\pi_y \rightarrow 4\sigma^*$ 97.2%

Source: Pritchina, E.A. and Gritsan, N.P., unpublished results

Abramovitch[34] studied in detail photolysis of nine *sec-* and *tert*-alkyl azides. Formation of imines, derived from 1,2-shifts of groups on the ipso carbon atom, was observed.

$$R^2 - \underset{\underset{R^3}{|}}{\overset{\overset{R^1}{|}}{C}} - N_3 \xrightarrow[-N_2]{h\nu} R^1R^2C = NR^3 + R^2R^3C = NR^1 + R^1R^3C = NR^2$$

$$R^1 = Ph, 2\text{-}PhC_6H_4, PhCH_2CH_2, t\text{-}Bu, n\text{-}Pr, n\text{-}Am$$

$$R^2 = Ph, Me, n\text{-}Pr, n\text{-}Am$$

$$R^3 = Ph, Me, H, n\text{-}Am$$

It was proposed that singlet excited alkyl azides eliminate nitrogen with concomitant rearrangement to form imine products, without the intervention of a nitrene intermediate.[34]

There are only a few examples of photolysis of alkyl azides that result in any process other than rearrangement to an imine.[33] Photolysis of highly fluorinated azide **1** in cyclohexane gave amide **3** at 18% yield after a hydrolytic workup, implicating a nitrene C–H insertion product **2**.[35]

$$CF_3CHFCF_2N_3 \xrightarrow[c\text{-}C_6H_{12}]{h\nu} CF_3CHFCF_2NHC_6H_{11} \xrightarrow[H_2O]{MeOH} CF_3CHF\overset{\overset{\displaystyle O}{\|}}{C}NHC_6H_{11}$$
$$\quad\quad\textbf{1} \quad\quad\quad\quad\quad\quad\quad \textbf{2} \quad\quad\quad\quad\quad\quad\quad \textbf{3}$$

Intramolecular cyclization via C–H insertion as a minor pathway was observed upon photolysis of a steroidal azide.[36]

Dunkin and Thomson[37] studied the photochemistry of *tert*-butyl azide (**4**) in an N_2 matrix at 12 K. Using IR spectroscopy, they detected the formation of only one product — imine **5**.

$$Me_3C \text{—} N_3 \xrightarrow[-N_2]{h\nu} Me_2C\text{=}NMe$$

$$\phantom{Me_3C \text{—} N_3}\mathbf{4}\mathbf{5}$$

The formation of highly strained bridgehead imines was observed on the irradiation of a series of matrix-isolated bridgehead azides.[38–41] The photochemistry of matrix-isolated 1-azidonorbornane (**6**) was studied using monochromatic irradiation; IR, UV, and ESR spectroscopy; and trapping with methanol and CO.[42] The azide photochemistry was very complicated, and the formation of two types of imines (**7** and **8**) and triplet nitrene **9** were observed.

The structure assignment of the nitrene **9** was based on its ESR signal at 8124 G ($E = 0$, $|D/hc| = 1.65$ cm^{-1}), a sharp UV peak at 298 nm, and photochemical trapping with CO in Ar at 36 K.

Recently[43] photolysis of perfluoromethyl azide **10** was studied in frozen Ar and in pentane at cryogenic temperatures. Photolysis (254 nm) of **10** in pentane at 6 to 10 K produced a persistent ESR spectrum, typical of a triplet nitrene, centered at 8620 G. The spectrum ($|D/hc| = 1.736$ cm^{-1}) is attributed to triplet CF$_3$-N (^{3}**11**) and is very similar to that of triplet NH ($|D/hc| = 1.863$ cm^{-1})[44] and CH$_3$-N ($|D/hc| = 1.720$ cm^{-1}).[45]

Azide **10** ($\lambda_{max} = 257$ nm) was also decomposed in an Argon matrix at 14 K by exposure to 254-nm light, with the

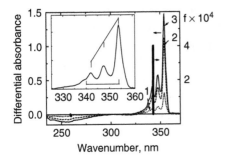

Figure 5.5 Differential electronic absorption spectra, recorded after irradiation of CF_3N_3 in an Ar matrix at 12 K with 254 nm light after 5-min (curve 1), 25-min (curve 2) and 45-min (curve 3). The calculated absorption bands (TD-B3LYP/6-31*) of triplet CF_3N (positive) and CF_3N_3 (negative) are depicted as solid vertical lines. Insert: differential electronic absorption spectrum detected after 25 min of irradiation.

concurrent formation of sharp, structured absorption bands at 342, 347.5, and 354 nm (Figure 5.5). The UV bands were assigned to the triplet nitrene [3]**11** on the basis of both the similarity to the spectrum of CH_3N and the results of time-dependent density functional theory (TD–B3LYP) calculations.[43] The product of formal nitrene rearrangement (CF_2NF, **12**) was also observed and its IR spectrum assigned on the basis of theoretical calculations.[46a] The formation of [3]**11** under direct irradiation of **10** indicates that singlet perfluoromethylnitrene [1]**11** was generated and is indeed an energy minimum on the CF_3N potential-energy surface.

Methylnitrene is the simplest of all alkylnitrenes. Photolysis of methyl azide (CH_3N_3, **13**) does not produce $^1NCH_3$ (^{1}**14**) as a trappable species, and even attempts to detect it by femtosecond flash photolysis have failed.[46] Triplet methylnitrene was not formed on direct irradiation[30,47] of **13**, and only methyleneimine (**15**), the product of formal isomerization of ^{1}**14**, was detected in cryogenic matrices.[48,49]

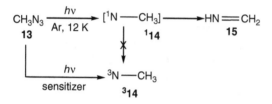

Triplet methylnitrene can be produced by sensitized photolysis[30,47] in the condensed phase or in a gas-phase corona discharge.[50-53] Barash et al.[30] and Wasserman[47] reported a value of the zero-field splitting parameter $|D/hc|$ of 1.595 cm^{-1} for matrix isolated triplet CH_3N. This earlier experiment is questionable,[51] and a revised value ($|D/hc| = 1.720$ cm^{-1}) was obtained for CH_3N using gas-phase spectroscopy.[45] The triplet-triplet absorption spectrum[54-57] of ^{3}**14** and its emission spectrum[50-53,56-58] are well known. The 0-0 transition in the absorption spectrum of ^{3}**14** was found to occur at 316.9 nm in an N_2 matrix[55] and at 314.3 nm in the gas phase,[56] similar to the spectrum of the parent NH (336 nm).[12,13]

The experimental results available in the literature for methylnitrene can be interpreted in two ways. One can posit that singlet methylnitrene is a discrete intermediate but that it does not live long enough, even at cryogenic temperatures, to relax to the lower-energy triplet state. Alternatively, it can also be proposed that singlet excited methyl azide eliminates nitrogen with concomitant rearrangement to form the methyleneimine **15**, without the intervention of a nitrene intermediate.[34]

The negative ion photoelectron spectrum of CH_3N^- contains features that were assigned to the lowest singlet state of **14**, and the singlet-triplet splitting was measured to be ΔE_{ST} = 1.352 ± 0.011 eV (31.2 ± 0.3 kcal/mol).[59] The features assigned to the singlet nitrene in the photoelectron spectrum of CH_3N^- were interpreted as belonging to a resonance, rather than to a true minimum on the singlet CH_3N potential-energy surface.[59] To further investigate this issue a series of quantum chemical calculations were performed.[60–69] The results of these calculations are discussed in the next section.

5.3.2 Computational Studies

5.3.2.1 Methyl Azide and Methylnitrene

Most calculations available in the literature are concerned with the simplest alkyl azide and alkylnitrene, CH_3-N_3 (**13**) and CH_3-N (**14**), respectively. The first *ab initio* theoretical study of the different electronic states of **14** was reported in 1974 by Yakony et al.[60] The geometries of the lowest $^3A_2$, ^1E, and $^1A_1$ electronic states of **14** were calculated at the self-consistent-field level of theory with a double ζ basis set. The ^1E and $^1A_1$ states were predicted to lie 14,200 and 27,700 cm^{-1} above the $^3A_2$ ground state. Yakony et al.[60] did not try to explain "why **14** has been so difficult to isolate in the laboratory."

In 1980 Demuyanck et al.[61] undertook the first *ab initio* study of the unimolecular rearrangement of triplet and singlet CH_3N to $CH_2=NH$. Geometries of the stationary points were determined and characterized at the double ζ (DZ) basis set self-consistent field (SCF) level of theory. At these geometries, correlation effects were evaluated using the CI procedure with all single and double excitations, and the DZ basis set was expanded to include polarization functions. The isomerization of triplet nitrene **14** to the triplet state of methyleneimine **15** (in a spin-allowed process) was predicted to be endothermic by 18 kcal/mol, with an activation energy of 48 kcal/mol.[61]

$$^3CH_3-N \rightarrow {}^3CH_2=NH$$

It was concluded that the ground-state triplet **14** is a stable species in the absence of collisions. This conclusion was very important at the time (1980), because the first unequivocal spectroscopic identification of triplet nitrene **14** was achieved only in 1984.[50]

It was found[61] that singlet $CH_2=NH$ lies 46 kcal/mol below the $^3A_2$ ground state of **14** and might be accessible if the spin-orbit coupling is substantial, since rearrangement of the triplet is a spin-forbidden process. The rearrangement of singlet nitrene **14** was also considered, using two-configuration SCF wave functions. Starting from the C_{3V} SCF equilibrium geometry of the ^1E state, no barrier was found for the isomerization of singlet methylnitrene to singlet imine **15**.[61]

In 1983, Pople et al.[62] recalculated geometries of ^{3}**14** and ^{3}**15**, using the unrestricted SCF level of theory in conjunction with the 6-31G* basis set. They used the MP4SDQ/6-31G** method for single-point energy calculations. The results of their calculations for the triplet nitrene rearrangement were very similar to those in previous paper.[61]

The authors[62] noted that ^1E state of **14** should be subject to Jahn–Teller distortion. At the HF/6-31G* level they found a local minimum at C_S symmetry that was 89.1 kcal/mol higher than the global minimum (**15**). However, after corrections were made the transition structure connecting this minimum to **15** was found to have lower energy and the authors concluded that there is no evidence for ^{1}**14** as a distinct species.[62]

A few years later[51] the first experimental evidence for the structure of ^{3}**14** was reported. It became clear then that the calculations available at that time[60–62] did not reproduce the experimental C–N bond length (1.411 A) of ^{3}**14**. Demuyanck et al.[61] obtained a value for the C–N bond length of 1.471 Å, and Pople et al.[62] obtained a value of 1.433 Å. An *ab initio* prediction of the C–N bond distance, accurate to within 0.01 Å of experiment, was finally achieved[64] by using configuration interaction, including all singlet and double excitations (CISD), in conjunction with a basis set of quality quadruple-ζ plus double polarization plus f functions on the C and N atoms (DZ + 2P + f).

Thus, in a period of only a few years (1980 to 1987), methylnitrene in its triplet ground state had gone from the "tenuous" category to what can be considered the best characterized of all known organic nitrenes. Nevertheless, the question of whether singlet methylnitrene ^{1}**14** is an intermediate in the thermal and photochemical decomposition of **13** remained unsolved.

As in the case of NH, at C_{3v} symmetry the closed-shell singlet and the open-shell singlet of ^{1}**14** have exactly the same energy and form the two components of a ^{1}E state. Demuyanck et al.[61] studied the isomerization of ^{1}**14** in a ^{1}E state constrained in C_{3v} symmetry. Pople et al.[62] pointed out that the ^{1}E state is subject to a Jahn–Teller distortion. The local minimum exhibits C_S symmetry, and they examined isomerization in the ^{1}A' (closed-shell) state. Nguyen[63] studied the isomerization of ^{1}**14** in both ^{1}A' and ^{1}A'' states using Restricted Hartree-Fock (RHF) and Unrestricted Hartree Fock (UHF) procedures with a 3-21G basis set for geometry optimization and MP4SDQ level of theory for single-point calculations. At this low level of the theory, Nguyen found that the ^{1}A'' state lies about 8 kcal/mol above the closed-shell ^{1}A' state, which is a saddle point on the potential-energy surface. The energy barrier for rearrangement in the ^{1}A'' state was calculated to be 9.5 kcal/mol.[63]

Richards et al.[65] examined the same rearrangement on the lowest-lying singlet surface at the CISD level of theory with a two-configuration (TC) reference wave function. A basis set up to triple zeta plus double polarization plus f functions (TZ2P + f) was used. The Jahn–Teller ^{1}A'-^{1}A'' energy splitting was found to be very small (<0.01 kcal/mol). Richards and coworkers[65] did, in fact, find a C_s energy maximum. The barrier to rearrangement on the ^{1}A' potential surface was computed to be 12.1 kcal/mol at the DZP TC-SCF level, whereas the TZ2P TC-SCF barrier is reduced to 10.6 kcal/mol. The two-reference CISD method caused the calculated barrier to fall to 1.9 kcal/mol with the DZP basis set and to 1.3 kcal/mol with the DZ2P basis set. The Davidson correction further reduced the barrier, and the TZ2P + f, TC-CISD + Q result was 0.89 kcal/mol. The exothermicity of the rearrangement was calculated to be 85 kcal/mol.

The authors apparently found that the energy of the C_s maximum decreases upon an asymmetric distortion of the geometry, because they concluded, "If there is a $^1A'$ transition state at this level of theory, it occurs in C_1 symmetry."[65] The authors proposed[65] that higher-level correlation effects and excursions into C_1 symmetry would eliminate the barrier entirely, in agreement with the previous, lower-level studies.[61,62]

The rearrangement of **14** on the $^1A''$ potential-energy surface significantly differs from that on the $^1A'$ surface.[65] The formation of the excited $^1A''$ state of **15** is expected; therefore, the exothermicity of the rearrangement is much more modest (~17 kcal/mol), and a barrier about 30 kcal/mol was computed for this state.[65]

A few years later Nguyen et al.[66] attempted to calculate the thermal decomposition of **13** using the MP2/6-31G(d,p) method for geometry optimization and QCISD(T)/6-311++G(3df,2p) level for single-point energy calculations. They concluded that the decomposition of **13** occurs through a concerted motion of N_2 elimination with hydrogen shift, giving **15**. A similar conclusion was made on the basis of DFT (B3LYP and B3PW91) calculations.[67] No singlet nitrene was found by this study.[66,67]

The process of thermal decomposition of **13** was also reinvestigated in 1999, using the CASSCF(10,8) method and Moller–Plesset theory (CAS/MP2) in conjunction with the 6-31G* basis set.[68] In contrast to the previous calculations,[66,67] this reaction was predicted to occur in two steps via a nitrene intermediate.[68] Therefore, the predicted reaction mechanism is drastically changed by the method used in the calculations. The authors concluded that the CASCCF predictions should be the more reliable, because this method has sufficient flexibility to describe in a balanced way most of the potential energy surface.[68]

A very shallow minimum for singlet methylnitrene in the $^1A'$ state was found, and its rearrangement to **15** was studied using the CAS(10,8)/MP2 procedure.[68] It was possible to localize a transition state of C_1 symmetry only 1.4 kcal/mol above the minimum of singlet nitrene.

Independently, Kemnitz and coworkers[69] undertook CASSCF and CASPT2 calculations on the singlet methylnitrene rearrangement. Their study was motivated by the fact that the closely related rearrangement of singlet methylcarbene to ethylene is computed to have a small but finite energy barrier[70,71], and methylcarbene-d_4 has, in fact, been trapped chemically.[72–75] Because the rearrangement of singlet methylcarbene to ethylene does have a barrier, it is not obvious why the rearrangement of singlet methylnitrene to methyleneimine should not.

To investigate the possibility that an energy barrier separates singlet methylnitrene from methyleneimine, CASSCF(12,11) and CASPT2 calculations were performed in which all 12 valence electrons and all 11 valence orbitals were included in the active space.[69] The CASSCF(12,11) geometry optimizations and the CASPT2 single-point calculations were carried out with Dunning's correction consistant polarized double ζ(cc-pVDZ) and cc-pVTZ correlation-consistent basis sets. The former was used for vibrational analyses, performed at CASSCF(12,11)/cc-pVDZ optimized geometries.

The CASSCF(12,11)/cc-pVTZ optimized geometries for the lowest singlet state of methylnitrene, methyleneimine, and the transition state that connects these two energy minima are shown in Figure 5.6. The CASSCF(12,11) and CASPT2 energies are given in Table 5.2.

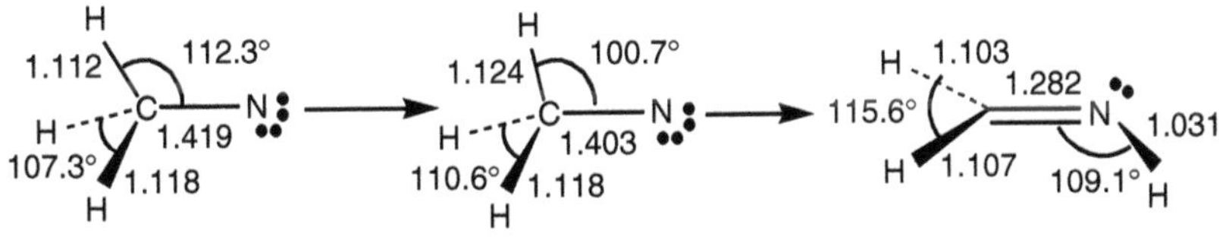

Figure 5.6 CASSCF(12,11)/cc-pVTZ geometries (bond lengths in angstroms, bond angles in degrees) of the stationary points along the lowest-energy pathway for the rearrangement of the lowest singlet state of methylnitrene (^{1}A' MN) to methyleneimine (MI).[69]

TABLE 5.2 Relative CASSCF(12,11) Zero-Point Energies (ZPEs) and CASSCF(12,11) and CASPT2(12,11) ZPE-Corrected Electronic Energies (kcal/mol) of $^1A'$ and $^3A_2$ Methylnitrene **14** and Methyleneimine **15** and of the Transition Structures Connecting $^1A'$ **14** to **15** (**TS1**) Computed with cc-pVDZ and cc-pVTZ Basis Sets[69]

	ZPE	E (CASSCF/ cc-pVDZ)	E (CASPT2/ cc-pVDZ)	E (CASSCF/ cc-pVTZ)	E (CASPT2/ cc-pVTZ)
$^1A'$ **14**	0	0	0	0	0
$^3A_2$ **14**	1.8	−41.1	−35.9	−39.7	−32.0
TS1	0.1	0.4	2.8	0.5	3.8
15	3.2	−91.7	−83.1	−92.0	−82.9

It was found, as in the previous calculations,[65,68] that the $^1A'$ state has one short and two long C–H bonds (Figure 5.6). The $^1A''$ state distorts from C_{3v} symmetry in the opposite sense. However, the distortions from C_{3v} symmetry of both components of ^1E are quite small; at their C_s optimized geometries, the $^1A'$ and $^1A''$ states differ in energy by less than 0.01 kcal/mol at the CASSCF(12,11) level of theory with both basis sets. As found by Xie et al.[64] and Richards et al.,[65] the C–N bond lengths in the two singlets are nearly the same, and both are slightly shorter than the C–N bond length in the lowest triplet state.

It is in the $^1A'$ state that migration of the unique hydrogen can most easily occur.[65] In fact, the lowest-frequency vibrational mode of 564 cm^{-1} in this state of methylnitrene corresponds to an a' motion that moves this hydrogen toward the nitrogen.

The zeroth-order wave function for $^1A'$ has the form $^1A' = c_1 |...7a'^2\rangle - c_2 |...2a''^2\rangle$. The 7a' Molecular Orbital (MO) is largely composed of the $2p_y$ Atomic Orbital (AO) on nitrogen, interacting in an antibonding fashion with the unique hydrogen; 2a'' is largely the $2p_x$ orbital on nitrogen, interacting with the remaining two hydrogens, also in an antibonding fashion. In the reactant, $c_1 \approx c_2$, so that the 7a' and 2a'' nonbonding (NB) MOs are each occupied by an average of about one electron.[69]

However, as hydrogen migration from carbon to nitrogen occurs, the electrons that form the unique C–H bond in the reactant are delocalized toward the $2p_y$ orbital nitrogen. Concomitantly, c_1 decreases and c_2 increases, so that the pair of electrons that are initially distributed nearly equally between $2p_x$ and $2p_y$ on nitrogen become increasingly localized in $2p_x$. Completion of the formation of methyleneimine can be thought of as requiring rotation around the C–N bond, so that this pair of electrons in $2p_x$ on nitrogen can delocalize into the now largely empty $2p_y$ orbital on carbon, thus forming the C–N π bonds in methyleneimine.

Product formation requires loss of the plane of symmetry that exists in the reactant. Therefore, to establish the existence of an energy barrier to product formation, it is not valid to simply find an energy maximum along a reaction pathway that preserves this symmetry plane. For a C_s energy maximum to be a true transition structure rather than a mountain top on the potential-energy surface for rearrangement of [1]**14** to **15**, the force constant for rotation around the C–N bond at the C_s energy maximum also must be positive.

The CASSCF(12,11)/cc-pVTZ geometry of the C_s energy maximum (Figure 5.6) is similar to the TC-CISD geometry found by Richards and coworkers[65] for this species. With both the cc-pVDZ and cc-pVTZ basis sets, the CASSCF(12,11) energy of this species is 0.3 to 0.4 kcal/mol above that of the reactant, and this energy difference increases by 0.1 kcal/mol with inclusion of the correction for differences in zero-point vibrational energies. At the CASPT2 level of theory, the C_s energy barrier increases to 2.8 kcal/mol with the smaller of the two basis sets and to 3.8 kcal/mol with the larger.[69]

Because CASPT2 usually overestimates the effect of including electron correlation because of excitations outside the valence space, the actual height of the barrier to passage over the C_s energy maximum is likely to be lower than 3.8 kcal/mol but closer to it than to the CASSCF(12,11)/cc-pVTZ value of 0.5 kcal/mol. A barrier height in the range of 2.5 ± 1.0 kcal/mol was proposed as being most reasonable.[69]

CASSCF(12,11) vibrational analyses with both the cc-pVDZ and cc-pVTZ basis sets found the C_s energy maximum to be a true transition state for rearrangement of [1]**14** to **15**. Except for the imaginary frequency for the a', symmetry-preserving vibration that corresponds to hydrogen migration, all other frequencies were found to be real. The lowest of these corresponds to a symmetry-breaking a" vibration, but this mode is computed at the CASSCF(12,11)/cc-pVTZ level to have a frequency of 854 cm^{-1}. CASPT2 calculations along this a" vibrational coordinate also found the force constant along it to be positive. Therefore, at both the CASSCF(12,11) and the CASPT2 levels of theory the C_s energy maximum does appear to be a true transition state on the potential-energy surface for rearrangement of singlet methylnitrene to methyleneimine.[69]

The tremendous exothermicity of the rearrangement of singlet [1]14 to 15, amounting to 83 kcal/mol at CASPT2, results in the C_s energy maximum occurring very early along the reaction coordinate before substantial motion of the hydrogen from carbon to nitrogen has occurred. In this C_s transition structure the unique C–H bond has lengthened by only 0.01 Å, and the C–N distance has decreased from that in the reactant by only 0.016 Å. Nevertheless, the occupation of the 7a' NBMO falls from 1.07 electrons in the reactant to 0.35 electrons in the transition structure.[69]

The CASPT2(12,11)/cc-pVTZ value for the singlet-triplet splitting was found to be $\Delta E_{ST} = 32.0$ kcal/mol,[69] in excellent agreement with the experimental value of $\Delta E_{ST} = 31.2 \pm 0.2$ kcal/mol.[59] The absence of the ν_6 vibration band from the singlet region of the negative ion photoelectron spectrum of CH_3N^- provides the only experimental evidence that singlet methylnitrene could be a transition structure, because one component of the ν_6 vibration leads toward migration of a hydrogen from carbon to nitrogen. The harmonic frequency of this vibration at the singlet energy minimum was calculated to be 564 cm^{-1}. However, there is no peak apparent in this energy region above the singlet origin.[59] One of the possible reasons for the absence of this band is that the first excited level of the ν_6 C-H rocking vibration is above the barrier for the rearrangement that this mode promotes. This would place an upper limit on the barrier to rearrangement of 564 cm$^{-1} = 1.6$ kcal/mol.[69]

Based on the results of these CASSCF/CASPT2 calculations, singlet methylnitrene is indeed predicted to be an energy minimum rather than a transition state; the barrier to rearrangement is most likely in the range of 2.5 ± 1 kcal/mol.[69] Thus, at least in principle, [1]14 should be an observable albeit very short-lived intermediate, because tunneling through the small barrier to hydrogen migration should be fast.

The triplet-triplet absorption spectrum of ^{3}**14** has a maximum at 316.9 nm in an N_2 matrix[55] and at 314.3 nm in the gas phase,[56] similar to the spectrum of the parent NH (336 nm).[12,13,16] In the latter case the spectra of the nitrene in the lowest triplet and singlet states ($X\ ^3\Sigma^-$ and $a^1\Delta$) have similar bands in the near UV region with maxima at 336 and 324 nm.[12,13] Therefore, singlet nitrene **14** is expected to absorb in the near-UV region as well.

The electronic absorption spectra of **14** in the lowest singlet and triplet states were calculated[75a] with the CASSCF(12,11)/CASPT2 procedure with the ANO-L basis set at the bond distances optimized previously.[69] All 12 valence electrons and all 11 valence orbitals were included in the active space.

The long wavelength transition of the triplet CH_3N was predicted at 307 nm (f = 1.4×10^{-2}), in very good agreement with experiment (314 to 317 nm).[52] The long wavelength transition of singlet CH_3N in the $^1A'$ state was calculated at 287 nm (f = 4.9×10^{-3}). As with the case of parent nitrene NH, both transitions were associated with electron promotion from a lone-pair orbital on the nitrogen to a non-bonding 2p orbital. The calculated value of ΔE_{ST} is 40.7 kcal/mol at the CASSCF level and 34.0 kcal/mol at the CASPT2 level, in good agreement with experiment[59] and with previous calculations with a different basis set.[69]

5.3.2.2 Calculations for Other Alkylnitrenes

Photolysis of perfluoromethyl azide **10** in an argon matrix at 14 K produced triplet nitrene **11** with sharp, structured absorption bands at 342, 347.5, and 354 nm (Figure 5.5).[43] The singlet-triplet splitting of **11** was calculated at the CASSCF(8,8)/6-31G* level of theory to be 43.7 kcal/mol.[43] This value is larger than the experimentally determined singlet–triplet splitting in **14** (31.2 kcal/mol).[59] However, at the CASSCF(12,11)/cc-pVDZ level of theory, the singlet-triplet splitting of **14** is calculated to be 41.1 kcal/mol, too high by

10 kcal/mol.[59] Thus, the singlet–triplet gap in **11**, calculated by CASSCF, is also likely to be too large by this amount.

Rearrangement of [1]**11** to imine **12** ($F_2C=NF$) was predicted by CASSCF(8,8)/6-31G* to be exothermic by 46 kcal/mol,[43] probably too high by ≈ 10 kcal/mol, due to the CASSCF error in calculating the energy of [1]**11**. Even so, the rearrangement of **11** to **12** is computed to be much less exothermic than the rearrangement of [1]**14** to **15** (82.9 kcal/mol),[69] due to the greater strength of the C–F bond relative to the C–H bond. Consequently, the barrier to isomerization is expected to be greater for [1]**11** than for [1]**14**.[43] This is presumably the reason intersystem crossing (ISC) of the perfluoromethyl singlet nitrene ([1]**11**) to [3]**11** can compete with rearrangement to **12** at cryogenic temperatures.

Recently, Tsao et al.[76] performed CASSCF/CASPT2 calculations of the properties of the cyclopropylnitrene (**16**). This study was motivated by the fact that cyclopropylcarbene can be chemically intercepted and has a lifetime of about 20 ns in solution at ambient temperature.[77,78] Recall that singlet methylcarbene[70,71] and singlet methylnitrene[68,69] have not been detected or chemically intercepted and are predicted to isomerize to, respectively, ethylene and imine over a very small barrier of 1 to 3 kcal/mol. Calculations were performed to estimate the lifetime of [1]**16** and its suitability for laser flash photolysis studies.[76]

Geometry optimizations for singlet and triplet nitrene **16** were performed with the standard 6-31G* basis set, using CASSCF(4,4) calculations with a four-electron and four-orbital active space.[76] This active space consisted of two 2p AOs on nitrogen and the highest σ and lowest unoccupied σ^* MOs, formed from the two β C–C bonds of the cyclopropyl group.

The CASSCF(4,4)/6-31G* optimized bond lengths in the lowest electronic states ($^3A''$, $^1A'$, and $^1A''$) of cyclopropylnitrene **16** are shown in Figure 5.7, and their relative energies are given in Table 5.3. The calculated singlet-triplet energy gap (ΔE_{ST}) of **16** is 27.9 kcal/mol at the CASPT2(8,8)/6-31G(2d,p) level with the triplet as the ground state,[76] ΔE_{ST} in **16** is smaller than that in methylnitrene **14** (31.2 kcal/mol).[59] Electron donation from the cyclopropyl ring sta-

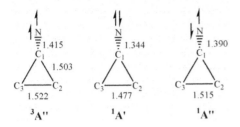

Figure 5.7 CASSCF(4,4)/6-31G* optimized bond lengths (in angstroms) of three low-lying spin states of cyclopropylnitrene **16**.[76]

TABLE 5.3 Relative Energies (kcal/mol) of Triplet and Singlet Cyclopropylnitrene[76a]

Method	³A"	¹A'	¹A"
CASSCF(4,4)/6-31G*	0.00	34.54	38.51
CASSCF(8,8)/6-31G*	0.00	30.68	39.4
CASSPT2(8,8)/6-31G*	0.00	28.59	33.66
CASSCF(8,8)/6-311G(2d,p)	0.00	29.17	39.00
CASSPT2(4,4)/6-311G(2d,p)	0.00	29.66	31.80
CASSPT2(8,8)/6-311G(2d,p)	0.00	27.88	32.26

ᵃEnergies were calculated with CASSCF(4,4)/6-31G* optimized geometries, including corrections for the differences in CASSCF(4,4)/6-31G* unscaled zero-point vibrational energies.

bilizes the closed-shell singlet state of the nitrene, relative to the triplet.

As expected, the closed-shell singlet state of **16** (¹A') has a shorter C–N bond length than the biradical-like open-shell singlet state (¹A"). The energy of the closed-shell singlet is calculated to be below that of the open-shell singlet at all levels of theory by 2.1 to 9.2 kcal/mol and 4.4 kcal/mol at highest level (Table 5.3). Note that in the methylnitrene **14** this calculated difference is less than 0.01 kcal/mol. (The opposite situation obtains in the case of vinylnitrene and phenylnitrene, as discussed in Sections 5.4 and 5.6.)

Thermolysis of azidopropane led to hydrogen cyanide (**17**), ethylene (**18**), and 1-azetine (**19**).[79]

Calculations were performed to provide some insight into these reactions.[76] Geometry optimization of the ${}^1A'$ state of **16** and the reaction products and location of possible transition states (**TS1** and **TS2**) were performed at the HF/6-31G* level of theory. Some calculated bond lengths for the intermediates and transition structure of the fragmentation reaction are shown in Figure 5.8. At the HF/6-31G* level of theory, the activation energy of the ring-expansion reaction (**16** → **TS1** → **19**) was 2.8 kcal/mol, and the activation energy of the fragmentation reaction (**16** → **TS2** → **17** + **18**) was 0.6 kcal/mol. The activation energy calculated for the ring-expansion reaction is smaller than the previously published value of 9.3 kcal/mol, which was obtained by HF/3-21G calculations without zero-point energy corrections.[80]

From single-point energies (Table 5.4) at correlated levels of theory (B3LYP, MP2, and CASPT2), ^{1}**16** is predicted to fragment spontaneously if one uses HF/6-31G* derived geometries.[76] In addition, transition state searches with (restricted) B3LYP and MP2 methods led to barrierless fragmentation.

The CASPT2(8,8)/6-31G(2d,p) calculations on structures along the intrinsic reaction coordinate (IRC) path at the HF/6-31G* level predict that the fragmentation reaction is barrierless, but the ring-expansion reaction has a small activation barrier ($\approx$2.4 kcal/mol). Therefore, the calculations indicate that, unlike cyclopropylcarbene, singlet cyclopropylnitrene will have a very short lifetime in solution at ambient temperature, and, like methylcarbene and methylnitrene, ^{1}**16** will be very difficult to detect.[76]

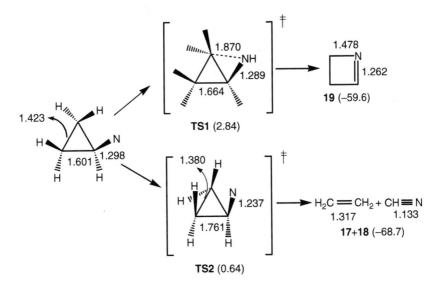

Figure 5.8 Intermediate and transition structures for the ring-expansion (via **TS1**) and fragmentation (via **TS2**) reactions of cyclopropylnitrene **16** at the HF/6-31G* level of theory.[76] Relative energies (in kcal/mol) are listed in parentheses at the HF/6-31G* level. Bond lengths are shown in Å.

TABLE 5.4 Relative Energies[a] (kcal/mol) of Intermediates and Stationary Points for Cyclopropylnitrene Rearrangement and Fragmentation[76]

Structure	HF	B3LYP	MP2	MP4(SDQ)	CASPT2(8,8)/ 6-311G(2d,p)
^{1}A'-16	0.00	0.00	0.00	0.00	0.00
^{3}A"-16	−45.16	−17.24	−20.49	−23.50	−27.88[b]
19	−59.56	−62.53	−70.44	−66.68	—[c]
17 + 18	−68.70	−63.05	−74.49	−71.84	—[c]
TS1	2.84	−1.43	−7.78	−1.72	−1.11
TS2	0.64	−6.44	−7.85	−4.95	−8.61

[a]Energies were calculated with a 6-31G* basis set and the HF/6-31G* optimized geometries, including corrections for difference in HF/6-31G* unscaled zero-point vibrational energies, unless noted otherwise.
[b]Using the CASCF(4,4)/6-31G* geometry.
[c]Not calculated.

5.4 PHOTOLYSIS AND THERMOLYSIS OF VINYL AZIDES: PROPERTIES OF VINYLNITRENES

5.4.1 Experimental Results

The simplest member of the class of vinyl azides, $H_2C=CH-N_3$, has been known for about 100 years. However, it was not until the late 1960s that vinyl azides became an important and synthetically useful class of organic compound.[81] The most interesting and important reaction of vinyl azides is the formation of azirines produced upon photolysis and thermolysis of vinyl azides.[81] Smolinsky[82] and Pryde[83] first observed azirine formation, together with a small amount of keteneimine, by gas-phase pyrolysis of α-aryl–substituted vinyl azides.

Hassner and Fowler[84,85] first discovered that several α-substituted (R≠H) vinyl azides produce 2-mono- or 2,3-disubstituted-1-azirines in high chemical yield (80 to 90%) upon photolysis.

Isolable 1-azirines were formed upon photolysis of α-unsubstituted (R=H) vinyl azides only at low temperature and underwent further decomposition upon heating.[86,87]

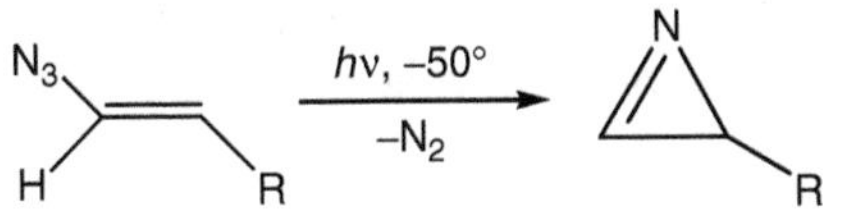

Three plausible mechanisms were postulated to explain these observations: the free nitrene route (pathway a), N_2 extrusion in concert with three-membered ring formation (pathway b), and cyclization to a 4H-1,2,3-triazole followed by rearrangement to azirine (pathway c).[81]

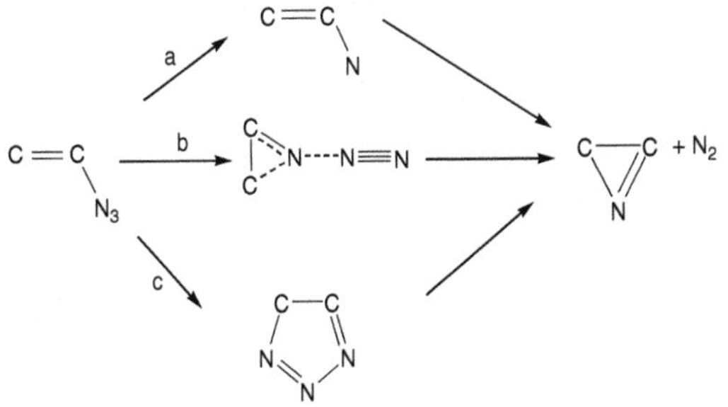

Scheme 5.0

The formation of 1-azirines (**AZ**) along with keteneimines (**K**) upon photolysis and thermolysis of vinyl azides was explained by invoking the intermediacy of singlet vinylnitrenes (^{1}**VN**) (pathway a).[88,89] Keteneimines can serve as a precursor to nitriles if R′ = H.

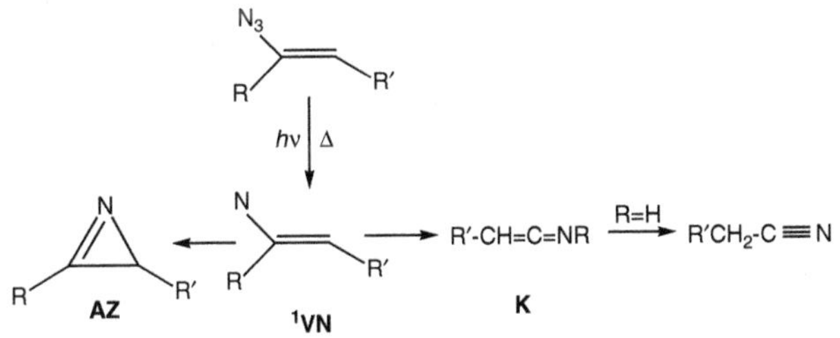

The intermediacy of a singlet nitrene was supported by the formation of indole during the thermolysis of β-azidostyrene.[90]

Note that in the early literature[81,85,88–90] vinylnitrene was thought to be a closed-shell dipolar species. Therefore, attempts were made to intercept this nitrene using 1,3-dipolarophiles.[85]

It was shown that 1-azirines can serve as precursors to vinylnitrenes.[91,92] A number of 1-azirines rearrange to indoles, most likely via vinylnitrene intermediates, upon heating. Two examples are known in which the nitrene formed during the thermolysis of azirines can be intercepted by tris(dimethylamino)phosphine[93] or triphenylphosphine.[94]

Triazole formation as a route to azirines (pathway c) was considered to be very unlikely.[81] Hassner,[81] and Fowler et al.[95] proposed that the concerted formation of azirines upon thermolysis of vinyl azides (pathway b) was more reasonable.

It is difficult to choose among the proposed mechanisms, based on the experimental data available, because both triplet and singlet vinyl nitrenes have still escaped direct observation. The formation of an azirine was observed upon photolysis of α-azidostyrene in an argon matrix at cryogenic temperature (8 K), but even under these conditions nitrene species were not observed.[96]

If a free singlet nitrene is an intermediate, produced upon decomposition of a vinylazide, then the activation energy of vinyl nitrene rearrangement to azirine must be very small. The situation is similar to the case of methylnitrene, produced upon photolysis of methyl azide (see Section 5.3).

Several computational methods have been used to evaluate the mechanism of azirine formation.

5.4.2 Computational Results on the Mechanism of Azirine Formation and Properties of Vinylnitrene

5.4.2.1 Semiempirical and Early *Ab Initio* Calculations

All calculations reported in the literature have been concerned with only the simplest vinyl azide and vinylnitrene, $CH_2=CH-N_3$ and CH_2-CH-N, respectively.[97–101] Until recently, all the theoretical studies on the vinyl azide to azirine transformation used semiempirical calculations[98] or *ab initio* calculations performed at relatively low levels of theory,[97,99] at least by current standards. Significantly, early calculations on the vinylnitrene-azirine interconversion have focused only on the closed-shell singlet excited state ($^1A'$) of vinylnitrene (**VN**).

According to MNDO calculations,[98] the concerted mechanism of azirine formation (b) was found to be the most reasonable process (see Scheme 5.0). The free nitrene route (a) for the thermal reaction was excluded on the basis of very high activation energy. A similar conclusion was made on the basis of HF, MP2, and CISD calculations with a minimal STO-3G basis set.[99] But as noted before, the authors of both publications[98,99] considered the formation and rearrangement of the closed-shell singlet nitrene (^{1}A'-**VN**).

Lohr et al.,[97] using a UHF procedure, were the first to discover that the lowest singlet state of **VN** has an open-shell electronic configuration with an energy 50.7 kcal/mol below that of the closed-shell singlet. However, the expectation value of S^2 for this singlet was 1.17, which indicated that there was substantial triplet character in the UHF wave function, and it was a poor representation of a true singlet. Therefore, these authors did not deal with the open-shell singlet and confined their attention to the rearrangement involving the closed-shell singlet. They found that there is little or no barrier to formation of 2*H*-azirine from the closed-shell singlet **VN**. They also found a pathway for the rearrangement of 2*H*-azirine to the more stable acetonitrile via the intermediate methylisocyanide.[97]

5.4.2.2 High-Level *Ab Initio* Calculations

Recently, Parasuk and Cramer[100] performed very accurate calculations of the electronic structure and energies of the lowest states of **VN**. Single-point calculations at the CASSCF(4,4) geometry were carried out at the MRCI and CASPT2 levels, using cc-pVDZ and cc-pVTZ basis sets. In the

MRCI calculations, single and double excitations of the reference wave functions were performed for all 16 valence electrons. The contribution of quadruple excitations was estimated using the method of Langhoff and Davidson.[102] The calculations predict **VN** to have a $^3A''$ ground state, and the lowest open-shell singlet ($^1A''$) and closed-shell singlet ($^1A'$) states to lie 15 and 40 kcal/mol higher in energy, respectively.

The C–N bond lengths for the three states of **VN** are 1.263 Å ($^1A''$), 1.309 Å ($^3A''$), and 1.368 Å ($^1A'$). The C=C bond length (1.346 Å) in the $^1A'$ state is typical for the carbon-carbon double bond. It is lengthened to 1.391 Å in the $^3A''$ state and to 1.461 Å in the $^1A''$ state. Therefore, the lowest open-shell singlet $^1A''$ state can be represented as a biradical.

In the $^1A''$ state, the electron in the π NBMO is predicted to be almost completely localized in the C1 carbon.

In $^1A''$, unlike the case in $^3A''$, the σ and π electrons have opposite spin; hence, they are not correlated by the Pauli exclusion principle. Localizing electrons of opposite spin to different regions of space — in this case the σ NBMO on nitrogen and the π NBMO of the C_1 carbon — minimizes their mutual Coulombic repulsion energy. Thus, ΔE_{ST} is much lower in **VN** (15 kcal/mol)[100] than in NH (36 kcal/mol)[16,17] or CH_3N (31 kcal/mol)[59] because the C=C substituent allows the π electron in the $^1A''$ state to become localized in a region of space that is disjoint from the region of space occupied by the σ electron. As will be discussed in Section 5.6, for the same reason the lowest singlet state in phenylnitrene is also the open-shell, $^1A''$ state.

In CASSCF(4,4)/6-31G* calculations on the rearrangement of the singlet **VN** to azirine, singlet vinylnitrene ($^1A''$-

VN) was found to be the transition state for interchange of the enantiotopic pair of hydrogens in 2*H*-azirine (**AZ**).[101] At the CASSCF optimized geometries, the CASSCF, CASPT2N, and MR-CISD (plus Davidson correction)[102] energies (in kcal/mol, with ZPE corrections) of [1]A"-**VN** relative to 2*H*-azirine are, respectively, 25.6, 29.8, and 31.9 with the 6-31G* basis set. With the 6-311G(2d,p) basis set the CASPT2N and MR-CISD+Q values are 27.9 and 31.3 kcal/mol.[101] With the larger basis set, it appears that CASPT2 overestimates the energy difference between the open-shell singlet nitrene and 2*H*-azirine by 3.4 kcal/mol.[101]

Although it is impossible to completely rule out the existence of the singlet **VN**, CASSCF(4,4)/6-31G* prediction, that the nitrene can cyclize to the azirine without any barrier[101] suggests that, if a barrier does exist, it is probably very small. This conclusion, based on the results of calculations, is wholly consistent with the fact, noted above, that the triplet and singlet vinyl nitrenes have escaped detection. However, further experimental studies, using very fast laser flash photolysis techniques, along with higher level *ab initio* calculations, are certainly warranted.

5.5 PHOTOLYSIS AND THERMOLYSIS OF ACYL AZIDES: PROPERTIES OF ACYLNITRENES

5.5.1 Early Experimental Results

Acyl azides and acylnitrenes have the following structures:

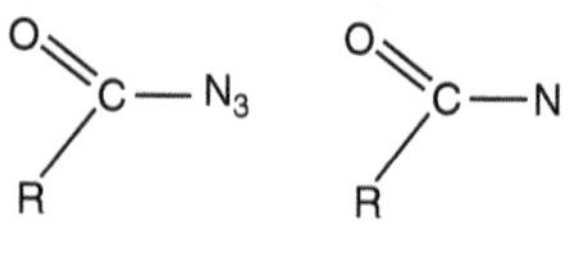

The substituent R can be an alkyl, aryl, alkoxy, aryloxy, or a hydroxyl group. This section will highlight the development of the theoretical description of the properties of acylnitrenes. Thus our coverage will be restricted to only those acyl azides and acylnitrenes for which calculations are available. We will not examine the carbamoyl and sulfonyl azides and nitrenes[103], because no theoretical data on these systems have yet been reported.

The most common thermal reaction of carbonyl azides is isocyanate formation. This reaction, known as the Curtius rearrangement, was reported as early as 1914.[104]

$$R - \overset{\displaystyle \overset{O}{\|}}{\underset{\displaystyle N_3}{C}} \quad \xrightarrow{\Delta} \quad R - N = C = O + N_2$$

$$R = \text{Alkyl, Aryl}$$

Photolysis of carbonyl azides gives rise to two types of reactions. The photo-Curtius rearrangement proceeds to form isocyanate. In addition, bimolecular trapping products, typical of the reactions of singlet carbonylnitrenes, are also observed.[103]

The mechanisms of the Curtius and photo-Curtius rearrangements have long been debated[103,105–110] Some authors have favored a concerted mechanism,[105–108] while others preferred a stepwise mechanism involving a carbonylnitrene intermediate.[109,110] In some cases carbonylnitrenes can be trapped in the photolytic, but not in the thermolytic, decomposition of carbonyl azides, which indicates that carbonylnitrenes are probably not intermediates in those particular thermal Curtius rearrangements.[103,111] For example, the yield of isocyanate, obtained upon thermolysis of pivaloyl azide (*t*-BuCON$_3$) in cyclohexane, is 99.4%.[111,112] No C-H insertion product was found, and it is now generally accepted[103] that loss of nitrogen and migration of R are concerted processes in the thermal Curtius rearrangement.

The yield of isocyanates, formed upon photolysis of a series of carbonyl azides (R–CO–N$_3$, R=*t*-butyl, aryl), remains constant in the presence and in the absence of a nitrene trap.[111–114] The yield of isocyanate, produced upon photolysis of pivaloyl azide, is about 40% in both inert and reactive alkene solvents.[111,112] The yields of the isocyanates were in the range of 40 to 50% upon photolysis of benzoyl azide and its *para*-methoxy, *para*-chloro, and *meta*-fluoro derivatives in both inert solvents and in solvents that intercept acylnitrenes.[113,114] Therefore, it was concluded that, if carbonylnitrenes (R–CO–N) are formed in these photolysis reactions, they rearrange to isocyanates (R–N=C=O) at a rate that is much lower than their capture by trapping agents.[103,112–114]

However, upon photolysis and thermolysis of azidoformates (RO–CO–N$_3$, R=alkyl, aryl), the major reaction path is the formation of products derived from capture of the nitrenes (RO–CO–N).[103, 115–119] Carbethoxynitrene (**21**) has been studied most extensively. This nitrene has been generated by thermolysis and photolysis of **20**[116,119] and by α-elimination of an arysulfonate ion from N-(p-nitrobenzenesulfonyloxy) urethane.[116,118]

The triplet multiplicity of the ground state of the nitrene **21** was first deduced on the basis of products analysis.[103,116,118,119] Nitrene **21** was generated by photolysis, thermolysis, and an α-elimination reaction, and the stereospecificity of its addition to *cis*- and *trans*-4-methylpentene-2 to form aziridines was investigated. It was found that the stereospecificity of the nitrene addition to cis alkenes decreases upon dilution of the alkene trap.

The thermolysis and α-elimination experiments can be quantitatively fit to a scheme in which all of the nitrene **21** is generated in the singlet state (**¹21**), which then decays to the ground triplet state (**³21**), in competition with stereospecific addition of **¹21** to olefins.[103,116,118,119] The triplet nitrene (**³21**) also reacts with olefins but nonstereospecifically, presumably through intermediate biradical formation.[103] The data for the photolytic generation of the nitrene were in accord with a scheme in which one third of the trappable nitrenes are produced in the triplet state, whereas are two thirds are generated in the singlet state.[103,119]

Insertion of singlet nitrene **21** into the C–H bond of alkanes and into the O–H bond of alcohols, addition of **21** to acetylenes, and reaction of **21** with benzene, followed by azepine formation, are all well documented.[103]

The proposed triplet multiplicity of the ground state of nitrene **21** was proven by ESR spectroscopy.[47] Wasserman[47] detected the ESR spectrum of ³**21** at low temperature in a rigid matrix with a $|D|$ value of 1.603 cm⁻¹. Some interaction of one of the spins with the carbonyl group is evident in the nonzero $|E|$ value of 0.0215 cm⁻¹. In subsequent studies, Autrey and Schuster[120] reproduced the ESR spectrum of ³**21** and, along with Sigman et al.,[121] recorded a very similar ESR spectrum ($|D|$ = 1.65 cm⁻¹, $|E|$ = 0.024 cm⁻¹) upon irradiation of a suspension of (4-acetylphenoxy)carbonyl azide (**22**) in Fluorolube at 77 K. This spectrum was assigned to triplet (4-acetylphenoxy)carbonylnitrene (**23**).

22   **23**

Wilde et al.[122] photolyzed methyl azidoformate in rare-gas matrices at 4 K. Methoxyisocyanate, formaldehyde, and isocyanic acid were identified as the photolysis products on the basis of infrared spectroscopy. No ESR observations of triplet carbmethoxynitrene were attempted.[122]

The photochemistry of benzoyl azide (**24**) has also been well studied. Early studies[113,114,123,124] of the photolysis of **24** in the presence of singlet nitrene traps (olefins, sulfides, etc.) demonstrated the formation of products typical of singlet nitrene reactions, along with a high yield (about 40%) of phenyl isocyanate, the product of a photo-Curtius rearrangement. Product distributions in the presence of trapping agents and photosensitizers suggested a triplet ground state for benzoylnitrene (**25**).[124] However, Inagaki and coworkers[125,126] have demonstrated that direct and triplet sensitized photolysis of **24** produces the same trapping products and that these products are characteristic of a singlet nitrene. Finally, no nitrene-like triplet ESR spectrum was detected after photolysis of **24** in glassy matrices.[28,47]

Comprehensive studies of the photochemistry of 2-naphthoyl and substituted benzoyl azides were undertaken to determine the multiplicity of the ground state of aroylnitrenes.[120,121,127,128] Irradiation (254 nm) of 2-naphthoyl azide (**26**) in cyclohexane at room temperature produces N-cyclohexyl-2-naphthamide (**27**, ~45%), 2-naphthyl isocyanate (**28**, ~50%), and a trace (<1%) of 2-naphthamide (**29**).

Irradiation of **26** in cyclohexane solution containing either cis- or trans-4-methyl-2-pentene forms aziridines with complete (>98%) retention of olefin stereochemistry.[120] Stereospecific aziridine formation is usually considered as evidence of a concerted, singlet nitrene addition reaction.

Triplet sensitized photolysis of **26** was also studied.[120] Both direct and triplet-sensitized photolysis generate products characteristic of the reactions of singlet 2-naphthoylnitrene **30**. The triplet nitrene was not detected in either chemical trapping or spectroscopic experiments. ESR signals attributable to triplet nitrene **30** were not observed after irradiation of **26** in fluorolube at 77 K. Therefore, the experimental data are most consistent with a singlet ground state for **30**.[120]

Similar results were obtained for a series of acetyl- and nitro-substituted aroyl azides (**31 –34**).[121,128]

Irradiation into the $\pi-\pi^*$ bands of aroyl azides **31** and **32** with deep-UV light leads to formation of the corresponding aroylnitrenes (**35** and **36**) in competition with photo-Curtius rearrangement to form isocyanates (23 and 40%, respectively).[121] Irradiation into the $n-\pi^*$ bands of azides **31** and **32** with near-UV light gives products derived from only the singlet aroylnitrenes (**35** and **36**). ESR signals attributable to triplet aroylnitrenes **35** and **36** were not detected after photolysis of **31** and **32** in glassy matrices.[121]

The triplet-excited states of azides **31** and **32** were detected chemically and by transient absorption spectroscopy.[121] It was concluded that nitrogen extrusion, after near-UV irradiation of the azides, occurs exclusively from the excited triplet states of the azides. However, products formed under these conditions are consistent with reactions originating from the singlet state of the aroylnitrenes.

As in the case of azides **31** and **32**, irradiation of nitro-substituted aroyl azides **33** and **34** leads both to photo-Curtius rearrangement (32 and 38% yields) and to the formation of products derived from the capture of singlet aroylnitrenes (**37** and **38**). The chemical and spectroscopic properties of nitrenes **37** and **38** indicate the singlet nature of their ground states.

Although, the authors were unable to determine the multiplicity of aroylnitrenes unambiguously by direct spectroscopic observation, the authors concluded that the sum total of their data left little doubt that aroylnitrenes have singlet ground states.[120,121,128] On the basis of this finding, they proposed acetyl- and nitro-substituted aroyl azides as potential photolabeling reagents.[127,128]

However, the opposite conclusion from that drawn for aroylnitrenes was made about ground-state multiplicity of aroyloxynitrene **23**.[121] An ESR spectrum attributable to triplet (4-acetylphenoxy) carbonylnitrene ([3]**23**) was detected[121] at 8 K after irradiation of a suspension of **22** in fluorolube at 77 K. In addition, it was found that the irradiation of **22** in benzene in the presence of both *cis*- and *trans*-pentenes (0.05 to 3 M) produces a mixture of aziridines, which is typical of the presence of both singlet and triplet nitrene cycloaddition reactions. The triplet-triplet absorption spectrum of azide **22** was detected, and it was found that naphthalene inhibits the photoreaction of **22** by quenching of its triplet state. It was concluded, therefore, that the ground state of nitrene **23** is a triplet and that it is formed exclusively upon sensitized photolysis of azide **22** through the triplet state of **22**.[121] No isocyanate was detected in the photolysis products of **22**.

Until recently singlet carbonylnitrenes, in a manner reminiscent of alkyl and vinyl nitrenes, had escaped direct spectroscopic detection. Only the ESR spectra of alkyloxy and aroyloxycabonyl nitrenes [RO–CO–N ([3]**21**, R=Et and [3]**23**, R=Ar)] have been detected.[47,121] ESR spectra of carbonyl nitrenes that do not have an oxygen atom attached to the carbonyl group [R–CO–N (R=alkyl, aryl)] have never been observed.[47,103,121]

The experimental results on acylnitrenes are clearly complicated, and calculations should be very useful for understanding and interpreting them.

5.5.2 Early Theoretical Results

Most of the calculations originally available in the literature were concerned with the simplest acyl azide and acylnitrene: formyl azide (HCO–N$_3$, **39**) and formylnitrene (HCO–N, **40**).[129,134] The first *ab initio* theoretical study of the lowest singlet and triplet electronic states of nitrene **40**, acetylnitrene (CH$_3$CO–N, **41**), and carbohydroxynitrene (HOCO–N, **42**), along with their corresponding amides, was reported in 1973.[129] The authors calculated the energies and electronic structures of the triplet and closed-shell singlet states of the nitrenes at the self-consistent-field (SCF) level of theory, using a very small basis set — STO-3G. For nitrenes **40** and **41**, which are models of alkanoylnitrenes, the triplet states were found to be more stable than the corresponding singlet states by 15 and 12 kcal/mol, respectively. The singlet state of **42**, a model of a carbalkoxynitrene, was found to be more stable than the triplet state by 23 kcal/mol. These results contradict the ESR experiments[47] for carbethoxynitrene **21**.

Harrison and Shalhoub[130] also performed calculations of the electronic structure and geometry of the low-lying states of a series of carbonylnitrenes (XCO–N [X = H, F, CH_3, and OCH_3]) using the small STO-3G basis set. However, they did include some configuration interaction, using single and double excitations. They found that for all carbonylnitrenes studied, the ground state is the triplet, with the next two states being closely spaced singlets. With nitrene **40**, the results obtained were highly reminiscent of those for alkylnitrenes, because the acylnitrene singlets ($^1A'$ and $^1A''$) were predicted to be approximately 35 and 40 kcal/mol above the triplet state ($^3A''$)[130] using a Dunning basis set.

Shortly thereafter, Rauk and Alewood[131] studied the thermal and photochemical decomposition and Curtius rearrangement of formyl azide **39** by using SCF calculations augmented with CI. They used both STO-3G and extended (9s5p/4s contracted to 4s2p/2s) basis sets and calculated the structures of the six isomers of the formula HNCO. Isocyanic acid HN=C=O, **43** the parent isocyanate and the product of the Curtius rearrangement of **39** was predicted to be the most stable isomer, 91 kcal/mol lower energy than nitrene **40** in the singlet $^1A'$ state. Figure 5.9 displays the geometry of the $^1A'$ state of **40** and of oxazirene **44**, which were calculated to be very close in energy (within 1 kcal/mol).[131] The authors[131] also attempted calculations of the potential-energy surface for the decomposition of the azide **39** in the ground and low-lying excited states.

Poppinger et al.[132] explored the singlet potential-energy surface of the CHNO system with the aid of the restricted HF method and with STO-3G, 4-31G, and 6-31G* basis sets. These authors found the structures of **40** and **44** to be minima on the singlet potential-energy surface by using an STO-3G basis

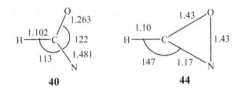

Figure 5.9 Structures of singlet nitrene and oxazirene calculated by Rauk and Alewood.[131] Bond lengths are in angstoms, angles in degrees.

set. The attempted optimization of **40** or **44** at the 4-31G level led to the HNCO global minimum **43**. The transition state for rearrangement to **43** was calculated to lie 23 kcal/mol above **44** at the STO-3G level, but 4-31G calculations at the STO-3G transition state placed it 9 kcal/mol below **44**. As a result, the authors[132] predicted that singlet formylnitrene **40** should collapse in a barrierless process to isocyanic acid **43** and that cyclic oxazirene **44** corresponds to a very shallow minimum or may not be a minimum at all.

In their next paper, Poppinger and Radom[133] studied the effect of substituents (R=H, Li, BH$_2$, CH$_3$, NH$_2$, OH, F, Cl, CN, C$_6$H$_5$, p-NO$_2$-C$_6$H$_4$, and p-NH$_2$-C$_6$H$_4$) on the structure and stability of the RNCO isomers at the previously applied level of theory.[132] On the basis of these calculations, they concluded that singlet acetylnitrene and chlorocarbonylnitrene both collapse to isocyanates without an activation barrier. However, they found hydroxycarbonylnitrene, the parent carbalkoxynitrene, and fluorocarbonylnitrene to be energy minima, although they were predicted to lie in relatively shallow potential wells. There was no theoretical evidence for the existence of a substituted oxazirene that would be sufficiently stable to observe directly.[133]

In 1980, Mavridis and Harrison[134] recalculated the electronic structures of the triplet and low-lying singlet states of formylnitrene **21** using self consistent field (SCF) and generalized valence bond (GVB) procedures with a double ζ plus polarization functions (DZ + P) basis set. The triplet ^{3}A'' state was found to be the ground state, whereas the first two excited states (^{1}A' and ^{1}A'') were calculated to be 36.8 and 39 kcal/mol above the triplet state. The geometries of the singlet states were not optimized, and the open-shell singlet $^1A''$ was assumed to have a geometry identical to its companion $^3A''$ state. In constructing the wave function for a closed-shell singlet, $^1A'$, all bond angles were taken to be 120°.

5.5.3 Recent Computational and Experimental Studies

Recently,[135] calculations of the properties of the singlet and triplet states of benzoyl and 2-naphthoylnitrenes (**25** and **30**) were perfomed at the B3LYP/6-31G* level of theory. It was found that the structure of the CON fragment in the lowest singlet state of **25** and **30** ($^1A'$) resembles that of a cyclic oxazirene, although the calculated N–O distance (~1.76 Å) is much longer than in a normal N–O single bond (about 1.5 Å in strained rings).[136,137] It is interesting to note that Cornell et al.[138] proposed a cyclic structure for carbalkoxylnitrenes about 40 years ago.

Starting from the geometry of either a closed-shell singlet nitrene or an oxazirene structure with normal N–O bond lengths, geometry optimizations led to the minimum with the 1.76 Å N–O bond length. The optimized structures of triplet and singlet benzoylnitrenes are presented in Figure 5.10. The energy difference between the triplet states of **25** and **30** and the corresponding singlet species, with cyclic structures but unusually long N–O bonds, was calculated to be small. However, the triplet state was still computed to be lower in energy by about 5 kcal/mol.[135]

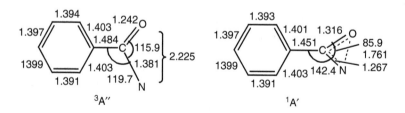

Figure 5.10 Bond lengths (in angstroms) and bond angles (in degrees) in the singlet A' and triplet A'' states of benzoylnitrene optimized at the B3LYP/6-31G* level.[135]

The reason for the dramatic stabilization of $^1A'$, relative to the $^3A''$ state of aroylnitrenes, appears to be a bonding interaction between a lone pair on the carbonyl oxygen atoms and an empty 2p orbital on nitrogen. The resulting weak sigma bond results in the structure of the singlet being intermediate between those of a normal nitrene and an oxazirene. Since B3LYP/6-31G* calculations might overestimate the value of ΔE_{ST}, the authors[135] conjectured that aroylnitrenes probably do have singlet ground states, in agreement with the experimental results of Autrey and Schuster[120] and Sigman et al.[121]

Shapley and Bacskay[139] performed high-level calculations on the structure and energy of triplet formylnitrene and a "cyclic" singlet. The geometries of these intermediates, calculated using the MP2/6-31G* and CASSCF(11,11)/cc-pVDZ methods, are very similar to those optimized by Gritsan and Pritchina[135] using the B3LYP/6-31G* procedure. The ΔE_{ST} values, calculated by Shapley and Bacskay,[139] are 3.6 kcal/mol and 0.7 kcal/mol, respectively, at the CASPT2(11,11) and QCISD(T) levels of theory with the cc-pVTZ basis set. Using data obtained with the G2 method,[139] ΔE_{ST} was calculated to be –2.5 kcal/mol; that is, the singlet was predicted to be the ground state.

To analyze the influence of the basis set and the level of the theory on the calculated value of ΔE_{ST}, the energies of the lowest singlet and triplet states of **40** were also computed at the coupled cluster level with single and double substitutions (CCSD) and with non-iteratively included triple substitutions (CCSD(T)) levels of theory.[140]. Dunning's correlation-consistent basis sets were used, ranging in size from double-zeta to quadruple-zeta. It was found that B3LYP/6-31G* provides very good geometries for the singlet and triplet states of formylnitrene. However, whereas the B3LYP calculations predict $\Delta E_{ST} > 8$ kcal/mol, benchmark CCSD(T)/cc-pVQZ//CCSD(T)/cc-pVTZ calculations give $\Delta E_{ST} = -0.13$ kcal/mol. Furthermore, basis set extrapolation[141–143]) leads to $\Delta E_{ST} = -0.72$ kcal/mol and an even clearer prediction of a singlet ground state.

The prediction of a singlet ground state should also be valid for the aroylnitrenes. A ΔE_{ST} value of about 5 kcal/mol was calculated for aroylnitrenes **25** and **30** by a B3LYP/6-31G* procedure,[135] but higher-level calculations show that B3LYP/6-31G* overestimates ΔE_{ST} for formylnitrene by about 9 kcal/mol. Assuming errors of similar size in the B3LYP calculations on **25** and **30**, the B3LYP values of ΔE_{ST} for these two nitrenes suggest that they too have singlet ground states.

Although the triplet ESR spectra of a carbonylnitrene with a structure of R–CO–N (R=Alkyl, Aryl) has never been detected,[47,103,121] the ESR spectra of triplet carbethoxynitrene [3]**21** and aroyloxynitrene [3]**23** (RO–CO–N, R=Et and Ar) have been obtained.[47,121] To understand the difference between these two types of nitrenes (R–CO–N and RO–CO–N), some preliminary calculations were performed on carbohydroxynitrene (HOCO–N, **42**), using it as a model for carbalkoxy and carbaryloxynitrenes.[143a] The geometries of the singlet and triplet states of **42** were optimized at the B3LYP/6-31G* level, and the energies of the optimized structures were recalculated at the CCSD(T) level with cc-pVDZ and cc-pVTZ basis sets.

Figure 5.11 compares the optimized structures of the lowest singlet states ($^1A'$) of formylnitrene and two rotamers of nitrene **42**. It was found at the B3LYP/6-31G* level that the ΔE_{ST} value is 16.1 kcal/mol for rotamer **42a** and 11.0 kcal/mol for rotamer **42b**. Recall that at this level of theory ΔE_{ST} = 8.1 kcal/mol for **40**. The value of ΔE_{ST} for **42a** at the CCSD(T)/cc-pVTZ level drops but only to 7.8 kcal/mol, providing computational stronger evidence for a triplet ground state in **42**.

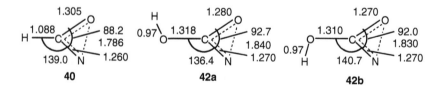

Figure 5.11 Bond lengths (in angstroms) and bond angles (in degrees) in the singlet $^1A'$ states of formylnitrene and carbohydroxynitrene optimized at the B3LYP/6-31G* level.

The difference in ΔE_{ST} between **42** and formylnitrene **40** is attributed to the smaller bonding interaction between the nitrogen and the oxygen atoms in **42**, which manifests itself in the longer N–O bond length in **42** than in formylnitrene (Figure 5.11). The replacement of the hydrogen in formylnitrene by the hydroxyl group in **42** apparently inhibits σ-bonding between the nitrogen and the carbonyl oxygen by increasing the strain in the three membered oxazirene ring.

In order to determine experimentally the multiplicity and the structure of the lowest state of benzoylnitrene (**25**), a study of the photochemistry of benzoyl azide (**24**) was performed in an Ar matrix at 12 K.[140] The formation of two species was observed upon exposure of **24** to 254 nm light. One of these species has an IR spectrum, which is consistent with that expected for phenyl isocyanate (**45**). The IR and UV spectrum of the second intermediate are in very good agreement with the calculated spectra of singlet nitrene in the ^{1}A' state (**25**). This intermediate undergoes isomerization to isocyanate (**45**) upon further exposure of the sample to 313-nm light.[140]

Structure 5.26

A definititve answer can now be given[135,140] to the question posed by Autrey and Schuster[120]: "Are aroylnitrenes ground-state singlets?" Calculations and experiments now agree that aroylnitrenes are indeed ground-state singlets, but calculations show that these singlets have electronic structures and geometries that are intermediate between those of nitrenes and oxazirenes.

5.6 PHOTOLYSIS OF ARYL AZIDES: PROPERTIES OF ARYLNITRENES

Before discussing recent experimental and theoretical studies on the reactivity and spectroscopy of arylnitrenes, we will describe some of the earliest experimental and theoretical work on these molecules. Our purpose is to give a brief overview in order to provide a context for discussion of more recent results. For detailed descriptions of the chemistry of arylnitrenes, we refer the reader to the many excellent reviews in this area[144–151]

5.6.1 Early Experimental and Theoretical Studies

5.6.1.1 Experiments

Phenylazide (**46**) was first synthesized by Greiss[152] in 1864. Wolf[153] first studied the thermal decomposition of phenylazide. In the presence of aniline, thermolysis of **46** leads to extrusion of molecular nitrogen, generation of a reactive intermediate C_6H_5N, and eventual formation of azepine **47**, a reaction elucidated by Huisgen and Vossius[154] and Huisgen.[155]

Similarly, Doering and Odum[156] demonstrated that photolysis of **46** leads to the evolution of molecular nitrogen, the formation of a diethylamine-trappable intermediate, and ultimately azepine **48**.

For almost 100 years; chemists have argued over the identity of the C_6H_5N species. Candidate structures for C_6H_5N have been singlet (^{1}**49**) and triplet (^{3}**49**) phenylnitrene, benzazirine (**50**), and cyclic ketenimine (**51**), a menagerie of species described by Schrock and Schuster[157] as "wonderfully complex."

The question of the intermediates formed by the photolysis of **46** is not just of academic interest, because aryl azides have important applications as photoresists in lithography,[8] in the formation of electrically conducting polymers,[158] in organic synthesis,[144,159] as photoaffinity labels,[4–6] and in the covalent modification of polymer surfaces.[7]

Chemical analysis of reaction mixtures has provided evidence for all these intermediates, **49** to **51**, under different conditions.

Triplet phenyl nitrene and other triplet arylnitrenes dimerize to form azo compounds.[147,160]

The products of nucleophilic trapping (**47, 48**), after decomposition of **46**, were initially rationalized as arising from benzazirine **50**.[154,155] This explanation was generally accepted in subsequent studies[161–163] and supported by calculations,[164] but in 1978, Chapman and LeRoux[165] detected 1-aza-1,2,4,6-cycloheptatetraene (**51**) using matrix isolation techniques. The existence of the cyclic ketenimine **51,** was confirmed by later spectroscopic studies in matrices[166] and in solution.[167–169] It was also established that ketenimine **51** is the species trapped by nucleophiles in solution to form azepines **47** and **48**.[169]

Chapman and LeRoux 's experiments produced no evidence for the intermediacy of **50**.[165] The strongest experimental evidence to date for the intermediacy of **50** is the observation that photolysis of **46** in ethanethiol affords o-thioethoxyaniline (**52**) in 39% yield, presumably from the nucleophilic trapping of **50**.[170]

Unfortunately, the major product obtained upon decomposition of phenylazide (and many, if not most, of its derivatives) in solution is polymeric tar.[158] Thus, progress in classical nitrene chemistry was much slower than with carbenes, which form robust, easily characterized adducts with most organic molecules and even with alkanes.[171]

Before moving on to the contribution of spectroscopic methods, we will note two early experiments. First, high dilution of phenylazide suppresses polymer formation and encourages the formation of the azo compound,[168,172,173] implying that a singlet intermediate, such as benzazirine or ketenimine, can serve as a reservoir for generation of triplet arylnitrenes, which subsequently dimerize.

Second, Leyva et al.[174] found that the solution phase photochemistry of phenylazide (**46**) was temperature dependent. Photolysis of **46** in the presence of diethylamine at ambient temperature yields azepine **48**, first prepared by Doering and Odum.[156] Lowering the temperature suppresses the yield of **48** and encourages the formation of the azo compound. Thus, high temperatures favor reactions of singlet state intermediates, whereas low temperatures favor reactions associated with triplet phenylnitrene.

Smolinsky and coworkers[175] obtained the EPR spectrum of triplet phenylnitrene (^{3}**49**), immobilized in a frozen glass. The temperature dependence of one EPR signal demonstrated that ^{3}**49** is lower in energy than ^{1}**49**.

Reiser's and Frazer[176] published an important series of papers beginning in 1965. They were the first to observe the low-temperature UV-Vis spectrum of triplet phenylnitrene ^{3}**49**.[177] A later study[174] of low-temperature glassy matrices revealed an additional long-wavelength band in the spectrum of ^{3}**49**, reported earlier,[177] showing that it was contaminated by the presence of ketenimine **51**. It was found that ^{3}**49** is extremely light sensitive and, upon photo excitation at 77 K, ^{3}**49** rapidly isomerizes to the isomeric **51**.[174]

Initial flash photolysis experiments involving **46** gave conflicting results, with different authors favoring the presence of either triplet phenylnitrene ^{3}**49**,[180,181] benzazirine

50,[162] or cyclic ketenimine **51**[157] as the carrier of the transient spectra.

The currently accepted spectroscopic assignments were obtained by a combination of techniques. Leyva et al.[174] applied matrix absorption and emission spectroscopy, along with flash photolysis techniques. Chapman and LeRoux[165] obtained the matrix IR spectrum of cyclic ketenimine **51**, and Hayes and Sheridan[182] obtained the matrix IR and UV-Vis spectrum of triplet phenylnitrene and cyclic ketenimine **51**. Shields et al.[168] and Li et al.[169] applied time-resolved IR and UV-Vis spectroscopy to demonstrate that cyclic ketenimine **51** is formed in solution and that this species absorbs strongly at 340 nm.

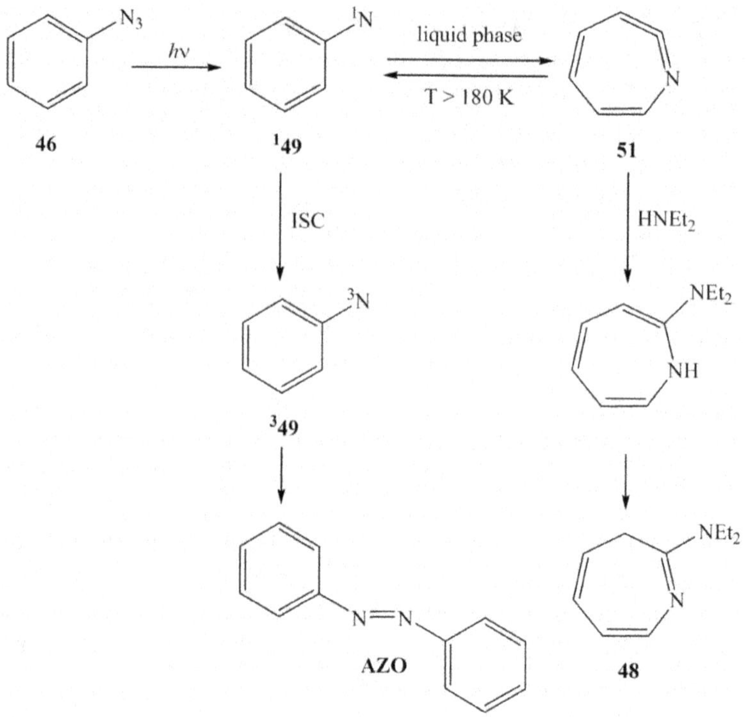

Scheme 5.1

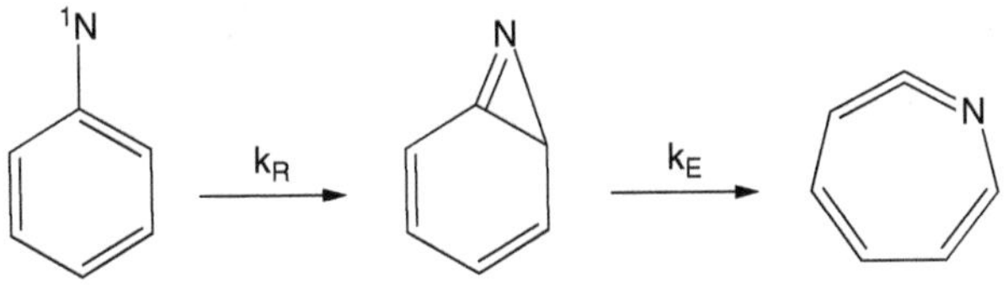

Scheme 5.2

By 1992, Schuster and Platz[149] could write Scheme 5.1, which economically explained much of the condensed-phase photochemistry of **46**. UV photolysis of **46** produces singlet phenylnitrene and molecular nitrogen. In the liquid phase, [1]**49** isomerizes over a small barrier to form cyclic ketenimine **51**. Later computational work from Karney and Borden[101] showed this to be a two-step process involving benzazirine **50** (Scheme 5.2), the species trapped by ethanethiol. At ambient temperature in the liquid-phase [1]**49** prefers rearrangement to intersystem crossing (ISC) to the lower-energy triplet state.

The rate of cyclization slows upon cooling but ISC is not expected to be an activated process so its rate is not expected to vary with temperature. The isokinetic temperature is about 180 K, measured with two precursors: **46** and sulfoximine **53**,[174,183] Below 180 K, ISC to triplet **49** predominates.

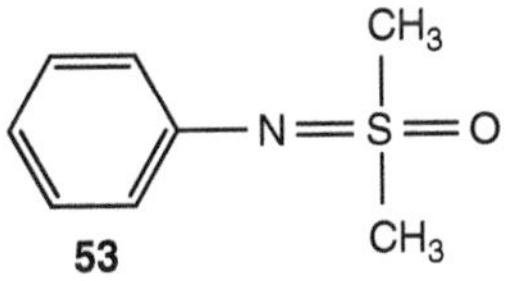

The key intermediate in Scheme 5.1 is singlet phenylnitrene ([1]**49**) — the only intermediate that by 1992 had not been detected directly or chemically intercepted. However, in 1997, Gritsan et al.[184] and the Born et al.[185] simultaneously reported that laser flash photolysis of **46** or of phenyl isocyanate **54** produces a previously undetected transient with λ_{max} = 350 nm and a lifetime of $\approx$1 ns at ambient temperature.

54

The transient decays at the same rate that cyclic keten-imine **51** is formed, suggesting that the newly detected transient is singlet phenylnitrene (^{1}**49**).[184] This assignment was secured with the aid of computational chemistry[186] and by studying the temperature dependence of the kinetics.[184,186] The barrier to rearrangement of ^{1}**49** has been determined to be 5.6 ± 0.3 kcal/mol, with an Arrhenius pre-exponential factor of $10^{13.1 \pm 0.3}$ s^{-1}.[186] A value for the rate constant for inter-system crossing was also extracted: k_{ISC} = 3.2 ± 0.3 × 10^6 s^{-1}, which is nearly four orders of magnitude smaller than k_{ISC} for phenylcarbene.

5.6.1.2 Semiempirical Calculations

The C_6H_5N potential-energy surface originally received less attention from theoreticians than from experimentalists. Until recently, species on the reaction pathway for ring expansion of ^{1}**49** and some of its simple derivatives had been studied using only semiempirical methods.[150,164,169,173,187] MNDO calculations by the Schuster group[169] predicted the intermediacy of azirine **50** and placed azepine **51** below **50** in energy. Their calculations found barriers of 12.4 and 3.6 kcal/mol for the first and second steps of the ring expansion, respectively.[169] The much lower barrier computed for the ring opening of **50** to **51** is consistent with the experimental finding that **51**, not **50**, is the species that is trapped in solution.[169]

Li et al.[169] also performed calculations on azacyclohep-
tatrienylidene (**55**), the planar carbene isomer of ketenimine
51. On the basis of their MNDO results, they proposed that
the experimentally observed thermal reversion of **51** to triplet
phenylnitrene (^{3}**49**) occurs not via singlet **49** but rather via a
triplet state of **55**.[169]

55

Gritsan and Pritchina[150,173] performed AM1 and MNDO
calculations of the electronic structure of phenylnitrene **49**
and its para-substituted derivatives **56** to **58**.

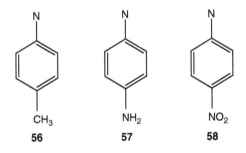

| CH₃ | NH₂ | NO₂ |

CH_3 NH_2 NO_2

56   **57**   **58**

They found that the triplet state is the ground state for all
arylnitrenes studied. The lowest singlet state of nitrene **49**
and its derivatives **56** to **58** was predicted to be the open-shell
singlet.[150,173] The singlet-triplet energy gap ΔE_{ST} was calculated to be 8.4 kcal/mol for **49** at the AM1 level, and both

electron-donor (NH_2) and electron-withdrawing (NO_2) substituents reduced ΔE_{ST} to 6.6 and 7.4 kcal/mol, respectively.[173] The next singlet state of nitrene **49** was calculated to be a closed-shell singlet, 12.7 kcal/mol above the open-shell singlet. The energy difference between open-shell and closed-shell singlets were predicted to be 9.5 kcal/mol for **57** and 14.3 kcal/mol for **58**.[173] Subsequent *ab initio* calculations revealed that these results were only qualitatively correct and significantly underestimated the value of ΔE_{ST}.

5.6.2 Phenylnitrene: Electronic Structure and Spectroscopy

In phenylnitrene **49**, a lone pair occupies a hybrid orbital, rich in 2s character; and, unlike the case in phenylcarbene (**59**), the two nonbonding electrons both occupy pure 2p orbitals. One of these is a 2p-π orbital, and the other a 2p orbital on nitrogen that lies in the plane of the benzene ring. The near-degeneracy of the two 2p orbitals gives rise to three low-lying spin states: a triplet ($^3A_2$), an open-shell singlet ($^1A_2$), and a closed-shell singlet ($^1A_1$). The orbital occupancies and CASSCF(8,8)/6-31G* geometries of these are shown in Figure 5.12.[101]

59

In the $^3A_2$ and $^1A_2$ states, the 2p-π orbital and the in-plane 2p orbital on N are both singly occupied. The $^1A_1$ state of **49** is a mixture of two dominant configurations — one in which the in-plane 2p orbital on N is doubly occupied and the

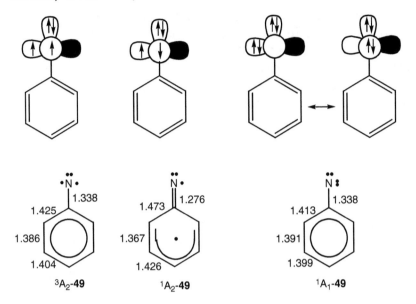

Figure 5.12 Orbital occupancies for nonbonding electrons and CASSCF(8,8)/6-31G* optimized geometries of the lowest triplet and singlet states of phenylnitrene.[49,101] Bond lengths are in angstroms.

2p-π orbital is empty, which is slightly lower in energy than the configuration in which these orbital occupancies are reversed.[188,189] In both the $^3A_2$ and $^1A_1$ states, the C–N bond is relatively long and the phenyl ring shows little bond-length alternation. In the $^1A_2$ state, however, strong delocalization of the electron in the nitrogen p-π orbital into the ring leads to a very short C–N bond (1.276 Å), and the bond lengths within the aromatic ring resemble those of a cyclohexadienyl radical.[188,189]

As discussed in Section 5.2, in NH $^1A_1$ and $^1A_2$ are degenerate and form the two components of a ^1E state. However, in phenylnitrene the $^1A_2$ state lies well below the $^1A_1$ state. In the $^1A_2$ state delocalization of the electron in the singly occupied π orbital into the benzene ring confines this electron and the opposite-spin electron in the in-plane 2p AO on nitrogen to different regions of space, thus minimizing their mutual Coulomb repulsion energy.[190–192] This strong delocalization of the unpaired π electron in $^1A_2$ accounts for the difference between the bond lengths in this state and those in $^3A_2$.[188,189]

The calculated[101,188,189,193,194] and experimentally determined[193,194] relative energies of the spin states of **49** are shown in Table 5.5. All levels of theory shown predict the lowest singlet state, ($^1A_2$) to lie about 18 kcal/mol higher in energy than the triplet ground state ($^3A_2$), in excellent agreement with the results of photoelectron[193] and electron photodetachment[194] spectroscopy. Predictions for the relative energy of the $^1A_1$ state vary substantially, with the DFT result of 29.5 kcal/mol[195] (relative to the $^3A_2$ state) apparently giving the best agreement with the experimental estimate.[193]

Both phenylnitrene **49** and the isoelectronic phenylcarbene **59** have triplet ground states; however, the lowest singlet

TABLE 5.5 Relative Energies (kcal/mol) of the Lowest Spin States of Phenylnitrene (**49**)

Method	Electronic state			Ref.
	$^3A_2$	$^1A_2$	$^1A_1$	
CISD+Q/DZ+d	0	18.3	32.4	188
σ-S, π-SDCI/6-31G*// CASSCF(8,8)/3-21G	0	18.3	38.7	189
SDCI/6-31G*// CASSCF(8,8)/3-21G	0	18.3	30.6	195
CASSCF(8,8)/6-31G*	0	17.5	42.2	101
CASPT2(8,8)/6-311G(2d,p)// CASSCF(8,8)/6-31G*	0	18.5	36.9	101
BLYP/cc-pVTZ// BLYP/cc-pVDZ	0	-	29.5	196
CCSD(T)/cc-pVDZ// CASSCF(8,8)/cc-pVDZ	0	-	35.2	196
Experiment	0	18 ± 2	30 ± 5	193
	0	18.3 ± 0.7	-	194

is the open-shell ($^1A_2$) state in **49** but the lower singlet is the closed-shell ($^1A'$) state in **59**. This difference between the carbene and the nitrene can be ascribed to the fact that in singlet **59** the two nonbonding electrons can occupy a hybrid AO, whereas in **49** the two nonbonding electrons occupy two pure 2p orbitals. Therefore, the near degeneracy of the nonbonding MOs in **49** is strongly lifted in **59**, and in the lowest singlet state of **59**, both nonbonding electrons preferentially occupy the lower-energy, hybridized σ-orbital rather than the 2p-π orbital.[197–200]

As a result, the energy difference between the lowest singlet and the triplet state is much smaller in **59** ($E_{ST} \approx 2$ kcal/mol[201]) than in **49** ($E_{ST} \approx 18$ kcal/mol[193,194]). The smaller value of E_{ST} in **59** than in **49** contributes to the much faster rate of intersystem crossing in phenylcarbene[202,203] relative to phenylnitrene.[186]

Another contributor is the difference between the nature of the lowest singlet states in **49** and **59**. The transition of an electron from a σ-orbital in the $^1A'$ state of **59** to a π orbital in the $^3A''$ state creates orbital angular momentum, and spin-orbit coupling facilitates the change in spin angular momentum that occurs in intersystem crossing in the carbene. In contrast, the $^1A_2$ and $^3A_2$ states of **49** have the same orbital occupancy, so there is no change in orbital angular momentum to facilitate intersystem crossing via spin orbit coupling.

5.6.2.1 UV-Vis Spectroscopy of Triplet Phenylnitrene ^{3}**49**

The calculations[101,188,189,193–196] predict that the ground state of phenylnitrene has triplet multiplicity ($^3A_2$) in accordance with the earlier EPR experiment of Smolinsky et al.[175] Figure 5.13 represents the electronic absorption spectrum of ^{3}**49** in an Ether-pentane-alcohol (EPA) matrix at 77 K.[174] There is a strong sharp band at 308 nm, a broad structured band at 370 nm, and a broad unstructured feature, which tails out to 500 nm.[174,177]

The π system of ^{3}**49** is closely related to that of the benzyl and anilino radicals. Thus, it is no surprise that the triplet absorption spectrum of ^{3}**49** is very similar to the experimental

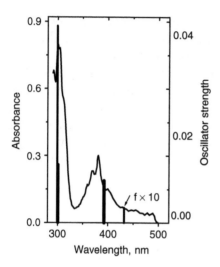

Figure 5.13 The absorption spectrum of triplet phenylnitrene in EPA glass at 77 K.[174] The computed positions and oscillator strengths (f, right-hand axes) of the absorption bands are depicted as solid vertical lines.[186] For very small oscillator strength, the value multiplied by 10 is presented (f × 10).

spectra of the benzyl ($C_6H_5CH_2$) and anilino (C_6H_5NH) radicals. Benzyl radical has a medium intensity feature at 316 nm and a very weak band at 452 nm.[204] The anilino radical has a medium intensity band at 308 nm and a weak transition at 400 nm.[205–207]

Using the INDO/S method, Shillady and Trindle[164] performed the first calculation of the electronic absorption spectrum of ^{3}**49**. The long wavelength maximum was calculated to be at 380 nm and was assigned to an n → π* transition. The long wavelength transition, observed at about 450 nm, was not reproduced by this calculation. The next transition was computed to be at 330 nm, and it was assigned to the excitation of a nitrogen lone-pair electron into the singly occupied nitrogen 2p orbital that lies in the molecular plane ($2p_y$). Note that the latter transition is of the same nature as those in triplet nitrenes NH and in CH_3N (Sections 5.2 and 5.3).

Later, Kim et al.[188] performed configuration interaction calculations of ground and excited states of ^{3}**49** with all single and double excitations, but they failed to reproduce the electronic absorption spectrum of triplet phenylnitrene quantitatively.

However, recently[186] the spectrum of ^{3}**49** was calculated using the CASPT2 level of theory. The computed spectrum was in very good agreement with experiment (Figure 5.13). The improved correspondence is a result of the combination of an improved description of the reference wave function and an adequate treatment of the dynamical electron correlation by the CASPT2 procedure.

CASPT2 calculations predict that the vertical excitation energy to the first excited state (1^3B_1) will be at 432 nm (f = 3.4×10^{-4}). This excited state consists principally of two electronic configurations: $\pi (1a_2) \rightarrow \pi (3b_1)$ and $\pi (3b_1) \rightarrow \pi_1^* (2a_2)$, where $\pi (3b_1)$ is the singly occupied π–orbital. The second excited state is 2^3A_2, and it has a vertical excitation energy of 393 nm (f = 9.4×10^{-3}), which is associated with the $\pi (2b_1) \rightarrow \pi (3b_1)$ and $\pi (3b_1) \rightarrow \pi_2^* (4b_1)$ transitions.

Triplet phenylnitrene has a very strong absorption band at 308 nm. The CASPT2 calculations predict that transitions to the 2^3B_1 (at 301 nm, f = 0.013) and 3^3B_1 (at 299 nm, f = 0.044) states contribute to this absorption. The electronic configurations for the 3^3B_1 state are the same as for the 1^3B_1 state. The main configuration involved in the $1^3A_2 \rightarrow 2^3B_1$ transition consists of excitation of an electron from the lone pair orbital (n_z) on nitrogen to the singly occupied nitrogen 2p orbital that lies in the molecular plane (p_y). This transition, $1^3A_2 \rightarrow 2^3B_1$ (around 300 nm), is very similar to those observed from the triplet ground states of the parent NH (336 nm),[12,13] methylnitrene (315 nm),[55,56] 1-norbornylnitrene (298 nm),[42] and perfluoromethylnitrene (354 nm).[43]

5.6.2.2 UV-Vis Spectroscopy of Singlet Phenylnitrene ^{1}**49**

The electronic absorption spectrum of ^{1}**49** was first detected in 1997.[184] Laser flash photolysis (266 nm, 35 ps) of **46** in pentane at 233 K produces a transient absorption spectrum, which shows two sharp bands with maxima at 335 and 352 nm (Figure 5.14). In later work[186] the spectrum of ^{1}**49** was

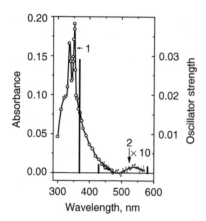

Figure 5.14 Transient spectrum of singlet phenylnitrene produced upon LFP of phenyl azide. Spectrum 1 was recorded 2 ns after laser pulse (266 nm, 35 ps) at 233 K. Long-wavelength band (spectrum 2) was recorded with an optical multichanal analyzer at 150 K (with a 100-ns window immediately after the laser pulse, 249 nm, 12 ns). The computed positions and oscillator strengths (f, right-hand axes) of the absorption bands are depicted as solid vertical lines.[186] For very small oscillator strengths a value multiplied by 10 is presented (f × 10).

reinvestigated and an additional very weak, long wavelength absorption band at 540 nm was observed. The transient spectrum in Figure 5.14 was assigned to singlet phenylnitrene in its lowest open-shell electronic configuration: $^1A_2$.[184,186]

This assignment was supported by the similarity of the spectrum (Figure 5.14) to that of the long-lived perfluorinated singlet arylnitrenes.[208] The decay of this transient absorption is accompanied by the formation of cyclic ketenimine **51**. Furthermore, the electronic absorption spectrum of ^{1}**49** in the $^1A_2$ state, calculated at the CASPT2 level, is in good agreement with the transient spectrum that is observed for ^{1}**49** (Figure 5.14).

Calculations on ^{1}**49** are more challenging than those on ^{3}**49**, because the $^1A_2$ state of **49** is an open-shell singlet state. The first two electronically excited singlet states of ^{1}**PN** are both of A_1 symmetry and are predicted computationally to be found at 1610 and 765 nm. Neither of these transitions has been detected, because not only do excitations from $^1A_2$ into both of these states have zero oscillator strength because of symmetry considerations, but these excited states are predicted to lie outside the wavelength range accessible to the spectrometer.[186]

The CASPT2 calculations predict a transition at 581 nm to a 1^1B_1 excited state, with a very small oscillator strength (1.6×10^{-4}). This transition could be assigned to a very weak band with a maximum around 540 nm (Figure 5.14). This state consists of the same electronic configurations as the long wavelength transition in $^3\mathbf{49}$.

The next excited state in the singlet manifold is the 2^1A_2 state. This transition has a small oscillator strength (2.1×10^{-3}) and an excitation energy of 429 nm. In the experimental spectrum (Figure 5.14), this band seems to be a shoulder on the tail of a strong band at 350 nm.

The only intense absorption band in the absorption spectrum of $^1\mathbf{49}$ is localized around 350 nm, which is a pronounced shift from the 308 nm band in $^3\mathbf{49}$. This band has a long tail out to 450 nm and displays some fine structure that may be associated with the vibrations of the phenyl ring in $^1\mathbf{49}$ (Figure 5.14). The strongest absorption band in $^1\mathbf{49}$, predicted by the CASPT2 method, is the transition to the 2^1B_1 exited state, which has a 368 nm excitation energy. The main configuration involved in this transition is similar to that of the 2^3B_1 state and consists of an electron from the lone-pair orbital on nitrogen (n_z), which is promoted to the singly occupied nitrogen 2p orbital that lies in the molecular plane (p_y). Therefore, the nature of this band is the same as that of the most intense band in the spectrum of $^3\mathbf{49}$. Note also the same origin as the UV bands of the triplet and singlet nitrene NH (Section 5.2) and triplet and singlet alkylnitrenes (Section 5.3).

Although, the electronic absorption spectra of $^1\mathbf{49}$ and $^3\mathbf{49}$ are very similar (Figure 5.13 and Figure 5.14). However, all of the calculated and experimental bands of $^1\mathbf{49}$ exhibit a red shift compared with those of $^3\mathbf{49}$.

5.6.3 Dynamics of Singlet Phenylnitrene

5.6.3.1 Recent Experiments

The decay of singlet phenylnitrene in pentane was monitored at 350 nm over a temperature range of 150 to 270 K, which allowed direct measurement of the rate constant for intersystem crossing (k_{ISC}) and of accurate barriers to cyclization.[186] The disappearance of ¹**49** at 298 K was faster than the time resolution of the spectrometer, and the lifetime (τ) of ¹**49** was estimated to be ~1 ns under these conditions.[184,186] This estimated lifetime of ¹**49** is consistent with that of about 0.6 ns, measured in CH_2Cl_2 at ambient temperature by Born and coworkers.[185]

The formation of the products (cyclic ketenimine **51** and triplet nitrene ³**49**) was monitored at 380 nm.[184] The decay of ¹**49** and growth of the products are first order and can be analyzed to yield an observed rate constant, k_{OBS}. An Arrhenius treatment of the k_{OBS} data (open circles) is presented in Figure 5.15.

The magnitude of k_{OBS} decreases with decreasing temperature until about 170 K, whereupon it reaches a limiting

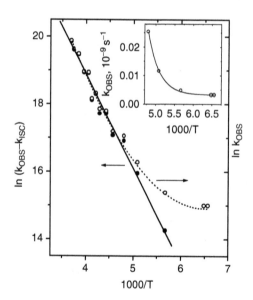

Figure 5.15 Arrhenius treatment of the k_{OBS} data (open circles) and k_R data (filled circles) for singlet phenylnitrene deduced upon assuming that k_{ISC} is independent of temperature.[186] Insert: temperature dependence of k_{OBS} data.

value of about 3.2×10^6 s^{-1}. Below this temperature, k_{OBS} remains constant.[186] The breakpoint in the Arrhenius plot is about 180 to 200 K and is in exactly the same temperature range where the solution-phase chemistry changes from trapping of ketenimine **51** with diethylamine to dimerization of ^{3}**49**.[174] Thus, the low-temperature data in Figure 5.15 is associated with the rate constant for intersystem crossing of ^{1}**49** to ^{3}**49** ($k_{ISC} = 3.2 \pm 0.3 \times 10^6$ s^{-1}) and the high-temperature data with k_R, the rate constant for rearrangement of ^{1}**49** (Scheme 5.1).

Because at any temperature $k_{OBS} = k_R + k_{ISC}$, by assuming that k_{ISC} does not change with temperature, it was possible to deduce values of k_R as a function of temperature and to obtain its Arrhenius parameters. Indeed, an Arrhenius plot of $k_R = k_{OBS} - k_{ISC}$ was linear (Figure 5.15, solid circles), giving an activation energy for rearrangement of $E_a = 5.6 \pm 0.3$ kcal/mol and pre-exponential factor $A = 10^{13.1 \pm 0.3}$.s^{-1}.[186]

5.6.3.2 Recent *Ab Initio* Calculations

Recent calculations on the ring expansion of the lowest singlet state of phenylnitrene ($^1A_2$-**49**) to azacycloheptatetraene (**51**) predict a two-step mechanism, which involves the bicyclic azirine intermediate **50**.[101] The CASPT2 energetics are depicted in Figure 5.16, and the CASSCF optimized structures of the stationary points are shown in Figure 5.17.

The first step, cyclization of ^{1}**49** to the azirine **50**, is predicted to be the rate-determining step. The CASPT2 calculated barrier of 9.2 kcal/mol is somewhat higher than the experimental barrier of 5.6 ± 0.3 kcal/mol.[186] The discrepancy between the calculated and the experimental barrier heights is due to the general tendency of the CASPT2 method to overstabilize open-shell species (in this case, $^1A_2$-**49**) relative to closed-shell species (in this case, all the other stationary points on the reaction path).[209]

This tendency is also seen in comparing the CASPT2 and multireference configuration interaction (MR-CISD + Q) values for the energy difference between open-shell singlet ($^1A''$) vinylnitrene and 2*H*-azirine.[100,101] The nitrene and azirine

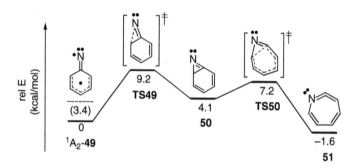

Figure 5.16 Energetics of the ring expansion of singlet phenylnitrene ($^1A_2$-**49**), calculated at the CASPT2(8,8)/6-
311G(2d,p)//CASSCF(8,8)/6-31G* level.[101]

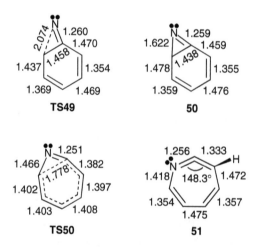

Figure 5.17 CASSCF(8,8)/6-31G* optimized geometries of stationary points in the ring expansion of singlet phenylnitrene ($^1A_2$-**49**).[101]
Bond lengths are in angstroms, and bond angles are in degrees.

serve as models for $^1A_2$-**49** and **50**, respectively. This comparison with MR-CISD + Q shows that CASPT2 underestimates
the energy of the open-shell nitrene reactant relative to the
closed-shell 2*H*-azirine product by 3.4 kcal/mol.[101] If the relative energy of $^1A_2$-**49** is also too low by a comparable amount,
then a better computational estimate of the barrier for the

first step in the ring expansion of **49** would be about 5.8 kcal/mol, in excellent agreement with experiment.[186]

The CASPT2/6-311G(2d,p) barrier for the process **50** → **51** is only about 3 kcal/mol, and this reaction is calculated to be exothermic by about 6 kcal/mol. These computational results are consistent with the failure of time-resolved IR experiments by Shields et al.[168] and Li et al.[169] to detect **50**. A 3-kcal/mol barrier implies rapid conversion of **50** to **51** at room temperature, and a 6-kcal/mol difference in energy between **50** and **51** means that at 25°C the equilibrium would overwhelmingly favor **51**. In addition, azirine **50** probably absorbs less strongly than cyclic ketenimine **51** in the infrared, rendering detection of the azirine even more difficult. Nevertheless, although **50** has not been observed spectroscopically, it apparently can be intercepted by ethanethiol to form an ortho-substituted aniline.[170]

If the MRCI-derived upward correction of 3.4 kcal/mol to the energy of $^1A_2$-**49** is made, the energy difference between $^1A_2$-**49** and **51** is estimated to be only 5 kcal/mol. Therefore, at equilibrium, a small amount of singlet **49** should be present at ambient temperatures. Intersystem crossing of singlet **49** to triplet **49** should then lead to the irreversible conversion of **51** to the triplet ground state of **49**. It has, in fact, been observed that **51** ultimately reverts to triplet **49** over time in inert solvents.[168,172,173]

5.6.4 Spectroscopy of Substituted Phenylnitrenes

Direct observations of singlet arylnitrenes before 1997 were exceedingly rare. Before the detection of the spectrum of ^{1}**49**[184,186] and its 2,4,6-tribromosubstituted analogue,[185] only the spectrum of singlet 1-pyrenyl nitrene (^{1}**60**), reported by Sumitani et al.[210] in 1976, was known. In 1985, Kobayashi et al.[211] studied p-(dimethylamino)phenyl azide by picosecond laser flash photolysis (LFP). They detected the transient spectrum of triplet nitrene (^{3}**61**) and that of a precursor, which had a lifetime of 120 ps. The authors did not specify the nature of this precursor to the triplet, but in our opinion this inter-

mediate must have been singlet p-(dimethylamino)phenylni-
trene (^{1}**61**).

60   61

Miura and Kobayashi[212] studied the photochemistry of
4,4'-biphenyl bisazide. They concluded that photolysis of this
diazide leads to extrusion of a single molecule of nitrogen with
the formation of a singlet-state species, "X," with a lifetime of
19 ns. Miura and Kobayashi did not specify the nature of X.
However, this species absorbs at 380 nm, and its spectrum
and lifetime are similar to those we found for singlet para-
biphenylnitrene.[213] Therefore, we conclude that X is singlet
4'-azido-4-biphenylnitrene.

X

After the spectrum and dynamics of parent phenylnitrene were measured, our group began a comprehensive study
of the influence of substituents on the spectroscopy and reactivity of simple substituted phenylnitrenes.[213-218]The spectra

of many *para*- and *ortho*-substituted singlet phenylnitrenes as well as *ortho,ortho*-disubstituted singlet phenylnitrenes were recorded (Table 5.6). The spectra of most singlet arylnitrenes reveal strong absorbtion bands in the near UV region with maxima in the range of 320 to 440 nm (Table 5.6). We detected the spectra of all singlet aryl nitrenes studied, with the notable exceptions of *para*-nitro- and *para*-cyanophenylnitrenes.

Figure 5.18 displays representative spectra of singlet *ortho*-fluoro, *ortho*-phenyl, and *ortho,ortho*-dicyanophenylnitrenes as well as that of parent singlet **49**. Figure 5.18 shows that the maximum of the o-fluorosubstituted [1]**49** is shifted slightly to the blue region, and the maxima of cyano-substi-

TABLE 5.6 Maxima (in nm) of the Most Intense Absorption Bands in Electronic Absorption Spectra of Substituted Singlet and Triplet Phenylnitrenes (Near UV and Visible)

Substituent	Singlet Nitrene	Ref.	Triplet Nitrene	Ref.
4-F	365	214	a	
4-Cl	360	—	a	
4-Br	361	—	a	
2,4,6-triBr	395	185	326, 340	174
4-I	328	214	a	
4-Me	365	—	315	219
4-CF$_3$	~320	—	a	
4-COCH$_3$	334	—	No	219
4-Ph	345	213	320	219
4-(4'-azido-phenyl)	380	212	a	
2-Me	350	215	a	
2,6-diMe	350	—	297, 310	174
2,4,6-triMe	348, 366	—	319	—
2-F	342	218	294, 315	
2,6-diF	331, 342	—	313	174
2,3,4,5,6-pentaF	330	184	315	—
2-CN	382	217	328	217
2,6-diCN	385, 405	—	341	—
2-Ph	409	213	344	213
2-Ph,4,6-diCl	437	b	355	b

[a]Spectrum was not detected.
[b]Tsao, M.L., Gtitsan, N.P., and Platz, M.S., unpublished data.

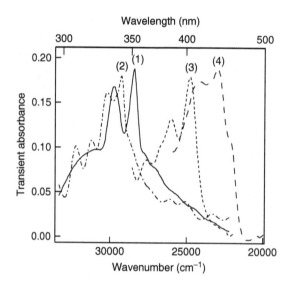

Figure 5.18 Electronic absorption spectra of selected singlet arylnitrenes: phenylnitrene (curve 1), 2-fluorophenylnitrene (curve 2) and 2-phenyl-4,6-dichlorophenylnitrene (curve 4) in pentane and 2,6-dicyanophenylnitrene (curve 3) in CH_2Cl_2.

tuted and especially of *ortho*-phenyl–substituted phenylnitrenes undergo a more pronounced shift to the red.

    In the case of [1]**49** the origin of the absorption band at 350 nm largely involves the excitation of an electron from the lone pair orbital on nitrogen (n) to the singly occupied nitrogen 2p orbital that lies in the molecular plane (p_y). In the case of [3]**49** two transitions ($T_0 \rightarrow T_3$, $T_0 \rightarrow T_4$) contribute to the absorption band around 300 nm, one of which is the same in nature as in the case of [1]**49** (n $\rightarrow p_y$). It is reasonable to assume that the same situation will be found with the substituted phenylnitrenes. Therefore, we can predict that the influence of the substituents on the maxima of the most intense absorption bands of substituted singlet and triplet phenylnitrenes will be similar. Indeed, Figure 5.19 shows that the maxima of triplet *ortho*-biphenylnitrenes and *ortho,ortho*-dicyanophenylnitrenes are shifted to the red just as with the singlet phenylnitrenes (Figure 5.18).

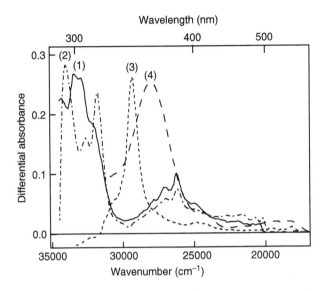

Figure 5.19 Electronic absorption spectra of selected triplet arylnitrenes: phenylnitrene (curve 1, EPA), 2-fluorophenylnitrene (curve 2, MCH), 2,6-dicyanophenylnitrene (curve 3, MCH), and 2-phenyl-4,6-dichlorophenylnitrene (curve 4, MCH) at 77 K.

In the case of triplet *ortho*-fluorophenylnitrene, it is most probable that two excited states (T_3 and T_4) are separated (~2200 cm^{-1}, Figure 5.19) in energy. A similar separation of T_3 and T_4 states was observed experimentally[174] and verified computationally[220] for triplet perfluorophenylnitrene.

Figure 5.20 displays the correlation between the maxima in the spectra of substituted triplet and singlet phenylnitrenes. A clear linear dependence of the maximum of the electronic spectra of various singlet nitrenes with the maxima in the spectra of the corresponding triplet nitrenes (slope = 1.36 ± 0.12 and correlation coefficient 0.98) is observed. Furthermore, analysis of the data of Table 5.6 verifies that *ortho*-substituents influence the absorption spectra of singlet and triplet phenylnitrenes more significantly than do *para*-substituents.

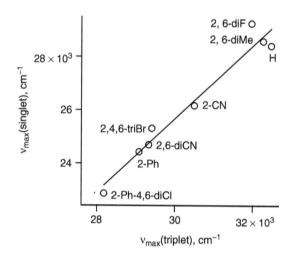

Figure 5.20 Linear correlation between maxima in the electronic absorption spectra of substituted singlet and triplet phenylnitrenes.

5.6.5 Substituent Effects on the Intersystem Crossing

Values of k_{OBS} in substituted singlet phenylnitrenes were measured by the decay of singlet nitrene absorptions at the wavelength of their maxima as a function of temperature.[213–218] As in the case of unsubstituted phenylnitrene (Figure 5.15), the magnitude of k_{OBS} decreases as the temperature decreases, until a limiting value is reached (Figure 5.21 and Figure 5.22). The temperature-independent rate constant observed at low temperature was associated with k_{ISC}. In the case of 4-bromo, 4-iodophenylnitrene (Figure 5.21), and 2,6-dimethylphenylnitrene (Figure 5.22), the values of k_{OBS} are independent of temperature over a very large temperature range (~120 to 200 K). This confirms that the rate constants for ISC in arylnitrenes are temperature independent in solution over typical temperature ranges.

5.6.5.1 *Para*-Substituted Derivatives of Phenylnitrene

Values of k_{ISC} for singlet *para*-substituted phenylnitrenes are given in Table 5.7. Table 5.7 also contains the k_{ISC} value for

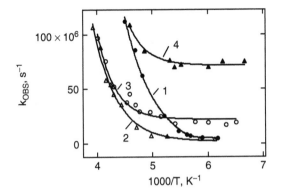

Figure 5.21 Temperature dependence of k_{OBS} values of *para*-fluoro (curve 1), *para*-chloro (curve 2), *para*-bromo (curve 3), and *para*-iodo (curve 4) singlet phenylnitrene in pentane.[214]

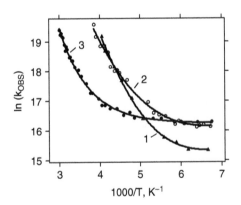

Figure 5.22 Arrhenius treatment of k_{OBS} values of singlet *para*-methyl (curve 1) and *ortho*-methyl (curve 2) phenyl nitrene in pentane and singlet *ortho, ortho*-dimethylphenylnitrene (curve 3) in hexane.[215]

singlet *para*-dimethylaminophenylnitrene[211] and a lower limit to k_{ISC} for para-nitrophenylnitrene, based on the data of Liang and Schuster.[172]

The rate constant of ISC for *para*-bromo singlet phenylnitrene is about seven times larger than that of parent [1]**49** and the *para*-fluoro and *para*-chloro analogs. This is easily attrib-

TABLE 5.7 Kinetic Parameters of *Para*-Substituted Singlet Aryl Nitrenes $(X–C_6H_4–N)$ in Pentane

Para-X	τ_{295K}, ns	k_{ISC} ($\times 10^6$ s^{-1})	E_a (kcal/mol)	Log A (s^{-1})	Ref.
H	~1	3.2 ± 0.3	5.6 ± 0.3	13.1 ± 0.3	186
CH$_3$	~1	5.0 ± 0.4	5.8 ± 0.4	13.5 ± 0.2	214
CF$_3$	1.5	4.6 ±0.8	5.6 ± 0.5	12.9 ± 0.5	—
C(O)CH$_3$	5.0	8 ± 3	5.3 ± 0.3	12.5 ± 0.3	—
F	~0.3	3.5 ± 1.4	5.3 ± 0.3	13.2 ± 0.3	—
Cl	~1	3.9 ± 1.5	6.1 ± 0.3	13.3 ± 0.3	—
Br	~3	17 ± 4	4.0 ± 0.2	11.4 ± 0.2	—
I	a	72 ± 10	a	a	—
OCH$_3$	<1	>500	a	a	—
CN	8 ± 4	6 ± 2	7.2 ± 0.8	13.5 ± 0.6	217
Ph	15 ± 2	12 ± 1	6.8 ± 0.3	12.7 ± 0.3	213
N(CH$_3$)$_2$[b]	0.12	8300 ± 200	a	a	211
NO$_2$[c]	<20	>50	a	a	172

[a]Not measured.
[b]In toluene.
[c]In benzene.

utable to a small heavy atom effect. The heavy atom effect of iodine is larger than that of bromine, as expected, and increases the rate of ISC by more than a factor of 20, relative to parent [1]**49**.

A very large increase in k_{ISC} is observed with *para*-methoxy and -dimethylamino substituents (Table 5.7). This result is consistent with the solution phase photochemistry of *para*-methoxy and para-dimethylaminophenyl azides,[169] which largely yield azobenzenes on photolysis.

The CH$_3$-, CF$_3$-, acetyl-, fluoro-, and chloro-substituents are not sufficiently strong π donors or acceptors to significantly influence the size of k_{ISC} (Table 5.7). However, the strong π-donating *para*-methoxy and *para*-dimethylamino groups do have a huge influence on k_{ISC}.

The electron-withdrawing substituents (CF$_3$, COCH$_3$, CN, and NO$_2$) have a smaller, but measurable, influence on

TABLE 5.8 Intersystem Crossing Rate Constants of *Ortho-* and *Meta* Substituted Phenylnitrenes

Substituent	Solvent	k_{ISC} ($\times 10^6$ s^{-1})	Ref.
2-Methyl	Pentane	10 ± 1	215
2,6-Dimethyl	Pentane	15 ± 3	—
—	CF$_2$ClCFCl$_2$	30 ± 8	—
2,4,6-Trimethyl	Pentane	29 ± 3	—
—	CF$_2$ClCFCl$_2$	20 ± 1	—
2-Fluoro	Pentane	3.3 ± 0.5	218
3,5-Difluoro	Pentane	3.1 ± 1.5	—
2,6-Difluoro	Hexane	2.4 ± 0.3	—
—	CCl$_4$	2.7 ± 0.3	—
2,3,4,5,6,-Pentafluoro	Pentane	3.3 ± 1.5	—
—	CH$_2$Cl$_2$	10.5 ± 0.5	208
2-Cyano	Pentane	2.8 ± 0.3	217
2,6-Dicyano	CH$_2$Cl$_2$	4.5 ± 0.5	
—	Pentane	6.2 ± 0.8	—
—	THF	5.9 ± 1.5	—
2-Pyrimidyl	CH$_2$Cl$_2$	800 ± 200	216

k_{ISC}. It is interesting to note that both donating and withdrawing substituents accelerate ISC.

5.6.5.2 *Ortho-* and *Meta* Substituted Derivatives of Phenylnitrene

Intersystem crossing rate constants of *ortho-* and *meta* substituted singlet phenyl nitrenes are presented in Table 5.8. Mono- and di-*ortho* fluorine substituents have no influence on ISC rate constants. No effect with *meta, meta* difluoro substitution is observed either. Pentafluoro substitution has no effect on k_{ISC} in pentane, although a modest acceleration is observed in the slightly more polar solvent, methylene chloride.

An increase of triplet nitrene absorption, relative to azepine absorption, was observed upon LFP of perfluorophenyl azide in methanol.[221] This effect was not observed in acetonitrile, which has a similar dielectric constant, nor in tetrahydrofuran, which contains an oxygen atom but cannot donate hydrogen bonds. It was proposed that the ISC rate constant for singlet perfluorophenylnitrene is dramatically increased in methanol, which explains much of the solution-phase organic chemistry of this nitrene in this solvent.[221]

An *ortho* cyano group has little influence on k_{ISC}, but two *ortho* cyano groups slightly accelerate intersystem crossing. Electron-withdrawing groups in the *para* positions of singlet arylnitrenes have a modest effect on increasing the rate constant of ISC.

An *ortho* methyl group accelerates intersystem crossing relative to singlet *para*-tolylnitrene. Two *ortho* methyl groups are more effective than one at accelerating intersystem crossing. Singlet 2,4,6-trimethylphenylnitrene undergoes intersystem crossing about as readily as 2,6-dimethylphenylnitrene. These results suggest that *ortho* methyl groups may accelerate intersystem crossing through a steric effect that forces the nitrogen out of the plane of the aromatic ring, thus facilitating vibronic mixing of the low-lying singlet excited states into the lowest singlet state.

5.6.5.3 Calculations of the Spin-Orbit Coupling

Recently Johnson et al.[222] calculated the electronic structures and energies of the triplet and three singlet states for phenylnitrene and 31 singly substituted derivatives. The spin–orbit coupling constants (SOC) were calculated as well to better understand the factors affecting rates of ISC.

All structures were optimized at the CASSCF(8,8) level with the cc-pVDZ basis set. For multireference calculations involving bromine and iodine, the Cowan–Griffin *ab initio* model potential with a relativistic effective core potential was used.[222] CASPT2 calculations were performed on all optimized CASSCF(8,8)/cc-pVDZ geometries, using the CASSCF wave functions as the reference wave functions. SOCs were computed by using the Pauli–Breit Hamiltonian.

The geometries and energies of the triplet and three singlet states of arylnitrenes were also calculated using DFT at the BPW91 level.[222] In the case of calculations of the lowest singlet states the unrestricted DFT was used, and the energies were computed using the sum method of Ziegler. et al.[223] and Cramer et al.[224]

Table 5.9 lists the electronic energies of the three singlet states S_1, S_2, and S_3 relative to the triplet as computed at the CASPT2 and BPW91 levels of the theory for 31 substituted phenylnitrenes studied. Analysis of Table 5.9 shows that for all substituents except those having a σ value more negative than -0.14, a near constant T_0–S_1 energy separation is predicted. Even in the case of substituents with the most negative σ values, the decrease of the energy separation does not exceed 2 to 3 kcal/mol.

It was noted previously that the strong π-donating *para*-methoxy and *para*-dimethylamino groups (Table 5.7) have a huge influence on k_{ISC}. For biradicals where two electrons are spatially proximate, as in nitrenes, SOC is the primary mechanism for ISC. For small amounts of SOC, relative rate constants k_{ISC} between two systems can be estimated from the Landau–Zener model[225] as

$$\frac{k'_{ISC}}{k_{ISC}} = \left(\frac{\langle S_1|H_{SO}|T_0\rangle'}{\langle S_1|H_{SO}|T_0\rangle}\right)^2 \left(\frac{\Delta E}{\Delta E'}\right)^2, \tag{5.1}$$

where H_{SO} is the spin-orbit coupling Hamiltonian and ΔE is the energy difference between T_0 and S_1. In phenylnitrenes, however, the SOC matrix elements between T_0 and S_1 states are required by symmetry to be zero (Table 5.10). SOC can be accomplished only dynamically, whereby S_2 and S_3 character is mixed into S_1 by geometric distortion. Johnson et al.[222] used a (2,2) CI model[225–227] to obtain a qualitative picture of SOC in the arylnitrenes. It was concluded that the vibrationally averaged SOC matrix elements are required for a more quantitative description of the ISC in arylnitrenes.[222]

5.6.6 Substituent Effects on the Rates and Regiochemistry of the Ring Expansion of Phenylnitrene

Cyclic ketenimine **51** is the major, trappable, reactive intermediate in solution when phenyl azide (at moderate concentrations) is decomposed photolytically at 298 K. The rate of decay of singlet phenylnitrene is equal to the rate of formation

TABLE 5.9 Hammett σ Values and CASPT2 and BPW91 Singlet Energies (kcal/mol) Relative to the Triplet State for R-Substituted Phenylnitrenes[222]

R	σ	S_1		S_2		S_3	
		CASPT2	BPW91	CASPT2	BPW91	CASPT2	BPW91
p-NHCH$_3$	-0.46	16.7	12.2	30.4	23.3	63.7	53.9
p-OH	-0.38	18.3	13.4	34.5	27.6	61.2	49.9
p-N(CH$_3$)$_2$	-0.32	16.3	12.2	30.5	24.4	63.2	52.4
p-NH$_2$	-0.30	17.1	12.5	31.2	23.9	63.5	53.4
p-CH$_3$	-0.14	18.8	13.8	36.9	32.2	58.6	44.5
p-OCH$_3$	-0.12	18.3	13.3	34.5	27.7	65.2	49.4
m-N(CH$_3$)$_2$	-0.10	19.3	14.1	36.7	32.7	57.4	43.0
m-NHCH$_3$	-0.10	19.6	14.3	36.8	33.1	57.0	42.8
m-NH$_2$	-0.09	19.2	14.1	37.1	33.3	57.2	43.2
m-CH$_3$	-0.06	19.5	14.3	37.3	33.6	57.6	43.1
H	0.00	19.3	14.3	37.4	33.9	57.8	43.0
m-OCH$_3$	0.10	20.0	14.7	37.4	33.9	57.1	43.1
m-OH	0.13	19.6	14.5	37.6	34.3	57.0	43.1
p-F	0.15	19.1	14.1	36.3	30.8	59.5	47.1
p-Cl	0.24	18.4	13.7	37.3	32.6	58.2	45.1
p-Br	0.26	18.4	13.8	37.5	33.0	57.9	44.5
p-I	0.28	18.2	13.7	37.7	33.1	57.4	43.7
m-F	0.34	19.4	14.4	37.9	34.3	57.5	43.9
m-I	0.34	19.3	14.2	37.6	32.2	57.2	41.8
m-COCH$_3$	0.36	19.8	14.7	37.7	35.1	57.1	39.2
m-Cl	0.37	19.4	14.3	37.8	a	57.5	43.3
m-Br	0.37	19.3	14.3	37.7	33.0	57.3	42.8

p-CO$_2$H	0.44	18.0	15.4	38.7	38.1	55.9	38.3
p-CO$_2$CH$_3$	0.44	18.0	13.4	38.6	36.7	56.0	37.7
m-CF$_3$	0.46	19.8	14.7	37.8	a	57.4	42.7
p-COCH$_3$	0.47	17.5	13.0	38.7	37.1	56.1	33.9
p-CHO	0.47	17.2	12.9	38.9	37.9	56.0	36.1
p-CF$_3$	0.53	19.1	14.2	38.3	40.7	62.8	36.1
m-CN	0.62	19.7	14.7	37.9	35.1	57.3	42.7
p-CN	0.70	17.5	13.2	39.5	37.0	56.7	40.6
m-NO$_2$	0.71	19.7	14.7	38.0	a	57.4	a
p-NO$_2$	0.81	18.1	13.5	38.8	38.8	55.5	34.1

[a]SCF convergence for this state was not achieved.

TABLE 5.10 Spin-Orbit Coupling Constants for the First Three
Singlet States of Four Para-Substituted Phenylnitrenes in cm^{-1}.[222]

State	R = NO$_2$	R = F	R = H	R = NHMe
S$_1$	0.0	0.0	0.0	0.0
S$_2$	11.9	16.6	15.5	18.8
S$_3$	44.3	43.5	43.5	41.8

of the cyclic ketenimine.[184] Nevertheless, the calculations of
Karney and Borden[101] reveal that this is a two-step process
(Scheme 5.2). The first step, cyclization to benzazirine **50**, is
rate determining, followed by fast electrocyclic ring opening
to cyclic ketenimine **51**. The predicted potential-energy surface is shown in Figure 5.16.

In the absence of nucleophiles, the cyclic ketenimine poly-
merizes. At high dilution it slowly reverts to benzazirine **50**
and from there to the singlet nitrene ¹**49**. Eventually the
singlet nitrene relaxes to the lower-energy triplet nitrene,
which dimerizes[167,169]

A study by Younger and Bell[228] nicely demonstrated the
interconversion of a disubstituted benzazirine and singlet
nitrene.

Structure 5.39

There is little direct experimental evidence for the intermediacy of **50** and its derivatives. Benzazirine **50**, has not been
detected by matrix IR.[165] However, fluorinated[229] and
naphthalenic[230] derivatives of **50** have been generated as persistent species in cryogenic matrices and characterized. Parent
benzazirine **50** has been intercepted with ethanethiol,[170] and
certain derivatives of **50** have been trapped with amines.[231–234]

For most substituted phenyl azides of interest[214–218] the rate constants of singlet nitrene decay and product formation (triplet nitrene and ketenimine) were found to be the same. With these arylnitrenes, cyclization to substituted benzazirines is the rate-determining step of the process of nitrene isomerization to ketenimine as in case of the parent phenylnitrene (Scheme 5.2). The only exception, o-fluorophenylnitrene, will be discussed in detail in Section 5.6.6.3.2.

The kinetics of rearrangement of substituted penyl nitrenes have been studied by laser flash photolysis. The temperature-independent observed rate constants are associated with k_{ISC}. Plots of ln ($k_{OBS}-k_{ISC}$) were linear (Figure 5.23 to Figure 5.25), and these plots were used to deduce the Arrhenius parameters for cyclization of the substituted singlet arylnitrenes (Table 5.7, Table 5.11 to Table 5.13).

5.6.6.1 Influence of *Para*-Substituents: The Electronic Effect

5.6.6.1.2 Recent Experiments

Activation parameters of cyclization of *para*-substituted singlet phenylnitrenes are presented in Table 5.7. It is readily

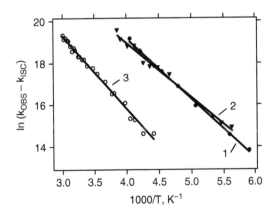

Figure 5.23 Arrhenius treatment of $k_R (= k_{OBS} - k_{ISC})$ data for singlet *para*-methyl **69d** (curve 1) and *ortho*-methyl **69a** (curve 2) phenylnitrene in pentane and for singlet *ortho,ortho*-dimethylpenylnitrene **69b** (curve 3) in hexane.[215]

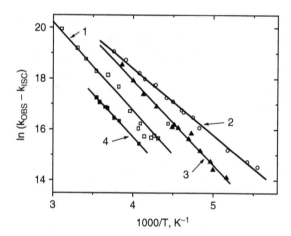

Figure 5.24 Arrhenius treatment of k_R (= k_{OBS} − k_{ISC}) data for singlet *para*-cyano **62a** (curve 1), *ortho*-cyano **62b** (curve 2) phenylnitrene in pentane, for singlet *ortho,ortho*-dicyanophenylnitrene **62c** in CH_2Cl_2 (curve 3), and for singlet *para*-biphenylnitrene (curve 4) in pentane.[213,217]

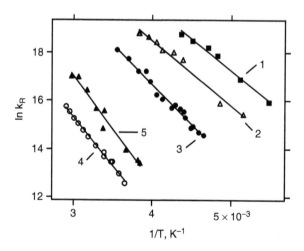

Figure 5.25 Arrhenius treatment of k_R(= k_{OBS} − k_{ISC}) data for singlet *para*-fluoro **70b** (curve 1), *meta,meta*-difluoro **70c** (curve 2), *ortho*-fluoromethylnitrene **70a** (curve 3) in pentane, and ortho,ortho-difluorophenylnitrene **70d** (curve 4) in CCl_4. Curve 5: Arrhenius treatment of the rate constant for the ring-opening reaction (k_E) for benzazirine**71a**.[218]

TABLE 5.11 Summary of Kinetic Results for Singlet Methyl-Substituted Phenylnitrenes[215]

Substituent	$\tau(295)$ (ns)	log A (s^{-1})	E_a (kcal/mol)	Solvent
2-methyl, **69a**	~1[a]	12.8 ± 0.3	5.3 ± 0.4	C$_5$H$_{12}$
2,6-dimethyl, **69b**	12 ± 1	13.0 ± 0.3	7.0 ± 0.3	C$_6$H$_{14}$
2,6-dimethyl, **69b**	13 ± 1	12.9 ± 0.3	7.5 ± 0.5	CF$_2$ClCFCl$_2$
2,4,6-trimethyl, **69c**	8 ± 1	13.4 ± 0.4	7.3 ± 0.4	CF$_2$ClCFCl$_2$
4-methyl, **69d**	~1[a]	13.2 ± 0.2	5.8 ± 0.4	C$_5$H$_{12}$

[a]Lifetime estimated by extrapolation of the data to 295 K.

TABLE 5.12 Summary of Kinetic Results for Singlet Cyano-Substituted Phenylnitrenes[217]

Substituent	$\tau(295)$ (ns)	log A (s^{-1})	Ea (kcal/mol)	Solvent
para-cyano, **62a**	8 ± 4	13.5 ± 0.6	7.2 ± 0.8	C$_5$H$_{12}$
ortho-cyano, **62b**	~2[a]	12.8 ± 0.3	5.5 ± 0.3	C$_5$H$_{12}$
2,6-dicyano, **62c**	~2.5[a]	13.3 ± 0.2	6.4 ± 0.3	CH$_2$Cl$_2$
2,6-dicyano, **62c**	~2.3[a]	13.5 ± 0.2	6.5 ± 0.4	C$_5$H$_{12}$
2,6-dicyano, **62c**	~2.3[a]	13.1 ± 1.0	6.0 ± 1.1	THF

[a]Lifetime estimated by extrapolation of the data to 295 K.

seen from the table that substituents such as *para* CH$_3$, CF$_3$, halogen, and acetyl have little influence on k$_R$. This is not very surprising, given that theory predicts emphatically that singlet phenylnitrene has an open-shell electronic structure.[188,189] Therefore, cyclization of singlet **49** requires only that the nitrogen bend out of the molecular plane, so that the singly occupied σ NBMO on it can interact with the singly occupied π NBMO.[101] Azirine formation is simply the cyclization of a quinoidal 1,3-biradical, which originally has two orthogonal, antiparallel spins. Consequently, large substituent effects are not anticipated.

It was impossible to study the effect of strong π-donor substituents on the rate of cyclization, because with *para*-methoxy and dimethylamino substituents the phenylnitrene underwent intersystem crossing to the triplet faster than cyclization at all temperatures.[214]

TABLE 5.13 Kinetic Parameters of Fluoro-Substituted Singlet Phenylnitrenes

Substituent	τ_{298} (ns)	log A (s^{-1})	E$_a$ (kcal/mol)	Solvent	Ref.
H	~1	13.1 ± 0.3	5.6 ± 0.3	C$_5$H$_{12}$	186
2-Fluoro, **70a**	8 ± 1	13.0 ± 0.3	6.7 ± 0.3	C$_5$H$_{12}$	218
	10 ± 2			CH$_2$Cl$_2$	218
	10 ± 2			CF$_2$ClCFCl$_2$	218
4-Fluoro, **70b**	~0.3	13.2 ± 0.3	5.3 ± 0.3	C$_5$H$_{12}$	218
3,5-Difluoro, **70c**	~3	12.8 ± 0.3	5.5 ± 0.3	C$_5$H$_{12}$	218
2,6-Difluoro, **70d**	240 ± 20	11.5 ± 0.5	7.3 ± 0.7	C$_6$H$_{14}$	218
	260 ± 20	12.0 ± 1.2	8.0 ± 1.5	CCl$_4$	218
2,3,4,5,6-Pentafluoro, **70e**	56 ± 4	12.8 ± 0.6	7.8 ± 0.6	C$_5$H$_{12}$	218
	32 ± 3	13.8 ± 0.3	8.8 ± 0.4	CH$_2$Cl$_2$	208
Perfluoro-4-biphenyl, **70f**	260 ± 10	13.2 ± 0.2	9.4 ± 0.4	CH$_2$Cl$_2$	208
	220 ± 10	12.5 ± 0.4	8.9 ± 0.3	CH$_3$CN	235
4-CONHC$_3$H$_8$- 2,3,5,6-tetrafluoro, **70g**	210 ± 20	13.2 ± 0.3	7.5 ± 0.3	CH$_3$CN	235

Two *para*-substituents, phenyl and cyano, depress k_R and retard the rate of cyclization significantly (Table 5.7). Phenyl and cyano are both radical stabilizing substituents. When attached to the carbon atom *para* to the nitrene nitrogen, these substituents concentrate spin density at this carbon and reduce the spin density at the carbons *ortho* to the nitrene nitrogen. The reduced spin density at carbons *ortho* to the nitrogen lowers the rate at which the 1,3-biradical cyclizes. The lifetimes of these singlet nitrenes at ambient temperature are 8 and 15 ns, respectively, and the activation barriers to cyclization are 7.2[217] and 6.8 kcal/mol,[213] respectively, compared with 5.6 kcal/mol for the unsubstituted parent. The longer lifetime of nitrene **62a** explains the high yield of hydrazine (~70%) observed upon photolysis of *para*-cyanophenyl azide in dimethylamine.[236]

5.6.6.1.2 *Recent Computational Studies*

Prior to these experiments, CASPT2/6-31G* calculations on the ring expansion reactions of *para*-cyanophenylnitrene (**62a**) were performed.[217] Table 5.14 shows that the barrier computed for cyclization of *para*-cyanophenylnitrene (**62a**) is more than 1 kcal/mol higher than that for parent phenylnitrene **49**. This prediction is in quantitative agreement with subsequent experiments (Table 5.7).

TABLE 5.14 (8/8)CASSCF and CASPT2/6-31G* Energies (kcal/mol),[a] Relative to the Reactants, for the Transition Structures and Products in the Cyclization Reactions of Singlet Phenylnitrene and of the *Para-*, *Ortho-*, *Meta-* and 2,6-dicyano Derivatives[217]

Substituent	Cyclization Mode[b]	Azirine	CASSCF		CASPT2	
			TS	Product	TS	Product
H	—	**50**	8.9	4.7	8.6	1.6
Para-cyano, **62a**	—	**63a=63a'**	9.4	5.0	9.8	3.3
Ortho-cyano, **62b**	Away from	**63b**	8.3	4.5	8.6	2.2
	Toward	**63b'**	8.4	2.6	7.5	0.3
2,6-Dicyano, **62c**	—	**63c = 63c'**	8.2	3.1	8.0	1.5
Meta-cyano, **62d**	Away from	**63d**	8.6	4.4	8.2	1.2
	Toward	**63d'**	8.1	2.9	7.6	–0.7

[a]Including zero-point energy (ZPE) corrections, which range from –0.3 to 0.1 kcal/mol for transition structures and from 0.9 to 1.4 kcal/mol for products.
[b]Mode of cyclization, toward or away from the substituted carbon.

Recently, Johnson et al.[222] performed calculations of the ring-expansion reaction for three *para*-substituted singlet phenylnitrenes with substituents ranging from highly electron donating ($NHCH_3$) to highly electron withdrawing (NO_2).

65a-d   66a-d   67a-d

The structures for the stationary points were optimized at the BPW91, BLYP, and CASSCF levels of theory. Table 5.15 gives data from only the most accurate calculations, those performed at the CASPT2/CASSCF level of theory.

The calculations showed that in the case of F and NO_2 substituents, similar to the case for the parent system, cyclization is weakly exothermic, and formation of azirine (**66**) is the rate-determining step. In the case of the highly electron-withdrawing NO_2 substituent, the barrier for cyclization was cal-

TABLE 5.15 CASPT2 Enthalpies (H_{298}, kcal/mol) of Stationary Points Relative to the Singlet Arylnitrenes (**65**) for Different *Para*-Substituents R[222]

R	TS1	66	TS2	67
$NHCH_3$, **65a**	12.3	8.5	13.3	1.7
H, **65b=49**	8.5	2.7	5.8	−1.9
F, **65c**	8.9	3.0	7.3	−1.6
NO_2, **65d**	9.5	3.7	4.8	−0.7

Note: All structures were optimized at the CASSCF(8,8)/cc-pVDZ level.

culated to be 1 kcal/mol higher than that for parent phenylnitrene **49**.[222] Unfortunately, it was impossible to measure the rate constant for cyclization of **65d** because of its very fast rate of ISC.[214] No trace of the 3*H*-azepine was reported among the products of p-nitrophenyl azide (**68**) photolysis in neat dimethylamine[236] or diethylamine.[169] At the same time, photolysis of **68** in diethylamine gives a hydrazine in 9% yield.[172] This indicates that the lifetime of **65d** is not very short. This finding is consistent with the absence of azepine only if the cyclization reaction is suppressed (Table 5.15).

For the NHMe-substituted case, ring expansion was predicted to be weakly endothermic, and electrocyclic opening of **66a** was predicted to be the rate-determining step.[222] The barrier to the cyclization of **65a** was found to be about 4 kcal/mol higher than that of the parent system **49**.[222] The predicted reduction of the reactivity, along with much faster ISC ($k_{ISC} = 8.3 \pm 0.2 \times 10^9$ s^{-1}) than in parent, accounts for the absence of 3H-azepines as products of *para*-dimethyaminophenyl azide photolysis in the presence of nucleophiles.[172]

5.6.6.2 The Influence of *Ortho*-Substituents: Steric and Electronic Effects

5.6.6.2.1 *Ortho-Methyl Substituents*

Ring expansion of alkyl-substituted phenylnitrenes has also been studied. Sundberg et al.[163] found that generation of several *ortho*-alkyl arylnitrenes (e.g., *ortho*-methylphenylnitrene, **69a**) in diethylamine affords nucleophilic trapping products that are consistent with initial cyclization to only the unsubstituted *ortho* carbon. Dunkin et al.[237] reported that matrix-isolated 2,6-dimethylphenylnitrene (**69b**) undergoes inefficient ring expansion, and Murata et al.[238] have observed the trapping of singlet mesitylnitrene (**69c**), as well as of its ring-expansion product, by tetracyanoethylene (TCNE).

69a   69b   69c

The results for methyl derivatives **69a–c** suggest that steric effects play a role in determining the barrier to ring expansion, as suggested by Dunkin et al.[237]

A recent kinetic study[215] demonstrated that a single *ortho* (**69a**) or *para* (**69d**) methyl substituent has no influence on the rate of cyclization of the singlet tolylnitrene to the azirine (Table 5.11). Spin localization effects are not observed, as they are with cyano and phenyl substitution.

In contrast to the case of **69a**, cyclization of **69b** or **69c** necessarily proceeds toward an *ortho* carbon bearing a substituent. The resulting steric effect raises the barrier to cyclization by 1.5 to 2.0 kcal/mol, in quantitative agreement with the results of calculations by Karney and Borden,[239], who predicted that the barrier to rearrangement of **69a** away from the methyl group is lower by 2 kcal/mol than cyclization toward the methyl group. The steric effect extends the lifetime of **69b** at ambient temperature to 13 ns in freon-113 and of **69c** to 8 ns, in the same solvent.

5.6.6.2.2 *Ortho-Cyano Substituents*

A cyano group is a smaller substituent than methyl and should help localize an unpaired electron at the carbon to which this substituent is attached. Thus, cyclization toward and away from the substituted *ortho*-carbon should be more evenly balanced for a cyano- than for a methyl substituent.

Consistent with this hypothesis, Lamara et al [240] found that singlet *ortho*-cyanophenylnitrene **62b** undergoes ring expansion to afford not only **63b**, the product formed by cyclization away from the cyano substituent, but also **63b'**, the product formed by cyclization toward the cyano group (Scheme 5.3). Similar results have been found in the ring expansion of singlet *ortho*-acetylphenylnitrene.[241]

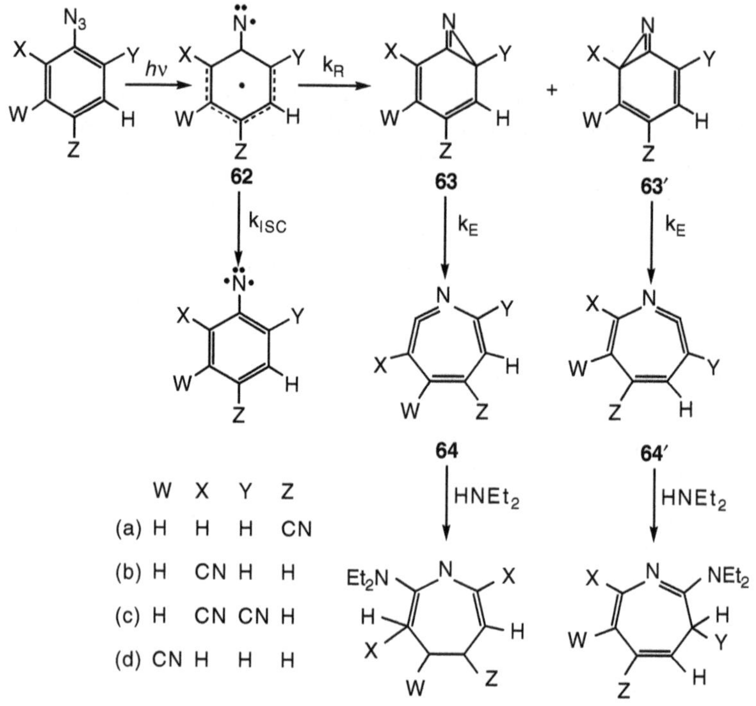

hν
k_R
62
63
63′
k_{ISC}
k_E
64
64′
HNEt₂
W X Y Z
(a) H H H CN
(b) H CN H H
(c) H CN CN H
(d) CN H H H

Scheme 5.3

Laser flash photolysis studies were performed on *para*-cyano (**62a**), *ortho*-cyano (**62b**), and *ortho,ortho*-dicyanophenyl (**62c**) azide.[217] The results are given in Table 5.12. In pentane the barrier to cyclization of *ortho*-cyanophenylnitrene is the same, within experimental error, as that of parent phenylnitrene. The barrier to cyclization of 2,6-dicyanophenylnitrene (**62c**) is about 1 kcal/mol larger than that of parent phenylnitrene (¹**49**). Variation of solvent has only a small effect on the kinetics of this nitrene.

Product studies demonstrated that *ortho*-cyanophenylnitrene (**62b**) prefers slightly to cyclize toward the carbon bearing the cyano-group (63:37) in pentane solvent.[217] Thus, the spin localization effect and the steric effect of cyano, relative to hydrogen, essentially cancel, and there is no net influence of the substitutent on the reaction barrier. The barrier to cyclization of 2,6-dicyanophenylnitrene **62c** increases, but the increase is smaller than that found for 2,6-dimethylphenylnitrene (**69b**).

5.6.6.2.3 Theoretical Analysis of the Influence of Cyano Substitution

The qualitative predictions and experimental findings in the case of cyano-substituents have been analyzed computationally by performing CASSCF(8/8) and CASPT2/6-31G* *ab initio* calculations.[217] Table 5.14 summarizes the results for the cyclization reactions of *ortho*-, *meta*-, *para*-, and 2,6-dicyanophenylnitrene (**62a–d**). The zero-point corrected energies of the two possible products, **63** and **63'**, are given, relative to the reactants. Also shown are the relative energies of the transition structures, TS (**62 → 63**) and TS (**62 → 63'**), leading to each of the products. For comparison, the CASSCF and CASPT2 relative energies for the cyclization reactions of unsubstituted phenylnitrene[101] are given as well.

CASSCF and CASPT2 calculations both overestimate the stability of the open-shell electronic structure of singlet nitrenes **62a–d** by about 3 kcal/mol,[217] as in the case of parent ¹**49**.[101] The ring opening is computed to require passage over a 2 to 3 kcal/mol lower energy barrier than reversion of the intermediates to the reactants. Therefore, cyclization is the rate-determining step in the ring-expansion reactions of cyano-substituted phenylnitrenes **62a–d** to derivatives of **64** and **64'** (Scheme 5.3).

Of particular interest in Table 5.14 are the results for cyclization of *ortho*-cyanophenylnitrene (**62b**). Cyclization toward the cyano-substituent is predicted to have a slightly lower barrier height than cyclization away from the cyano group, which is calculated to have the same barrier height as cyclization of **49** at the CASPT2 level of theory.

This prediction is very different from the computational[239] and experimental results for cyclization of *ortho*-methylphenylnitrene (**69a**)[163,215] and *ortho*-fluorophenylnitrene,[218,242] where cyclization away from the *ortho*-substituent is strongly preferred over cyclization toward the substituent.

In the cyclization of *m*-cyanophenylnitrene (**62d**), where the cyano group is located on a carbon at which the π NBMO in the reactant has a node, it seems unlikely that radical stabilization will influence the direction in which **62d** cyclizes. In fact, the barrier heights connecting **62d** to either of the two possible azirines are quite comparable, and the small kinetic preference predicted for cyclization toward the cyano group may well be a consequence of the slightly lower energy computed for the linearly conjugated product, relative to the cross-conjugated product.

The barrier to cyclization of *para*-cyanophenylnitrene (**62a**) is more than 1 kcal/mol higher than that for either of the other cyanophenylnitrenes. In addition, the cyclization of **62a** is more endothermic than any of the other cyclizations shown. Both facts are attributable to ~3 kcal/mol lower energy of **62a** compared with either **62b** or **62d**.

5.6.6.3 Ring Expension of Fluoro-Substituted Singlet Phenylnitrenes

Abramovitch et al.[243,244] and Banks' group[245–249] discovered that, unlike most arylnitrenes, polyfluorinated arylnitrenes have bountiful bimolecular chemistry. Polyfluorinated arylnitrenes are useful reagents in synthetic organic chemistry,[250] in photoaffinity labeling,[251–258] and for the covalent modification of polymer surfaces.[7] The effects of the number and positions of fluorine substituents on the ring expansion of phenylnitrene have been extensively investigated by members of the Platz group.[250, 259–263] Using the pyridine ylide probe method,[264] they found that, whereas both pentafluorophenylnitrene (**70e**) and 2,6-difluorophenylnitrene (**70d**) give nitrene ylides, 4-fluorophenylnitrene (**70b**) yields only the ketenimine ylide, and 2,4-difluorophenylnitrene affords a mixture of nitrene ylide and ketenimine ylide.[260–261] They concluded that fluorine substitution at both *ortho* positions is required to inhibit ring expansion effectively.

To understand the fluorine effect quantitatively, we studied the kinetics of fluoro-substituted phenylnitrenes (Scheme 5.4, **70a–e**) and interpreted the data with the aid of molecular orbital calculations.[218,239]

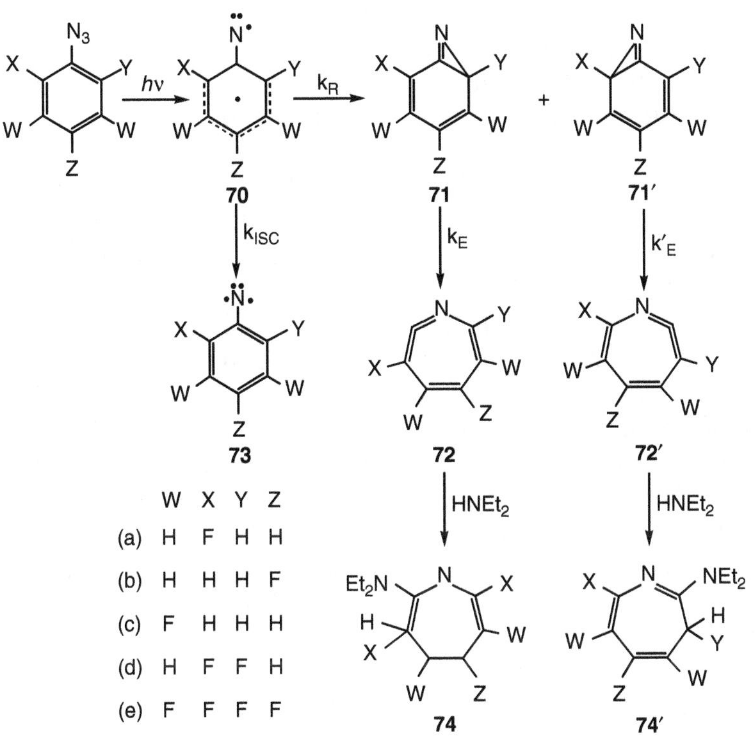
hν
k_R
k_ISC
k_E
k′_E
HNEt2
HNEt2
70
71
71′
73
72
72′
74
74′
W X Y Z
(a) H F H H
(b) H H H F
(c) F H H H
(d) H F F H
(e) F F F F

Scheme 5.4

Laser flash photolysis of a series of fluorinated aryl azides produced the transient spectra of the corresponding singlet nitrenes.[218] With the exception of singlet 2-fluorophenylnitrene (**70a**), the rate of decay of the singlet nitrene was equal to the rate of formation of the reaction products, for example, didehydroazepines (**72**) and triplet nitrenes (**73**). The temperature dependence of k_{OBS}, typical for nitrenes, was again observed. The data were interpreted in the usual manner to give k_{ISC} and the Arrhenius parameters for azirine formation. The latter are summarized in Table 5.13.

5.6.6.3.1 Influence of Fluorine Substituents on the Cyclization Reaction

Singlet 2,6-difluorophenylnitrene (**70d**) and singlet perfluorophenylnitrene (**70e**) react with hydrocarbon solvents by insertion into C–H bonds.[259–261,265] Therefore, in the case of nitrenes **70d** and **70e** in hydrocarbon solvent, k_{OBS} is

$$k_{OBS} = k_{ISC} + k_R + k_{SH}[SH]$$

where the latter term reflects the contribution of the reaction of the singlet nitrene with solvent. In these cases, the slope of a plot of $\log(k_{OBS} - k_{ISC})$ versus $1/T$ is not simply related to the barrier to cyclization. Thus, values of E_a in hydrocarbon solvents are smaller than those measured in the less reactive solvents CH_2Cl_2, CCl_4, and $CF_2ClCFCl_2$ (Table 5.13). On the basis of on product studies,[259, 266] we are confident that the activation energy barriers determined in the latter solvents can be associated with the cyclization of the singlet nitrene.

Placement of fluorine substituents at both *ortho* positions (**70d**) raises the barrier to cyclization by about 3 kcal/mol relative to the unsubstituted system (Table 5.13). The lifetimes of singlet phenylnitrene (**¹49**) and 4-fluorophenylnitrene (**70b**) are about 1 ns or less at 298 K. The lifetime of 3,5-difluorophenylnitrene (**70c**) is about 3 ns at 298 K, but that of 2,6-difluorophenylnitrene (**70d**) is 260 ns, in CCl_4. Because a *para*-fluoro group fails to exert an electronic influence on the cyclization process, it is tempting to attribute the effect of two *ortho*-fluorine substituents on the singlet nitrene lifetime to a simple steric effect.

This interpretation is consistent with the calculations of Karney and Borden,[239] who found that cyclization away from an *ortho*-methyl or an *ortho*-fluorine group is favored by 2 to 3 kcal/mol relative to cyclization toward the substituent (Table 5.16).

Leyva and Sagredo[242] demonstrated that cyclization of the singlet nitrene **70a** proceeds away from the fluorine sub-

TABLE 5.16   Calculated Relative Energies (kcal/mol)[a] for Species Involved in the First Step of the Ring Expension of Fluorosubstituted Phenylnitrenes[239]

Substituent		Mode	CAS/6[b]	PT2/6[c]	PT2/cc[d]
H	**49**	—	0	0	0
	TS		8.9	8.6	9.3
	50	—	4.7	1.6	3.5
	70a		0	0	0
2-F	**TS1**	Away	9.5	9.5	9.9
	71a	—	6.1	3.6	4.8
	TS2	Toward	13.6	12.3	13.0
	71a'	—	0.7	−2.4	−0.3
	70b		0	0	0
4-F	**TS**		7.9	8.5	9.1
	71b		3.3	1.6	3.3
	70c		0	0	0
3,5-diF	**TS**		8.5	7.9	8.6
	71c		3.2	−0.7	1.1
	70d		0	0	0
2,6-diF	**TS**		13.9	13.0	13.4
	71d		2.1	−0.5	1.0

[a]Energies pertain to CASSCF(8,8)/6-31G* optimized geometries.
[b]CASSCF(8,8)/6-31G* energy
[c]CASPT2N/6-31G* energy
[d]CASPT2N/cc-pVDZ energy

stituent. The steric argument predicts that a single *ortho*-fluorine substituent will have little influence on the rate of conversion of **70a** to **71a**, because cyclization occurs at the unsubstituted *ortho* carbon.

However, the barrier to this process is larger (outside experimental error) than that of the parent system (Table 5.13). In fact, the lifetime of singlet 2-fluorophenylnitrene (**70a**) at 298 K is 8 to 10 times longer than that of the parent (^{1}**49**) and 20 to 30 times longer than that of 4-fluorophenylnitrene (**70b**). Therefore a single *ortho*-fluorine atom exerts a small but significant bystander effect on remote cyclization that is not simply steric in origin. This result is in good quantitative agreement with the computational data of Karney and Borden,[239] who predicted that the barrier to cyclization of **70a** away from the fluorine substituent is about 1 kcal/mol higher than that for parent system ^{1}**49** (Table 5.16).

To understand this substituent effect, the atomic charges for the different centers were computed[218] using the CASSCF(8,8)/6-31G* wave functions and the natural population analysis (NPA) method of Reed et al.[266] It was found that fluorine substitution makes the adjacent carbon very positively charged (+0.48 e). In the transformation of 2-fluorophenylnitrene (**70a**) to TS1 (away from F) or TS2 (toward F), there is an increase in positive character at the (ipso) carbon bearing the nitrogen. The increased activation barrier to cyclization for 2-fluorophenylnitrene (**70a**) relative to ^{1}**49** or 4-fluorophenylnitrene (**70b**) is due to the build up of positive charges on the neighboring *ortho* and *ipso* carbons in TS1 for cyclization of **70a**.

For insertion toward F in TS2, there is an even greater amount of Coulombic repulsion between the *ortho* and *ipso* carbons than in TS1, and this effect is responsible, in part, for a higher activation barrier for insertion toward F than away from F. Therefore, the origin of the influence of *ortho,ortho*-difluoro substitution on prolonging the lifetime of singlet arylnitrene and increasing the activation energy for cyclization is due to a combination of the steric effect and the extraordinary electronegativity of the fluorine atom. In this case, the electronic and steric effects reinforce each other. This is the opposite of the case for *ortho,ortho*-dicyanophenylnitrene, where the electronic and steric effects oppose and nearly cancel each other.

5.6.6.3.2 *Ortho-Fluorophenyl Azide*

Unique kinetic results were obtained upon LFP of ortho-fluorophenyl azide **70a**.[218] In this case, the decay of **70a** was much faster than the formation of products (ketenimine **72a** and triplet nitrene **73a**) at temperatures above 230 K. For all the other substituted singlet phenylnitrenes described in this review, the rates of nitrene decay were equal to the rates of product formation.

Between 147 and 180 K, **70a** behaves normally. The rate constant for the growth of the triplet is equal to the rate constant for the disappearance of **70a**, and both rate constants are temperature-independent and close to the value of k_{ISC} for parent ^{1}**49** (Figure 5.26). In this temperature range (147 to 180 K), singlet nitrene **70a** cleanly relaxes to the lower-energy triplet nitrene **73a**.

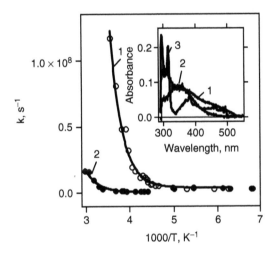

Figure 5.26 The temperature dependencies of the rate constant of decay of singlet 2-fluorophenylnitrene **70a** (curve 1) and the apparent rate constant of formation of triplet 2-fluorophenylnitrene **73a** and ketenimine **72a** (curve 2). Solid lines (1) and (2): results of nonlinear global fit of the data to an analytical solution according Scheme 5.5.[218] Insert: transient absorption spectra produced by LFP (249 nm) at 295 K of 2- fluorophenyl azide in pentane, detected 500 ns after the laser pulse (curve 1) and 4-fluorophenyl azide detected 50 ns after the laser pulse (curve 2) and persistent spectrum of triplet 2-fluorophenylnitrene (**73a**) in methylcyclohexane at 77 K (curve 3).

Above 180 K, **70a** decays by both intersystem crossing (k_{ISC}) and cyclization (k_R), with the latter process gaining relative to the former as the temperature increases. The data in Figure 5.26 for temperatures above 180 K were explained by positing that singlet nitrene **70a** and azirine **71a** (Scheme 5.5) interconvert under the experimental conditions.[218] This hypothesis was supported by analysis of the transient spectrum (Figure 5.26, insert) obtained upon LFP of 2-fluorophenyl azide. The small ratio of k_{ISC}/k_{OBS} can be reconciled with both the appearance of triplet nitrene **73a** in the transient spectrum at ambient temperature and the large chemical yield of triplet-derived azo product **74a**, if benzazirine **71a** serves as a reservoir for singlet 2-fluorophenylnitrene **70a**. The singlet nitrene eventually relaxes to the triplet, which is observed spectroscopically and dimerizes to form **74a**.

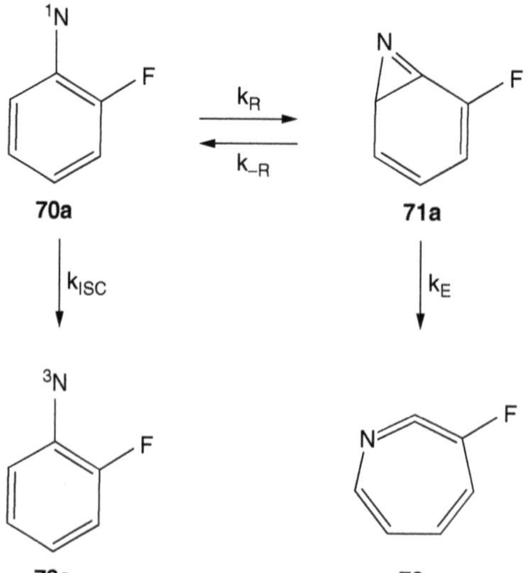

70a

k_R

k_{-R}

71a

k_{ISC}

k_E

73a

72a

Scheme 5.5

Kinetic data for singlet *ortho*-fluoronitrene **70a** were analyzed[218] following Scheme 5.5. The equilibrium constant K_R ($K_R = k_R/k_{-R}$) was estimated to be ~0.5 and ΔG to be ~350 cal/mol. Thus, **70a** and **71a** are very close in energy. The rate constant, k_E, for the ring opening reaction was measured and the Arrhenius parameters for it were found to be A = $10^{13.5 \pm 0.4}$ M^{-1} s^{-1} and E_a = 9.0 ± 0.5 kcal/mol.

To test computationally the proposed explanation of the unique kinetics observed for **70a**, a series of *ab initio* and DFT calculations on the second step of the ring expansion (electrocyclic ring opening of azirines **71a–e** to form the corresponding cyclic ketenimines **72a–e**) were performed.[218] The CASPT2/cc-pVDZ//CASSCF(8,8)/6-31G* and B3LYP/6-31G* energies are given in Table 5.17. The CASPT2 results are also depicted graphically in Figure 5.27 in a way that permits energetic comparisons of isomeric species.

As shown in Figure 5.27, in all cases except the "away" ring expansion of 2-fluorophenylnitrene (**70a**), the transition state for the second step of the ring expansion (**71** → **72**) is computed to be lower in energy than that for the first step (**70** → **71**) at the CASPT2 level of theory. This is consistent with the experimental finding that, for nitrenes **70b–d**, the nitrene decays at the same rate at which the corresponding ketenimine is formed, whereas for nitrene **70a**, nitrene decay is faster than ketenimine growth. Of the five electrocyclic ring-opening reactions investigated, the opening of **71a** to form **72a** is predicted to be the least exothermic (∆E = –1.6 kcal/mol). The exothermicity of this step for the other systems ranges from ∆E = –2.7 kcal/mol to ∆E = –6.0 kcal/mol.

The CASPT2 results are supported qualitatively by B3LYP/6-31G* calculations (Table 5.17). The DFT calculations predict that benzazirine **71a** has the second-highest barrier (after **71e**) to rearrangement to a ketenimine (**72a**). It also has the lowest barrier to reversion to the corresponding singlet nitrene (**70a**). Thus, **71a** reverts to the corresponding singlet nitrene (**70a**) more readily than does the parent system **50**. The DFT calculations also predict that, of the four fluoro-substituted arylnitrenes considered, the ring opening of **71a** → **72a** is the least exothermic. Thus, both CASPT2 and B3LYP calculations correctly predict that **71a** is the benzazirine most likely to revert to the corresponding singlet nitrene **70a**.

TABLE 5.17 Relative Energies (in kcal/mol) and Zero-Point Vibrational Energies of Azirines 71, Ketenimines 72, and the Transition States (TS) Connecting Them[218].[a]

Method	Azirine		Transition State		Ketenimine	
	E	ZPE	rel E	ZPE	rel E	ZPE
	50		**TS**		**51**	
CASPT2/cc-pVDZ [b]	−285.41815	60.9	2.5	59.8	−4.5	60.8
B3LYP/6-31G* [c]	−286.27659	57.7	4.7	56.7	−5.1	57.6
	71a		**TSa**		**72a**	
CASPT2/cc-pVDZ [b]	−384.47462	55.7	7.0	54.6	−1.6	55.8
B3LYP/6-31G* [c]	−385.50238	52.8	8.1	51.7	−3.0	52.8
B3LYP/6-311+G(2d,p) [c]	−385.62115		7.0		−5.5	
CCSD(T)/6-31+G* [c]	−384.48202		11.2			
	71a'		**TSa'**		**72a'**	
CASPT2/cc-pVDZ [b]	−384.48308	55.9	2.4	54.8	−6.0	55.7
B3LYP/6-31G* [c]	−385.51174	52.7	4.1	51.8	−6.8	52.8
	71b		**TSb**		**72b**	
CASPT2/cc-pVDZ [b]	−384.48077	55.6	4.5	54.5	−4.3	55.5
B3LYP/6-31G* [c]	−385.50865	52.6	5.7	51.5	−5.5	52.6
	71c		**TSc**		**72c**	
CASPT2/cc-pVDZ [b]	−483.52115	50.5	5.2	49.3	−2.7	50.3
B3LYP/6-31G* [c]	−484.74392	47.7	6.4	46.6	−3.9	47.6

	71d		TSd		72d	
CASPT2/cc-pVDZ [b]	−483.51566	50.7	5.7	49.6	−3.7	50.6
B3LYP/6-31G* [c]	−484.73747	47.8	6.8	46.8	−5.1	47.9

	71e		TSe		72e	
B3LYP/6-31G* [c]	−382.41002	32.6	8.5	31.5	−6.0	32.6

[a] Azirine energies are absolute energies, in hartree. Energies for transition states and keteneimines are relative energies, compared to azirine, and are corrected for differences in zero-point vibrational energy.
[b] Obtained using CASSCF(8,8)/6-31G* optimized geometry and zero-point vibrational energy.
[c] Obtained using B3LYP/6-31G* optimized geometry and zero-point vibrational energy.

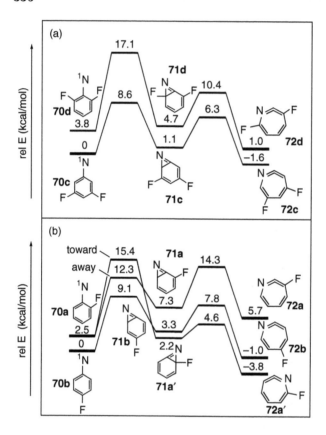

Figure 5.27 Relative energies (in kcal/mol) of species involved in the ring expansions of singlet fluoro-substituted phenylnitrenes calculated at the CASPT2/cc-pVDZ//CASSCF(8,8)/6-31G* level. (a) Difluorinated phenylnitrenes. (b) Monofluorinated phenylnitrenes.[218,239]

Significantly, this is the only system in which the rates of formation and disappearance of the benzazirine are predicted to be comparable, and this is the only case where there is compelling experimental evidence for formation of an intermediate between the singlet nitrene and its ketenimine isomer in solution.[218]

The experimental data reveal that singlet nitrene **70a** and benzazirine **71a** are very close in energy ($\Delta G \approx 350$ cal/mol). This value is much smaller than that predicted by

CASPT2 calculations (4.8 kcal/mol, Figure 5.27). However, as discussed above, the CASPT2 method typically underestimates the energies of open-shell species, such as **70a**, relative to closed-shell molecules, such as **71a**. Applying, as in the case of [149], an upward correction of ~3 kcal/mol to the energy of the singlet nitrene brings the computed energy difference between **70a** and **71a** into much better agreement with experiment.

Why is the barrier for ring opening of **71a** to **72a** so much higher than the corresponding barriers for **50** and **71b–d**? As shown in Figure 5.27 the energy of the transition state for this step roughly parallels the energy of the ketenimine product (**72**). The marked instability of ketenimine **72a**, relative to **72a'** and **72b**, is consistent with recent computational results, which predict that fluorine substitution destabilizes ketenimines.[267] This destabilization can be attributed to Coulombic repulsion between the two carbons of the ketenimine moiety. Because both of these carbons are attached to more electronegative atoms (one to N, one to F), both bear a partial positive charge, and severe electrostatic repulsion results.

The effect of Coulombic repulsion described for ketenimine **72a** can also be used to rationalize the higher energy of nitrene **70a** relative to **70b**, as well as the higher energy of azirine **71a** compared with **71a'** and **71b**. It is possible that the changes in relative orientation of the C=N and C–F bond dipoles that occur for **70a → 71a** and **71a→ 72a** are at least partly responsible for this process being predicted to be substantially more endothermic (in the case of **70a → 71a**) or less exothermic (in the case of **71a** → **72a**) than the other systems studied.

Curiously, the addition of a second *ortho*-fluorine substituent (i.e., in benzazirine **71d**) raises the barrier to reversion to singlet nitrene **70d**, relative to the mono *ortho*-fluoro system (Figure 5.27). This is partly due to steric hindrance by fluorine in the transition state for cyclization but also to the stabilization of **71d** by the fluorine attached directly to the azirine ring (*vide supra*).[239] The addition of the second fluorine substituent (benzazirine **71d**) slightly decreases the barrier to conversion of azirine **71d** to ketenimine **72d** (Figure 5.27). This is related to the more favorable thermodynamics of conversion in the case of **71d** compared with **71a**, due to the slight stabilization of ketenimine **72d** by the fluorine adjacent to nitrogen. The barrier for **71d** → **72d** is still predicted to be ~2.5 kcal/mol higher than the barrier for **50** → **51** at the same level of theory,[101] which helps explain why Morawietz and Sander[229] successfully detected benzazirine **71d** in their matrix isolation experiments.

5.7 CONCLUSIONS

Theoretical methods have been used extensively to understand properties of the simplest nitrene, NH, also known as imidogen. The experimental bond lengths and the spectroscopy of the singlet and triplet states of imidogen and the energy separation of these states are in excellent agreement with the predictions of theory. Imidogen is easy to study by experimental and theoretical methods because it is small and because it is not rapidly consumed by intramolecular rearrangement. Alkyl, vinyl, and acyl nitrenes are difficult to study experimetally because they have facile unimolecular decay pathways available. The key question with these singlet nitrenes is whether or not they are true intermediates, which are minima on the potential-energy surface and have finite lifetimes under any experimental conditions. More theoretical and experimental work is needed before this question can be definitively answered for singlet alkyl, vinyl, and acyl nitrenes.

In earlier reviews,[150,151] two of the authors wrote that the most important tasks remaining in the field of aryl azide photochemistry were the direct observation of singlet aryl nitrenes, direct study of the kinetics of singlet nitrene reactions, and high-level quantum chemical calculations of the nitrene potential-energy surface. This review illustrates that great progress has been made in these areas over the past few years. Singlet phenylnitrene has been detected directly and its spectrum analyzed with the aid of modern computational methods. The rate constant of intersystem crossing to the lower energy triplet state has been measured, and is much smaller than the corresponding values observed with aryl carbenes. This is a consequence of the open-shell electronic structure of phenylnitrene. The activation parameters for cyclization of singlet phenylnitrene to benzazirine have been obtained and are in good agreement with modern quantum chemical calculations.

The effect of substituents on the spectra and kinetics of singlet aryl nitrenes has been examined systematically. Groups that act as strong π donors dramatically accelerate the rate of ISC. Electron-withdrawing groups also slightly accelerate ISC. Para-substituents such as cyano and phenyl reduce spin-density ortho to the nitrene-bearing carbon and reduce the rate of cyclization to the corresponding benzazirine. Ortho substituents generally direct cyclization away from the substituted center because of unfavorable steric interactions. Ortho fluorine substituents retard cyclization toward and even away from the substituted carbon because of the development of unfavorable Coulombic interactions between the aryl carbons bearing the nitrogen and the fluorine substituents. The kinetics of cyclization of singlet ortho-fluorophenylnitrene reveal that cyclization to the corresponding benzazirine is reversible. All of these experimental observations are in agreement with the results of electronic structure calculations.

In the early 1990s, singlet aryl nitrenes had still avoided detection, and the influence of structure on reactivity was only dimly appreciated. Laser flash photolysis studies, in combination with calculations, have provided insight and have led to a comprehensive theory of substituent effects. We are proud to have been part of the progress that has been made in aryl nitrene chemistry during the past decade.

ACKNOWLEDGMENTS

The authors are deeply indebted to their graduate students (Dean Tigelaar, Monica Cerro Lopez, Meng-Lin Tsao, and Dmitrii Polshakov) and postdoctoral researchers (Bill Karney, Anna Gudmundsdottir, Carl Kemnitz, David Hrovat, Elena Pritchina, and Zhendong Zhu) whose efforts led to this review article. Support of this work by the National Science Foundation and the Russian Foundation for Basic Research is gratefully acknowledged.

REFERENCES

1. Lwowski, W., Ed., *Nitrenes,* Wiley, New York, 1970.

2. Scriven, E.F.V., Ed., *Azides and Nitrenes: Reactivity and Utility,* Academic Press, New York, 1984.

3. Singh, A., Thornton, E.R., and Westheimer, F.H., *J. Biol. Chem.,* 237, 3006–3008, 1962.

4. Bayley, H. and Knowles, J.R., *Methods Enzymol.,* 46, 69–114, 1977.

5. Bayley, H., *Photogenerated Reagents in Biochemistry and Molecular Biology,* Elsevier, New York, 1983.

6. Bayley, H. and Staros, J., Photoaffinity labeling and related techniques, in *Azides and Nitrenes: Reactivity and Utility,* Scriven, E.F.V., Ed., Academic Press, New York, 1984, pp. 434–490.

7. Cai, S.X., Glenn, D.R., and Keana, J.F.W., *J. Org. Chem.,* 57, 1299–1304, 1992.

8. Breslow, D.S., Industrial application, in *Azides and Nitrenes: Reactivity and Utility,* Scriven, E.F.V., Ed. Academic Press, New York, 1984, pp. 491–522.

9. Borden, W.T., Gritsan, N.P., Hadad, C.M., Karney, W.L., Kemnitz, C.R., and Platz, M.S., *Acc. Chem. Res.,* 33, 765–711, 2000.

10. Karney, W.L. and Borden, W.T., *Adv. Carbene. Chem.,* 3, 205–251, 2001.

11. Gritsan, N.P. and Platz, M.S., *Adv. Phys. Org. Chem.*, 36, 255–304, 2001.

12. Okabe, H. and Lenzi, M., *J. Chem. Phys.*, 47, 5241–5246, 1967.

13. Okabe, H., *J. Chem. Phys.*, 49, 2726–2733, 1968.

14. McDonald, J.R., Rabalais, J.W., and McGlynn, S.P., *J. Chem. Phys.*, 52, 1332–1340, 1970.

15. Okabe, H., *Photochemistry of Small Molecules*, Wiley, New York, 1978, pp. 232–234.

16. Masanet, J., Gilles, A., and Vermeil, C., *J. Photochem.*, 3, 417–429, 1974/75.

17. Engelking, P.C. and Lineberger, W.C., *J. Chem. Phys.*, 65, 4323–4324, 1976.

18. McDonald, J.R., Miller, R.G., and Baronavski, A.P., *Chem. Phys. Lett.*, 51, 57–60, 1977.

19. Baronavski, A.P., Miller, R.G., and McDonald, J.R., *Chem. Phys.*, 30, 119–131, 1978.

20. Rohrer, F. and Stuhl, F., *J. Chem. Phys.*, 88, 4788–4799, 1988.

21. Drozdovski, W.S., Baronavski, A.P., and McDonald, J.R., *Chem. Phys. Lett.*, 64, 421–425, 1979.

22. Kajimoto, O., Yamamoto, T., and Fueno, T., *J. Chem. Phys.*, 83, 429–435, 1979.

23. Stephenson, J.C., Casassa, M.P., and King, D.S., *J. Chem. Phys.*, 89, 1378–1387, 1988.

24. Alexander, M.H., Werner, H.-J., and Dagdigian, P.J., *J. Chem. Phys.*, 89, 1388–1400, 1988.

25. McDonald, J.R., Miller, R.G., and Baronavski, A.P., *Chem. Phys.*, 30, 133–145, 1978.

26. Kajimoto, O. and Fueno, T., *Chem. Phys. Lett.*, 80, 484–487, 1981.

26a. Pritchina, E.A. and Gritsan, N.P., unpublished results.

27. Huber, K.P. and Herzberg, G., *Constants of Diatomic Molecules*, Van Nostrand, New York, 1979.

28. Wasserman, E., Smolinsky, G., and Jager, W.A., *J. Am. Chem. Soc.*, 86, 3166–3167, 1964.

29. Wasserman, E., Jager, W.A., and Smolinsky, G., *J. Am. Chem. Soc.*, 84, 3220–3226, 1962.

30. Barash, L., Wasserman, E., and Jager, W.A., *J. Chem. Phys.*, 89, 3931–3935, 1967.

31. Wagner, P.J. and Scheve, B.J., *J. Am. Chem. Soc.*, 101, 378–383, 1979.

32. Gudmundsdottir, A.D., Mandel, S.M., Christman, R., Ault, B., and Krause Bauer, J.A., *Solid State Photolysis of Azidoarylketones. Book of Abstracts of International Conference on Reactive Intermediates and Reaction Mechanisms, Ascona, Switzerland,* 2002, p. 27.

33. Kyba, E.P., Alkyl azides and nitrenes, in *Azides and Nitrenes: Reactivity and Utility,* Scriven, E.F.V., Ed., Academic Press, New York, 1984, pp. 1–34.

34. Kyba, E.P. and Abramovitch, R.A., *J. Am. Chem. Soc.*, 102, 735–740, 1980.

35. Banks, R.E., Berry, D., McGlinchey, M.J., and Moore, M.J., *J. Chem. Soc. C*, 1017–1023, 1970.

36. Pancrazi, A. and Khuong-Huu, Q., *Tetrahedron*, 31, 2049–2056, 1975.

37. Dunkin, I.R. and Thomson, P.C.P., *Tetrahedron Lett.*, 21, 3813–3816, 1980.

38. Michl, J., Radziszewski, J.G., Downing, J.W., Wiberg, K.B., Walker, F.H., Miller, R.D., Kovacic, P., Jawdosiuk, M., and Bonai-Kouteck, V., *Pure Appl. Chem.*, 55, 315–321, 1983.

39. Dunkin, I.R., Shields, C.J., Quasi, H., and Seiferling, B., *Tetrahedron Lett.*, 24, 3887–3890, 1983.

40. Sheridan, R.S. and Ganzer, G.A., *J. Am. Chem. Soc.*, 105, 6158–6160, 1983.

41. Radziszewski, J.G., Downing, J.W., Wentrup, C., Kaszynski, P., Jawdosiuk, M., Kovacic, P., and Michl, J., *J. Am. Chem. Soc.*, 107, 594–603, 1985.

42. Radziszewski, J.G., Downing, J.W., Wentrup, C., Kaszynski, P., Jawdosiuk, M., Kovacic, P., and Michl, J., *J. Am. Chem. Soc.,* 107, 2799–2801, 1985.

43. Gritsan, N.P., Likhotvorik, I., Zhu, Z., and Platz, M.S., *J. Phys. Chem. A,* 105, 3039–3041, 2001.

44. Wayne, F.D. and Radford, H.E., *Mol. Phys.,* 32, 1407–1423, 1976.

45. Ferrante, R.F., *J. Chem. Phys.,* 86, 25–32, 1987.

46. Glowina, J.H., Misewich, J., and Sorokin, P.O., in *Supercontinuum Laser Sources,* Alfano, R.R., Ed., Springer Verlag, New York, 1989, p. 337.

46a. Gritsan, N.P., Likhotvorik, I., Zhu, Z., and Platz, M.S., unpublished results.

47. Wasserman, E., *Prog. Phys. Org. Chem.,* 8, 319–336, 1971.

48. Milligan, D.E., *J. Chem. Phys.,* 35, 1491–1497, 1961.

49. Jaxcox, M.E. and Milligan, D.E., *J. Mol. Spectrosc.,* 56, 333–356, 1975.

50. Carrick, P.G. and Engelking, P.C., *J. Chem. Phys.,* 81, 1661–1665, 1984.

51. Carrick, P.G., Brazier, C.R., Bernath, P.F., and Engelking, P.C., *J. Am. Chem. Soc.,* 109, 5100–5102, 1987.

52. Chappell, E.D. and Engelking, P.C., *J. Chem. Phys.,* 89, 6007–6016, 1988.

53. Brazier, C.R., Carrick, P.G., and Bernath, P.R., *J. Chem. Phys.,* 96, 919–926, 1992.

54. Franken, T., Perner, D., and Bosnali, M.W., *Z. Naturforsch. A,* 25, 151–153, 1970.

55. Ferrante, R.F., *J. Chem. Phys.,* 94, 4678–4679, 1991.

56. Shang, H., Yu, C., Ying, C., and Zhao, X., *J. Chem. Phys.,* 103, 4418–4426, 1995.

57. Ying, L., Xia, Y., Shang, H., and Tang, Y., *J. Chem. Phys.,* 105, 5798–5805, 1996.

58. Shang, H., Gao, R., Zhao, X., and Tang, Y., *Chem. Phys. Lett.,* 267, 345–350, 1997.

59. Travers, M.J., Cowles, D.C., Clifford, E.P., Ellison, G.B., and Engelking, P.C., *J. Chem. Phys.*, 111, 5349–5360, 1999.

60. Yakony, D.R., Schaefer, H.F., and Rothenberg, S., *J. Am. Chem. Soc.*, 96, 5974–5977, 1974.

61. Demuyanck, J., Fox, D.J., Yamaguchi, Y., and Schaefer, H.F., III, *J. Am. Chem. Soc.*, 102, 6204–6207, 1980.

62. Pople, J.A., Raghavachari, K., Frisch, M.J., Binckley, J.S., and Schleyer, P.V.R., *J. Am. Chem. Soc.*, 105, 6389–6398, 1983.

63. Nguyen, M.T., *Chem. Phys. Lett.*, 117, 290–294, 1985.

64. Xie, Y., Scuseria, G.E., Yates, B.F., Yamaguchi, Y., and Schaefer, H.F., III, *J. Am. Chem. Soc.*, 111, 5181–5185, 1989.

65. Richards, C., Jr., Meredith, C., Kim, S.J., Quelch, G.E., and Schaefer, H.F., III, *J. Chem. Phys.*, 100, 481–489, 1994.

66. Nguyen, M.T., Sengupta, D., and Ha, T.-K., *J. Phys. Chem.*, 100, 6499–6503, 1996.

67. Arenas, J.F., Otero, J.C., Sánchez-Gálvez, A., Soto, J., and Viruela, P., *J. Phys. Chem.*, 102, 1146–1151, 1998.

68. Arenas, J.F., Marcos, J.I., Otero, J.C., Sánchez-Gálvez, A., and Soto, J., *J. Chem. Phys.*, 111, 551–561, 1999.

69. Kemnitz, C.R., Ellison, G.B., Karney, W.L., and Borden, W.T., *J. Am. Chem. Soc.*, 122, 1098–1101, 2000.

70. Evanseck, J.D. and Houk, K.N., *J. Am. Chem. Soc.*, 112, 9148–9156, 1990.

71. Ma, B. and Schaefer, H.F., III, *J. Am. Chem. Soc.*, 116, 3539–3542, 1994.

72. Kramer, K.A. and Wright, A.N., *Tetrahedron Lett.*, 3, 1095–1097, 1962.

73. Frey, H.M., *J. Chem. Soc.*, 2293–2297, 1962.

74. Seburg, R.A. and McMahon, R.J., *J. Am. Chem. Soc.*, 114, 7183–7189, 1992.

75. Modarelli, D.A. and Platz, M.S., *J. Am. Chem. Soc.*, 115, 470–475, 1993.

75a. Pritchina, E.A. and Gritsan, N.P., unpublished results.

76. Tsao, M.-L., Hadad, C.M., and Platz, M.S., *Tetrahedron Lett.*, 43, 745–748, 2002.

77. Modarelli, D.A., Platz, M.S., Sheridan, R.S., and Ammann, J.R., *J. Am. Chem. Soc.*, 115, 10440–10441, 1993.

78. Huang, H. and Platz, M.S., *J. Am. Chem. Soc.*, 120, 5990–5999, 1998.

79. Harnisch, J. and Szeimies, G., *Chem. Ber.*, 112, 3914–3933, 1979.

80. Sun, H., Liu, C., Zhao, L., and Deng, L., *Chem. Phys. Lett.*, 228, 268–272, 1994.

81. Hassner, A., Vinyl azides and nitrenes, in *Azides and Nitrenes: Reactivity and Utility*, Scriven, E.F.V., Ed., Academic Press, New York, 1984, pp. 35–93.

82. Smolinsky, G., *J. Org. Chem.*, 27, 3557–3559, 1962.

83. Smolinsky, G. and Pryde, C.A., *J. Org. Chem.*, 33, 2411–2416, 1968.

84. Hassner, A. and Fowler, F.W., *Tetrahedron Lett.*, 1545–1548, 1967.

85. Hassner, A. and Fowler, F.W., *J. Am. Chem. Soc.*, 90, 2869–2875, 1968.

86. Isomura, K., Okada, M., and Taniguchi, H., *Tetrahedron Lett.*, 4073–4077, 1969.

87. Bauer, W. and Hafner, K., *Angew. Chem. Int. Ed. Engl.*, 8, 772–773, 1969.

88. Boier, J.H., Krueger, W.E., and Mikol, G.J., *J. Am. Chem. Soc.*, 87, 5504–5505, 1967.

89. Boier, J.H., Krueger, W.E., and Modler, R., *J. Org. Chem.*, 34, 1987–1989, 1969.

90. Isomura, K., Kobayashi, S., and Taniguchi, H., *Tetrahedron Lett.*, 3499–3503, 1968.

91. Taniguchi, H., Isomura, K., and Tanaka, T., *Heterocycles*, 6, 1563–1568, 1977.

92. Hemetsberger, H., Knittel, D., and Neidmann, H., *Montash Chem.*, 101, 161–165, 1970.

93. Padwa, A., Smolanoff, J., and Tremper, A., *J. Org. Chem.*, 41, 543–549, 1976.

94. Nishiwaki, T. and Fujiyama, F., *J. Chem. Soc. Perkin Trans.*, 1, 1456–1464, 1972.

95. Fowler, F.W., Hassner, A., and Levy, L.A., *J. Am. Chem. Soc.*, 89, 2077–2082, 1967.

96. Chapman, O.L. and Le Roux, J.-P., *J. Am. Chem. Soc.*, 100, 282–285, 1978.

97. Lohr, L.L., Jr., Hanamura, M., and Morokuma, K., *J. Am. Chem. Soc.*, 105, 5541–5547, 1983.

98. Bock, H., Dammel, R., and Aygen, S., *J. Am. Chem. Soc.*, 105, 7681–7685, 1983.

99. Yamabe, T., Kaminoyama, M., Minato, T., Hori, K., Isomura, K., and Taniguchi, H., *Tetrahedron*, 40, 2095–2099, 1984.

100. Parasuk, V. and Cramer, C.J., *Chem. Phys. Lett.*, 260, 7–14, 1996.

101. Karney, W.L. and Borden, W.T., *J. Am. Chem. Soc.*, 119, 1378–1387, 1997.

102. Langhoff, S.R. and Davidson, E.R., *Int. J. Quantum Chem.*, 7, 999–1019, 1973.

103. Lwowski, W., Acyl azides and nitrenes, in *Azides and Nitrenes: Reactivity and Utility*, Scriven, E.F.V., Ed., Academic Press, New York, 1984, pp. 205–246.

104. Curtius, T., *Z. Angew Chem.*, 27, 111–114, 1914.

105. Smith, P.A.S., in *Organic Reactions*, Vol. 3, Wiley, New York., 1946, pp. 337–449.

106. Hauser, C.R. and Kantor, S.W., *J. Am. Chem. Soc.*, 72, 4284–4285, 1950.

107. Hine, J., *Physical Organic Chemistry*, 2nd ed., McGraw-Hill, New York, 1962, p. 336.

108. Gould, E.S., *Mechanism and Structure in Organic Chemistry*, Henry Holt, New York, 1959, p. 623.

109. Roberts, J.D. and Caserio, M.C., *Basic Principles of Organic Chemistry*, WA Benjamin, New York, 1964, p. 656.

110. Fieser, L.F. and Fieser, M., *Advanced Organic Chemistry*, Reinhold, New York, 1961, p. 502.

111. Tisue, G.T., Linke, S., and Lwowski, W., *J. Am. Chem. Soc.*, 89, 6303–6307, 1967.

112. Linke, S., Tisue, G.T., and Lwowski, W., *J. Am. Chem. Soc.*, 89, 6308–6310, 1967.

113. Eibler, E. and Sauer, J., *Tetrahedron Lett.*, 2569–2572, 1974.

114. Semenov, V.P., Studenikov, A.N., Bespalov, A.D., and Ogloblin, K.A., *Zh. Organ Khim.* (Russ.). 13, 2202–2207, 1977.

115. Lwowski, W. and DeMauriac, R., *Tetrahedron Lett.*, 3285–3288, 1964.

116. Lwowski, W. and Woerner, F.P., *J. Am. Chem. Soc.*, 87, 5491–5492, 1965.

117. Breslow, D.S., Prosser, T.J., Marcantonio, T.J., and Genge, C.A., *J. Am. Chem. Soc.*, 89, 2384–2390, 1967.

118. McConaghy, J.S. and Lwowski, W., *J. Am. Chem. Soc.*, 89, 2357–2364, 1967.

119. McConaghy, J.S. and Lwowski, W., *J. Am. Chem. Soc.*, 89, 4450–4456, 1967.

120. Autrey, T. and Schuster, G.B., *J. Am. Chem. Soc.*, 109, 5814–5820, 1987.

121. Sigman, M.E., Autrey, T., and Schuster, G.B., *J. Am. Chem. Soc.*, 110, 4297–4305, 1988.

122. Wilde, R.E., Srinivasan, T.K.K., and Lwowski, W., *J. Am. Chem. Soc.*, 93, 860–863, 1971.

123. Puttner, R. and Hafner, K., *Tetrahedron Lett.*, 3119–3125, 1964.

124. Hayashi, Y. and Swern, D., *J. Am. Chem. Soc.*, 95, 5205–5210, 1973.

125. Inagaki, M., Shingaki, T., and Nagai, T., *Chem. Lett.*, 1419–1422, 1981.

126. Inagaki, M., Shingaki, T., and Nagai, T., *Chem. Lett.*, 9–12, 1982.

127. Woelfle, I., Sauerwein, B., Autrey, T., and Schuster, G.B., *Photochem. Photobiol.*, 47, 497–501, 1988.

128. Melvin, T. and Schuster, G.B., *Photochem. Photobiol.*, 51, 155–160, 1990.

129. Alewood, P.F., Kazmaier, P.M., and Rauk, A., *J. Am. Chem. Soc.*, 95, 5466–5475, 1964.

130. Harrison, J.F. and Shalhoub, G., *J. Am. Chem. Soc.*, 97, 4172–4176, 1975.

131. Rauk, A. and Alewood, P.F., *Can. J. Chem.*, 55, 1498–1510, 1977.

132. Poppinger, D., Radom, L., and Pople, J.A., *J. Am. Chem. Soc.*, 99, 7806–7816, 1977.

133. Poppinger, D. and Radom, L., *J. Am. Chem. Soc.*, 100, 3674–3685, 1978.

134. Mavridis, A. and Harrison, J.F., *J. Am. Chem. Soc.*, 99, 7651–7655, 1980.

135. Gritsan, N.P. and Pritchina, E.A., *Mendeleev Commun.*, 11, 94–96, 2001.

136. Kitaigorodski, A.I., Zorki, P.M., and Belski, V.K., *Structure of the Organic Compounds: Data of Structure Study. 1929-1970*, Science, Moscow, 1980, p. 628.

137. Kitaigorodski, I., Zorki, P.M., and Belski, V.K., *Structure of the Organic Compounds: Data of Structure Study. 1971-1973*, Science, Moscow, 1982, p. 511.

138. Cornell, D.W., Berry, R.S., and Lwowski, W., *J. Am. Chem. Soc.*, 87, 3626–3629, 1965.

139. Shapley, W.A. and Bacskay, G.B., *J. Phys. Chem.*, 103, 4514–4524, 1999.

140. Pritchina, E.A., Gritsan, N.P., and Bally, T., Autrey, T., Lui, Y., Wang, Y. and Toscano, J.P., *Phys. Chem. Chem. Phys.*, 5, 1010–1018, 2003.

141. Dunning, T.H., Jr., *J. Chem. Phys.*, 90, 1007–1023, 1989.

142. Feller, D.A., *J. Chem. Phys.*, 96, 6104–6114, 1992.

143. de Lara-Castells, M.P., Krems, R.V., Buchachenko, A.A., Delgado-Barrio, G., and Villarreal, P., *J. Chem. Phys.*, 115, 10438–10449, 2001.

143a. Gritsan, N.P., unpublished results.

144. Smith, P.A.S., in *Nitrenes,* Lwowski, W., Ed., Wiley-Interscience, New York, 1970, pp. 99–162.

145. Scriven, E.F.V., in *Reactive Intermediates,* Vol. 2, Abramovitch, R.A., Ed., Plenum, New York, 1982, pp. 1–14.

146. Wentrup, C., *Reactive Molecules,* Wiley-Interscience, New York, 1984, pp. 1–333.

147. Smith, P.A.S., in *Azides and Nitrenes: Reactivity and Utility,* Scriven, E.F.V., Ed., Academic, New York, 1984, pp. 95–204.

148. Platz, M.S. and Maloney, V.M., in *Kinetics and Spectroscopy of Carbenes and Biradicals,* Platz, M.S., Ed., Plenum, New York, 1990, pp. 303–320.

149. Schuster, G.B. and Platz, M.S., *Adv. Photochem.,* 17, 69–143, 1992.

150. Gritsan, N.P. and Pritchina, E.A., *Russ. Chem. Rev.,* 61, 500–516, 1992.

151. Platz, M.S., *Acc. Chem. Res.,* 28, 487–492, 1995.

152. Greiss, P., *Philos. Trans. R. Soc. London,* 13, 377, 1864.

153. Wolf, L., *Ann.* 394, 59–68, 1912.

154. Huisgen, R. and Vossius, D., *Chem. Ber.,* 91, 1–12, 1958.

155. Huisgen, T., *M. Appl. Chem. Ber.,* 91, 12–21, 1958.

156. Doering, W.V.E. and Odum, R.A., *Tetrahedron,* 22, 81–93, 1966.

157. Schrock, A.K. and Schuster, G.B., *J. Am. Chem. Soc.,* 106, 5228–5234, 1984.

158. Meijer, E.W., Nijhuis, S., and Von Vroonhoven, F.C.B.M., *J. Am. Chem. Soc.,* 110, 7209–7210, 1988.

159. Smith, P.A.S. and Brown, B.B., *J. Am. Chem. Soc.,* 73, 2435–2437, 1951.

160. Swenton, J.S., Ikeler, T.J., and Williams, B.H., *J. Am. Chem. Soc.,* 92, 3103–3109, 1970.

161. DeGraff, B.A., Gillespie, D.W., and Sundberg, R.J., *J. Am. Chem. Soc.,* 96, 7491–7496, 1974.

162. Sundberg, R.J., Brenner, M., Suter, S.R., and Das, B.P., *Tetrahedron Lett.,* 11, 2715–2718, 1970.

163. Sundberg, R.J., Suter, S.R., and Brenner, M., *J. Am. Chem. Soc.*, 94, 513–520, 1972.

164. Shillady, D.D. and Trindle, C., *Theor. Chim. Acta*, 43, 137–144, 1976.

165. Chapman, O. and Le Roux, J.P., *J. Am. Chem. Soc.*, 100, 282–285, 1978.

166. Donnelly, T., Dunkin, I.R., Norwood, D.S.S., Prentice, A., Shields, C.J., and Thomson, P.C.P., *J. Chem. Soc. Perkin Trans.*, 2, 307–310, 1985.

167. Schrock, A.K. and Schuster, G.B., *J. Am. Chem. Soc.*, 106, 5228–5234, 1984.

168. Shields, C.J., Chrisope, D.R., Schuster, G.B., Dixon, A.J., Poliakoff, M., and Turner, J.J., *J. Am. Chem. Soc.*, 109, 4723–4726, 1987.

169. Li, Y.Z., Kirby, J.P., George, M.W., Poliakoff, M., and Schuster, G.B., *J. Am. Chem. Soc.*, 110, 8092–8098, 1988.

170. Carroll, S.E., Nay, B., Scriven, E.F.V., Suschitzky, H., and Thomas, D.R., *Tetrahedron Lett.*, 18, 3175–3178, 1977.

171. Baron, W.J., DeCamp, M.R., Henric, M.E., Jones, M., Jr., Levin, R.H., and Sohn, M.B., in *Carbenes,* Vol. 1, Jones, M., Jr., and Moss, R.A., Eds. John Wiley, New York, 1973, pp. 1–152.

172. Liang, T.Y. and Schuster, G.B., *J. Am. Chem. Soc.*, 109, 7803–7810, 1987.

173. Gritsan, N.P. and Pritchina, E.A., *J. Inf. Rec. Mat.*, 17, 391–404, 1989.

174. Leyva, E., Platz, M.S., Persy, G., and Wirz, J., *J. Am. Chem. Soc.*, 108, 3783–3790, 1986.

175. Smolinsky, G., Wasserman, E., and Yager, Y.A., *J. Am. Chem. Soc.*, 84, 3220–3221, 1962.

176. Reiser, A. and Frazer, V., *Nature* (London), 208, 682–683, 1965.

177. Reiser, A., Bowes, G., and Horne, R., *Trans. Faraday Soc.*, 62, 3162–3169, 1966.

178. Reiser, A., Terry, G.C., and Willets, F.W., *Nature* (London), 211, 410, 1966.

179. Reiser, A., Wagner, H.M., Marley, R., and Bowes, G., *Trans. Faraday Soc.*, 63, 2403–2410, 1967.

180. Waddell, W.H. and Feilchenfeld, N.B., *J. Am. Chem. Soc.*, 105, 5499–5500, 1983.

181. Feilchenfeld, N.B. and Waddell, W.H., *Chem. Phys. Lett.*, 98, 190–194, 1983.

182. Hayes, J.C. and Sheridan, R.S., *J. Am. Chem. Soc.*, 112, 5879–5881, 1990.

183. Marcinek, A., Leyva, E., Whitt, D., and Platz, M.S., *J. Am. Chem. Soc.*, 115, 8609–8612, 1993.

184. Gritsan, N.P., Yuzawa, T., and Platz, M.S., *J. Am. Chem. Soc.*, 119, 5059–5060, 1997.

185. Born, R., Burda, C., Senn, P., and Wirz, J., *J. Am. Chem. Soc.*, 119, 5061–5062, 1997.

186. Gritsan, N.P., Zhu, Z., Hadad, C.M., and Platz, M.S., *J. Am. Chem. Soc.*, 121, 1202–1207, 1999.

187. Glaiter, R., Rettig, W., and Wentrup, C., *Helv. Chim. Acta*, 57, 2111–2124, 1974.

188. Kim, S.J.I., Hamilton, T.P., and Schaefer, H.F., III, *J. Am. Chem. Soc.*, 114, 5349–5355, 1992.

189. Hrovat, D., Waali, E.E., and Borden, W.T., *J. Am. Chem. Soc.*, 114, 8698–8699, 1992.

190. Borden, W.T. and Davidson, E.R., *J. Am. Chem. Soc.*, 99, 4587–4594, 1977.

191. Borden, W.T., in *Diradicals*, Borden, W.T., Ed., Wiley-Interscience, New York, 1982, pp. 1–72.

192. Borden, W.T., *Mol. Crystl. Liq. Crystl.*, 323, 195–218, 1993.

193. Travers, M.J., Cowles, D.C., Clifford, E.P., and Ellison, G.P., *J. Am. Chem. Soc.*, 114, 8699–8701, 1992.

194. McDonald, R.N. and Davidson, S.J., *J. Am. Chem. Soc.*, 115, 10857–10862, 1993.

195. Castell, O., Carefa, V.M., Bo, C., and Caballol, R., *J. Comput. Chem.*, 17, 42–48, 1996.

196. Smith, B.A. and Cramer, C.J., *J. Am. Chem. Soc.*, 118, 5490–5493, 1996.

197. Matzinger, S., Bally, T., Patterson, E.V., and McMahon, R.J., *J. Am. Chem. Soc.*, 118, 1535–1542, 1996.

198. Wong, M.W. and Wentrup, C., *J. Org. Chem.*, 61, 7022–7029, 1996.

199. Schreiner, P., Karney, W., Schleyer, P.V.R., Borden, W.T., Hamilton, T., and Schaeffer, H.F., III, *J. Org. Chem.*, 61, 7030–7039, 1996.

200. Cramer, C.J., Dulles, F.J.J., and Falvey, D.E., *J. Am. Chem. Soc.*, 116, 9787–9788, 1994.

201. Admasu, A., Gudmundsdottir, A.D., and Platz, M.A., *J. Phys. Chem. A*, 101, 3832–3840, 1997.

202. Sitzmann, E.V., Langen, J., and Eisenthal, K.B., *J. Am. Chem. Soc.*, 106, 1868–1869, 1984.

203. Grasse, P.B., Brauer, B.E., Zupancic, J.J., Kaufmann, K.J., and Schuster, G.B., *J. Am. Chem. Soc.*, 105, 6833–6845, 1983.

204. Huggenberger, C. and Fischer, H., *Helv. Chim. Acta*, 64, 338–353, 1981.

205. Porter, G. and Ward, B., *J. Chim. Phys.*, 61, 1517–1522, 1964.

206. Porter, G. and Wright, F.J., *Trans. Faraday Soc.*, 51, 1469, 1955

207. Leyva, E., Platz, M.S., Niu, B., and Wirz, J., *J. Phys. Chem.*, 91, 2293–2298, 1987.

208. Gritsan, N.P., Zhai, H.B., Yuzawa, T., Karweik, D., Brooke, J., and Platz, M.S., *J. Phys. Chem. A*, 101, 2833–2840, 1997.

209. Anderson, K., *Theor. Chim. Acta*, 91, 31–46, 1995.

210. Sumitani, M., Nagakura, S., and Yoshihara, K., *Bull. Chem. Soc. Jpn.*, 49, 2995–2998, 1976.

211. Kobayashi, T., Suzuki, K., and Yamaoka, T., *J. Phys. Chem.*, 89, 776–779, 1985.

212. Miura, A. and Kobayashi, T., *J. Photochem. Photobiol. A*, 53, 223–231, 1990.

213. Tsao, M.L., Gritsan, N.P., James, T.R., Platz, M.S., Hrovat, D. and Borden, W.T., J. Am. Chem. Soc. 125, 9343–9358, 2003.

214. Gritsan, N.P., Tigelaar, D., and Platz, M.S., *J. Phys. Chem. A,* 103, 4465–4469, 1999.

215. Gritsan, N.P., Gudmundsdóttir, A.D., Tigelaar, D., and Platz, M.S., *J. Phys. Chem. A,* 103, 3458–3461, 1999.

216. Cerro-Lopez, M., Gritsan, N.P., Zhu, A., and Platz. M.S., *J. Phys. Chem. A,* 104, 9681–9686, 2000.

217. Gritsan, N.P., Likhotvorik, I., Tsao, M.L., Çelebi, N., Platz, M.S., Karney, W.L., Kemnitz, C.R., and Borden, W.T., *J. Am. Chem. Soc.,* 123, 1425–1433, 2001.

218. Gritsan, N.P., Gudmundsdóttir, A.D., Tigelaar, D., Zhu, A., Karney, W.L., Hadad, C.M., and Platz, M.S., *J. Am. Chem. Soc.,* 123, 1951–1962, 2001.

219. Smirnov, V.A. and Brichkin, S.B., *Chem. Phys. Lett.,* 87, 548–551, 1982.

220. Kozankiewicz, B., Deparasinska, I., Zhai, H.B., Zhu, Z., and Hadad, C.M., *J. Phys. Chem.,* 103, 5003–5010, 1999.

221. Poe, R., Schnapp, K., Young, M.J.T., Grayzar, J., and Platz, M.S., *J. Am. Chem. Soc.,* 114, 5054–5067, 1992.

222. Johnson, W.T.G., Sullivan, M.B., and Cramer, C.J., *Int. J. Quant. Chem.,* 85, 492–508, 2001.

223. Ziegler, T., Rauk, A., and Baerends, E.J., *Theor. Chim. Acta,* 43, 261, 1977.

224. Cramer, C.J., Dulles, F.J., and Giesen, D.J., *Chem. Phys. Lett.,* 245, 165–170, 1995.

225. Michl, J. and Bonacic-Koutecky, V., *Electronic Aspects of Organic Photochemistry,* Wiley, New York, 1990, pp. 367.

226. Bonacic-Koutecky, V., Koutecky, J., and Michl, J., *Angew. Chem.,* 99, 216–36, 1987.

227. Michl, J. and Havlas, Z., *Pure Appl. Chem.,* 69, 785–790, 1997.

228. Younger, C.G. and Bell, R.A., *J. Chem. Soc. Chem. Commun.,* 21, 1359–1361, 1992.

229. Morawietz, J. and Sander, W., *J. Org. Chem.,* 61, 4351–4354, 1996.

230. Dunkin, I.R. and Thomson, P.C.P., *J. Chem. Soc. Chem., Commun.*, 9, 499–500, 1980.

231. Carroll, S.E., Nay, B., and Scriven, E.F.V., *Tetrahedron Lett.*, 38, 943–946, 1997.

232. Rigaudy, J., Igier, C., and Barcelo, J., *Tetrahedron Lett.*, 31, 3845–3848, 1975.

233. Hilton, S.E., Scriven, E.F.V., and Suschitzky, H., *J. Chem. Soc. Chem. Commun.*, 3, 853–854, 1974.

234. Leyva, E. and Platz, M.S., *Tetrahedron Lett.*, 28, 11–14, 1987.

235. Polshakov, D.A., Tsentalovich, Y.P., and Gritsan, N.P., *Russ. Chem. Bull.*, 49, 50–55, 2000.

236. Odum, R.A. and Aaronson, A.M., *J. Am. Chem. Soc.*, 91, 5680–5681, 1969.

237. Dunkin, I.R., Donnelly, T., and Lockhart, T.S., *Tetrahedron Lett.*, 26, 359–362, 1985.

238. Murata, S., Abe, S., and Tomioka, H., *J. Org. Chem.*, 62, 3055–3061, 1997.

239. Karney, W.L. and Borden, W.T., *J. Am. Chem. Soc.*, 119, 3347–3350, 1997.

240. Lamara, K., Redhouse, A.D., Smalley, R.K., and Thompson, J.R., *Tetrahedron*, 50, 5515–5525, 1994.

241. Berwick, M.A., *J. Am. Chem. Soc.*, 93, 5780–5786, 1971.

242. Leyva, E. and Sagredo, R., *Tetrahedron*, 54, 7367–7374, 1998.

243. Abramovitch, R.A., Challand, S.R., and Scriven, E.F.V., *J. Am. Chem. Soc.*, 94, 1374–1376, 1972.

244. Abramovitch, R.A., Challand, S.R., and Scriven, E.F.V., *J. Org. Chem.*, 40, 1541–1547, 1975.

245. Banks, R.E. and Sparkes, G.R., *J. Chem. Soc. Perkin Trans. I*, 1, 2964–2970, 1972.

246. Banks, R.E. and Prakash, A., *Tetrahedron Lett.*, 14, 99–102, 1973.

247. Banks, R.E. and Prakash, A., *J. Chem. Soc. Perkin Trans. I*, 3, 1365–1371, 1974.

248. Banks, R.E. and Venayak, N.D., *J. Chem. Soc. Chem. Commun.*, 9, 900–901, 1980.

249. Banks, R.E. and Sparkes, G.R., *J. Chem. Soc. Perkin Trans. I*, 1, 2964–2970, 1972.

250. Poe, R., Grayzar, J., Young, M.J.T., Leyva, E., Schnapp, K.A., and Platz, M.A., *J. Am. Chem. Soc.*, 113, 3209–3211, 1991.

251. Crocker, P.J., Imai, N., Rajagopalan, K., Kwiatkowski, S., Dwyer, L.D., Vanaman, T.C., and Watt, D.C., *Bioconjugate Chem.*, 1, 419–424, 1990.

252. Drake, R.R., Slama, J.T., Wall, K.A., Abramova, M., D'Souza, C., Elbein, A.D., Crocker, P.J., and Watt, D.S., *Bioconjugate Chem.*, 3, 69–73, 1992.

253. Pinney, K.C. and Katzenellenbogen, J.A., *J. Org. Chem.*, 56, 3125–3133, 1991.

254. Pinney, K.C., Carlson, K.E., Katzenellenbogen, S.B., and Katzenellenbogen, J.A., *Biochemistry*, 30, 2421–2431, 1991.

255. Kym, P.R., Carlson, K.E., and Katzenellenbogen, J.A., *J. Med. Chem.*, 36, 1111–1119, 1993.

256. Reed, M.W., Fraga, D., Schwartz, D.E., Scholler, J., and Hinrichsen. R.D., *Bioconjugate Chem.*, 6, 101–108, 1995.

257. Kapfer, I., Hawkinson, J.E., Casida, J.E., and Goeldner, M.P., *J. Med. Chem.*, 37, 133–140, 1994.

258. Kapfer, I., Jacques, P., Toubal, H., and Goeldner, M.P., *Bioconjugate Chem.*, 6, 109–114, 1995.

259. Poe, R., Schnapp, K., Young, M.J.T., Grayzar, J., and Platz, M.S., *J. Am. Chem. Soc.*, 114, 5054–5067, 1992.

260. Schnapp, K.A., Poe, R., Leyva, E., Soundararajan, N., and Platz, M.S., *Bioconjugate Chem.*, 4, 172–177, 1993.

261. Schnapp, K.A. and Platz, M.A., *Bioconjugate Chem.*, 4, 178–183, 1993.

262. Marcinek, A. and Platz, M.S., *J. Phys. Chem.*, 97, 12674–12677, 1993.

263. Marcinek, A., Platz, M.S., Chan, S.Y., Floresca, R., Rajagopalan, K., Golinski, M., and Watt, D., *J. Phys. Chem.*, 98, 412–419, 1994.

264. Jackson, J.E., Soundararajan, N., Platz, M.S., and Liu, M.T.H., *J. Am. Chem. Soc.,* 110, 5595–5596, 1988.

265. Young, M.J.T. and Platz, M.S., *J. Org. Chem.,* 56, 6403–6406, 1991.

266. Reed, A.E., Weinstock, R.B., and Weinhold, F.A., *J. Chem. Phys.,* 83, 733–746, 1985.

267. Sung, K., *J. Chem. Soc. Perkin Trans.,* 2, 1169–1173, 1999.

6

Semiempirical MR-CI Calculations
for Organic Photoreactions

MARTIN KLESSINGER

Organisch-Chemisches Institut der
Westfälischen Wilhelms-Universität
Münster, Germany

CONTENTS

6.1 INTRODUCTION

Semiempirical methods are generally quite useful in exploratory work, especially on larger systems.[1] This is true not only
for the ground-state chemistry, but also for excited-state phenomena such as photochemical reaction mechanisms, provided
the semiempirical method used is flexible enough for a description of excited states. Excited-state potential-energy surfaces
(PESs) can be determined with little computational effort, and
stationary points such as minima, barriers, and, in particular,
conical intersections[2] can be located quite reliably by semiempirical MR-CI (multireference configuration interaction) calculations even for larger systems.[3] Thus, such calculations may
be quite useful in investigating organic photoreactions, as will
be shown using the photochemistry of cycloheptatriene (CHT),
cyclooctatetraene (COT), and 2*H*-azirine as examples to demonstrate various aspects of the procedure.

   This chapter will be organized as follows. First, in Section
6.2 the requirements for an appropriate description of excited
states will be discussed very briefly and the corresponding
semiempirical method will be described. In Section 6.3 a
detailed example will be given for a straightforward application of this method in establishing the mechanism of a photochemical reaction. Next, the problem of conical intersections
will be addressed, as it has been known for quite some time
that organic singlet photoreactions generally proceed via conical intersections.[4] Usually, the location of the conical intersection, that is, the determination of its structure, is the
crucial and most difficult problem in determining a photo-

chemical reaction mechanism. This will be addressed in Sections 6.4 and 6.5, first by discussing a very useful method proposed by Zilberg and Haas[5] for qualitatively predicting the location of conical intersections and then by showing how this method can be used to determine the structure of the corresponding conical intersections on the basis of semiempirical MR-CI calculations. Section 6.6 will deal with the necessary verification of the results thus obtained by the *ab initio* method. In Section 6.7 the treatment of organic triplet photoreactions will be briefly covered, and Section 6.8 will provide a summary of the procedures discussed in this chapter together with an outlook for future applications. An appendix will present some of the data necessary to repeat the calculations discussed in the text in order to enable newcomers in the field to get some experience with excited-state calculations before turning to their own problems.

6.2 THE QUANTUM-CHEMICAL TREATMENT OF EXCITED STATES

Recent experimental results and new computational methods have provided a new mechanistic picture of photochemical reactions,[2] that can be understood in terms of the evolution of an excited-state species toward a conical intersection that acts as a photochemical decay channel. Thus, a theoretical discussion of the mechanism of a photochemical reaction will have to be based on excited-state PESs and on the location of conical intersections.

In general, a qualitatively correct description of the ground state of a closed-shell molecule is provided by a single Slater determinant. This is why semiempirical (one-determinant) self-consistent field (SCF) methods can be applied quite successfully to the determination of ground-state properties such as geometries, vibrational frequencies, and relative energies. Many electronically excited states, however, contain more then one dominant configuration state function. The simplest description of an excited state is the orbital picture where one electron has been moved from an occupied to an unoccupied orbital. The lowest level of theory for a qualitative description of excited states is therefore a CI including only singly excited determinants, denoted by CIS.[6] CIS gives wave functions of roughly HF quality for excited states, because no orbital optimization is involved. For valence excited states such as some of the $\pi \rightarrow \pi^*$ excited states of conjugated hydrocarbons, this may be a reasonable description. There are, however, normally quite low-lying states, too, that essentially correspond to a double excitation, as, for example, the lowest benzene $\pi \rightarrow \pi^*$ excited states, and they require at least the inclusion of doubles as well, that is, CISD (**C**onfiguration **I**nteraction including **S**ingles and **D**oubles). Furthermore, if an excited-state PESs is to be considered, it is most likely that the dominant configuration of the wave function will be different in different areas of the same surface. In particular, at a conical intersection, which is an intersection of PESs of different states described by different wave functions, the wave function must change. As a consequence, CIS methods will not be suited for calculations on the mechanism of a photochemical reaction.

A more balanced description thus requires multiconfiguration self-consistent field (MCSCF)–based methods,[7] where the orbitals are optimized for each particular state or optimized for a suitable average of the desired states (state-averaged MCSCF). In semiempirical methods, however, an MCSCF procedure is normally not required due to the limited flexibility of the minimal valence atomic orbital basis commonly used in these methods. Instead, a multireference CI method including a limited number of suitably chosen configurations will be appropriate.

Based on the work of Thiel,[8] such a semiempirical MR-CI was developed some time ago.[9] This MNDOC-CI method may be characterized as follows. For each state under consideration, one or a few reference configurations are used, and from these reference configurations single and double excitations within a properly chosen active space (AS) are considered in constructing spin-adapted configuration state functions (CSFs). To specify the active space and the refer-

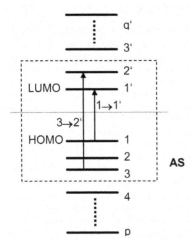

Figure 6.1 Numbering of molecular orbitals and definition of the active space (AS) and the excitation index (EI). For the singly excited configuration $\Phi_{3\rightarrow 2'}$ the excitation index is $m = 5$, and for the doubly excited configuration $\Phi_{3\rightarrow 2',1\rightarrow 1'}$ $n = 7$.

ence configurations, the occupied ground-state orbitals are numbered 1, 2,... starting at the HOMO, whereas the virtual orbitals are numbered 1', 2', ... starting at the LUMO. Typically, the number of CSFs is limited to a few hundred. If the active space has to be chosen such that not all singly and doubly excited configurations could be included, the CI is further truncated by the use of excitation indices (EI) m, n, which specify the largest single and double excitations to be included. For singly and doubly excited configurations $\Phi_{i\rightarrow k'}$ and $\Phi_{i\rightarrow k', j\rightarrow l'}$ the excitation indices are $m = (i + k')$ and $n = [(i + k') + (j + l')]$, respectively (see Figure 6.1). The truncation of the CI leads to inaccuracies in the calculated heats of formation but yields reliable geometries and comparable accuracies for the ground state and the various excited states.

Furthermore, the method includes the possibility for optimizing and characterizing the lowest point of a conical intersection.[10] Such an optimization procedure requires the local evaluation of the gradients, that is, the use of analytic gradients, for numerical gradients will cause difficulties because, strictly speaking, the gradient does not exist at the tip of a cone and because of the orbital crossing at this point. An efficient procedure for calculating analytic gradients in the framework of semiempirical MR-CI methods has been developed, and the same algorithm has been implemented to locate the lowest energy point of a conical intersection by the direct method of Bearpark et al.[10] Results for conical intersections obtained by this method are in very good agreement with *ab initio* data.[11] The MNDOC-CI method with analytic gradients represents a very powerful and efficient tool for the determination of excited-state PESs and, in particular, for conical intersections. To obtain absolute values for excitation energies, barrier heights, and energies of conical intersections, however, sophisticated *ab initio* calculations are required, for which the semiempirical results represent a good starting point.

Finally, a formalism for calculating spin–orbit coupling (SOC) effects in organic molecules based on Rumer spin eigenfunctions and the second quantization approach has been derived and implemented for the semiempirical MNDOC-CI model.[12] It allows for a straightforward determination of SOC surfaces within the context of MR-CI calculations at the same level of theory as PESs.

As the combined analysis of the PE and SOC surfaces is particularly suited for determining the triplet state reactivity, this approach extends the applicability of semiempirical methods to discuss triplet photoreactions as well.

In *ab initio* calculations on excited states, the static electron correlation is usually taken into account using the complete active space SCF (CASSCF) method[13] or a general MCSCF procedure.[7] Methods for calculating the lowest energy point of a conical intersection based on the CASSCF approach are available[10] and have been included in the GAUSSIAN program package.[14] Dynamic correlation, which is essential for calculating excitation energies or barrier heights, may then be included by perturbational methods such as

CASPT2.[15] (For details of *ab initio* calculations on excited states, see Chapter 2.)

6.3 THE PHOTOCHEMICAL [1,7] HYDROGEN SHIFT IN 1,3,5-CYCLOHEPTATRIEN

In ground-state reactions, the PESs of all states involved in a photochemical reaction are usually quite useful as a first step in establishing the reaction mechanism; the location of minima and barriers will give hints as to which states are involved in the photoreaction and where conical intersections are likely to be found. Because of the low computational efforts, semiempirical methods are particularly suited for such preliminary investigations.

As an example, the suprafacial [1,7] sigmatropic H-shift will be considered that occurs on irradiation of 1,3,5-CHT with formation of bicyclo[3.2.0]hepta-2,6-diene (BHD) as a minor product.[16] In 1-substituted CHT this [1,7]-H shift is regioselective; in 1-cyano-cycloheptatriene (CN-CHT) the hydrogen atom moves exclusively to the unsubstituted terminal carbon of the heptatriene moiety, whereas in 1-methyl-cycloheptatriene (Me-CHT) only 2% of the product exhibits this regiochemistry and 98% corresponds to a hydrogen shift toward the substituted carbon, as shown in Scheme 6.1.[17] Different models have been proposed to explain this regioselectivity.[18]

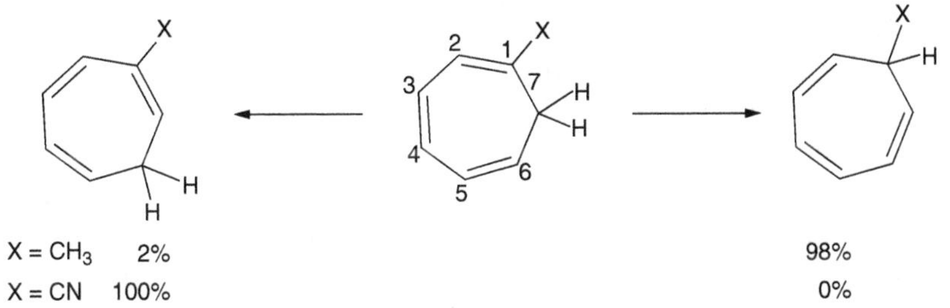

X = CH$_3$ 2%   98%
X = CN   100%   0%

Scheme 6.1

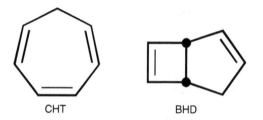

CHT   BHD

One of the major problems in determining the PES is the choice of appropriate coordinates, which allow the important features and characteristic points to show up in the PES. In the case of the hydrogen-shift reaction from C_1 to C_7 in CHT, the position of the migrating hydrogen H_m can be described by its distance R from the center of the C_1–C_7 bond and the angle θ, which describes the rotation around an axis through C_4 and the center of the C_1–C_7 bond (i.e., in the molecular plane and perpendicular to the C_1–C_7 bond) as indicated in Figure 6.2.

Using the semiempirical MNDOC-CI method, PESs of the two lowest singlet states were calculated subject to the condition that all atoms except H_m, H_1, and H_7 are coplanar; the other internal coordinates were optimized for the lowest excited state S_1 and for the ground state S_0.[18] The results are displayed on the left ($\theta = 30$ to $90°$) and on the right ($\theta = 90$ to $150°$) sides of Figure 6.3, respectively. Consequently, the diagram does not exhibit a plane of symmetry with respect to the transition state for the ground-state reaction at $\theta = 90°$. Vertical excitation occurs at the planar ground-state structure **1** in the right half of the diagram. However, the photochemical hydrogen-shift reaction starts on the S_1 surface at the relaxed excited-state minimum **2** then crosses a potential-energy barrier at the transition state **3** and reaches the pericyclic geometry **4**, where the two states S_1 and S_0 are degenerate. This region thus corresponds to a conical intersection from which either the reactant or the product may be reached. Energies of the characteristic points along this reaction path calculated at fully optimized geometries are collected in Table 6.1. These results show that the 1A" excited

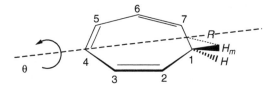

Figure 6.2 [1,7]-H shift reaction of CHT: definition of the coordinates R and θ for calculating the potential energy surfaces.

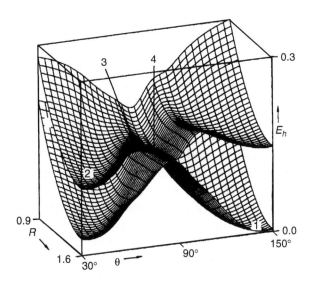

Figure 6.3 [1,7]-H shift reaction of CHT: S_0 and S_1 PESs. On the right-hand side (θ = 90 to 150°) geometries are optimized for the ground state S_0 subject to the condition that all atoms except H_m, H_1, and H_7 are coplanar; on the left-hand side (θ = 30 to 90°) they are optimized for the lowest excited state S_1.

state lies in the Franck–Condon region 0.4 kcal mol^{-1} below the 2A' state, but the minimum **2** lies on the totally symmetric PES (2A'), with the minimum of the spectroscopic 1A" state being 5.5 kcal mol^{-1} higher in energy. The transition state **3** between the 2A' minimum and the pericyclic geometry **4** is calculated to be 23.3. kcal mol^{-1} higher in energy.

TABLE 6.1   The [1,7]-H Shift Reaction of Cycloheptatriene[a]

Geometry		*ab initio*		
	MNDOC-CI	CASSCF/6-31G*	CASPT2/ANO[b]	
1 (S_0 min.)	65.9 (−69.1)	−269.76265 (−98.5)	−270.66516 (−78.1)	
2 (S_1 min.)	135.0 (0.0)	−269.69570 (0.0)	−270.54078 (0.0)	
3 (S_1 TS)	158.3 (23.3)	−269.58404 (13.6)	−270.53470 (3.8)	
4 (S_1/S_0 con. int.)	146.0 (11.0)	−269.61503 (3.0)	−270.54664 (−3.7)	

[a] Energies of stationary points: MNDOC-CI values (heats of formation) in kcal/mol, *ab initio* values (absolute energies) in au (1 au = 627.5 kcal/mol), and relative energies (in parentheses) in kcal/mol.
[b] ANO basis set from MOLCAS 4.0.[37]

The existence of an excited-state barrier on the 2A′ surface would be in agreement with the observation that in solution ground-state CHT appears only 26 picoseconds (ps) after photolysis.[19] More recent gas-phase experiments, however, indicate an appearance time of less then 100 femtoseconds (fs).[20] Thus, the calculated value of the barrier is definitely too high. Full optimization of the stationary points shows that in agreement with experiment[21] planarization of the CHT ring begins after excitation. Whereas the equilibrium ground-state conformation is boat-like, the excited-state minimum **2**, the transition structure **3**, and the pericyclic structure **4** are all calculated to be planar. The transition structure no longer exhibits C_s symmetry, as the migrating hydrogen has moved slightly toward C_7 with distances of 1.23 Å for $r(C_1$-H) (as compared with 1.12 Å for the ground-state equilibrium structure) and 1.56 Å for $r(C_7$-H). As a consequence of the incipient hydrogen-shift reaction, the unsaturated carbons become polarized and exhibit alternating charges, with the accepting carbon being positively charged. Thus, an electron-donating or electron-withdrawing substituent on the carbon adjacent to the methylene group will stabilize or destabilize, respectively, the transition state for migration of H_m to the substituted carbon, in agreement with the experimentally observed regioselectivity. Finally, at the pericyclic geometry, both C_1–H and C_7–H bond distances are equal to 1.41 Å. Because the excited state S_1 and the ground-state S_0 are degenerate at this geometry, a fast transition occurs to the ground-state PES, which slopes down steeply toward the product or back to the reactant.

Although for the [1,7]-sigmatropic shift reaction of CHT the semiempirical calculations yield a reaction mechanism that is in good qualitative agreement with all experimental observations, the quantitative results, and in particular the value of 23 kcal mol^{-1} for the excited-state barrier, require redetermination on the basis of more reliable *ab initio* methods. For details, see Section 6.6.

6.4 ZILBERG AND HAAS'S METHOD FOR LOCATING CONICAL INTERSECTIONS

In general, PESs such as those discussed in the previous section indicate where a conical intersection is likely to be found only if appropriate coordinates are used. The potential energy has the form of a double cone in the region of degeneracy only if the energy is plotted against two special variables, x_1 and x_2, which correspond to the gradient difference vector and the nonadiabatic coupling vector, respectively, and span the branching space (for details, see Chapter 2). Therefore, it is important to have methods available that allow prediction of the existence of a conical intersection and help locate it computationally.

Zilberg and Haas[5] recently proposed such a method based on the phase-change rule of Longuet-Higgins,[22] according to which a conical intersection necessarily arises within a region of space enclosed by a loop along which the total electronic wave function changes sign. Zilberg and Haas proposed (a) to construct a loop on the ground-state surface and chose three points, A, B, and C, which they call anchors, corresponding to the reactant A, the desired product B, and a third species C, which is chemically distinct from both A and B; and (b) to look for phase changes on the closed loop A-B-C-A formed by the reaction coordinates leading from A to B, from B to C, and from C to A. Two mechanisms by which a phase change on the ground-state surface can take place are considered: the orbital overlap mechanism, which creates a negative overlap (or an odd number of negative overlaps) between two adjacent

TABLE 6.2 The Phase Change of Hückel and Möbius Reactions

Number of Electron-Pair Exchanges	Reaction Type	
	Hückel	Möbius
Even (0,2,...)	Phase-inverting (i)	Phase-preserving (p)
Odd (1,3,...)	Phase-preserving (p)	Phase-inverting (i)

atomic orbitals during the reaction, which is termed Möbius-type; and the permutational mechanism, according to which Hückel-type reactions (in which an even number of new nodes will be formed along the reaction coordinate) are phase-inverting if an even number of electron pairs is exchanged in the reaction. This corresponds to an antiaromatic transition state, whereas Möbius-type reactions are phase-inverting if an odd number of electron pairs is exchanged, as summarized in Table 6.2.[23]

Following Haas and Zilberg,[24] we will consider as a first example the electrocyclic ring-closure reaction of butadiene. For the three anchors, we choose the reactant butadiene (A), the desired product cyclobutene (B), and as a chemically distinct third anchor bicyclobutane or the corresponding biradical (C), as shown in Figure 6.4. Only the electrons involved in the reaction, that is, the four π electrons of butadiene, are considered; they are represented schematically by the contour of the p-orbital, and the sign of the electronic wave function is indicated. The conrotatory reaction is phase preserving (the transition state corresponds to a $4n$ Möbius system, which is aromatic), whereas the disrotatory ring-closure reaction of butadiene to cyclobutane is phase-inverting (the transition state is antiaromatic). Ring closure to form the cyclopropane ring may or may not lead to a new node, depending on whether it is conrotatory or disrotatory. Because conrotatory and disrotatory cyclopropane formation yields different products, the reaction must have the same phase properties starting either

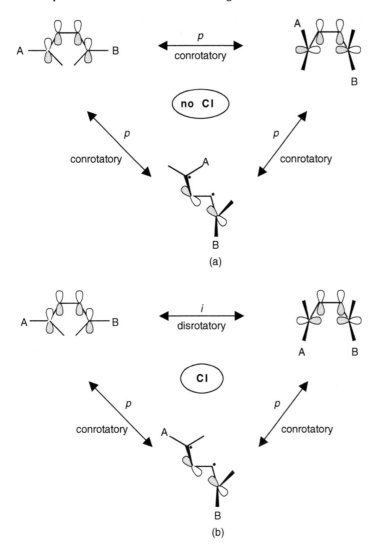

Figure 6.4 The butadiene ↔ cyclobutene ring closure reaction. (a) For the conrotatory reaction, the loop constructed from butadiene, cyclobutene, and bicyclobutane is p^3 (or pi^2), and no conical intersection is enclosed. (b) For the disrotatory reaction, the loop is ip^2 (or i^3) and encloses a conical intersection.

from butadiene or cyclobutene. Thus, if the butadiene $\leftrightarrow$ cyclobutene ring closure is conrotatory, the loop is either p^3 or pi^2, and no conical intersection can exist in the region enclosed by the loop. However, if the cyclobutene ring closure is conrotatory, the loop is either i^3 or ip^2, and it must enclose a conical intersection.

Alternatively, we may look at the spin pairing of the electrons involved in the three anchors, which is 1–2 and 3–4 in butadiene, {12,34}, 1–4 and 2–3 or {14,23} in cyclobutene, and 1–3 and 2–4 or {13,24} in bicyclobutane, where the electrons are numbered by the atoms whose atomic orbitals they occupy. The transition between any two of the three anchors thus involves spin exchange between two electron pairs. As conrotatory ring-closure reactions are of Möbius-type, the loop constructed from butadiene, cyclobutene, and bicyclobutene is according to Table 6.2 p^3; if conrotatory formation of the cyclopropane ring is considered, all three transitions between the three anchors are phase-preserving, or it is pi^2 if disrotatory formation of the cyclopropane ring is considered, and no conical intersection is included in the loop as shown in Figure 6.4a. If, however, the disrotatory butadiene $\leftrightarrow$ cyclobutene reaction is considered, an ip^2 or an i^3 loop results, which encloses a conical intersection (Figure 6.4b) in agreement with the Woodward–Hoffmann rules and the fact that this reaction proceeds photochemically.

Next we will consider some examples from the photochemistry of COT. Although the photochemistry of COT has not been studied in great detail and little is known about the mechanistic details of the various photoprocesses, a number of photoproducts have been identified under various conditions: benzene and acetylene among other hydrocarbons, as well as 3,6-dihydropentalene and its isomers, semibullvalene and cuneane, while photolysis of tricyclo[3.3.0.0²,⁶]octa-3,7-diene led to semibullvalene and COT in a 1:2 ratio.[31] We will deal with semibullvalene (**5**), tricyclo[3.3.0.0²,⁶]octa-3,7-diene (**6**), cuneane (**7**), and 3,6-dihydropentalene (**8**) as summarized in Scheme 6.2a. In addition, we will consider the hypothetical photocyclization of *syn*-tricyclo[4.2.0.0^{2,5}] octa-3,7-diene (**9**) to cubane (**10**), which does not take place (Scheme 6.2b).

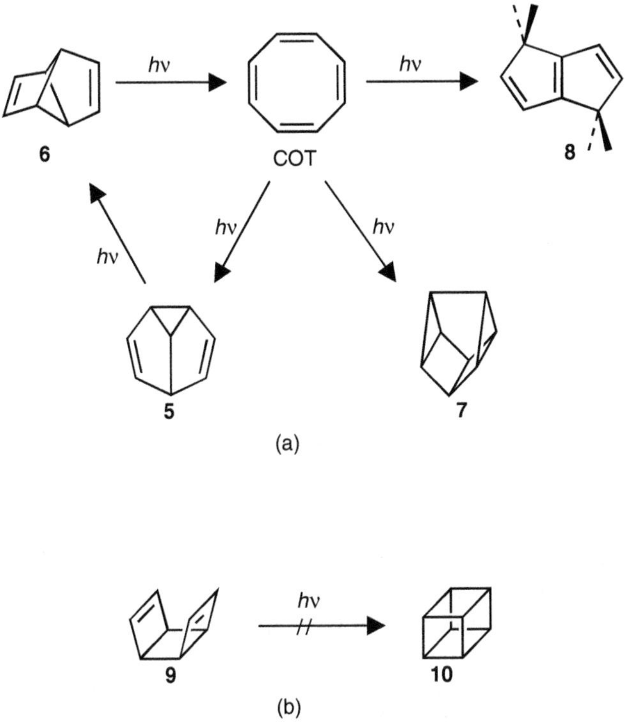

(a)

(b)

Scheme 6.2

Due to steric reasons, all ring-closure reactions shown in Scheme 6.2 have to proceed in a disrotatory way, and all transitions between the anchors shown in Figure 6.5 and Figure 6.6 are Hückel-type reactions. The transition from one of the Kekulé forms of COT to the other one involves spin exchange between four pairs of electrons and is thus phase-inverting, whereas the formation of semibullvalene from both Kekulé forms involves spin exchange between three pairs of electrons and is thus phase-preserving, resulting in an ip^2

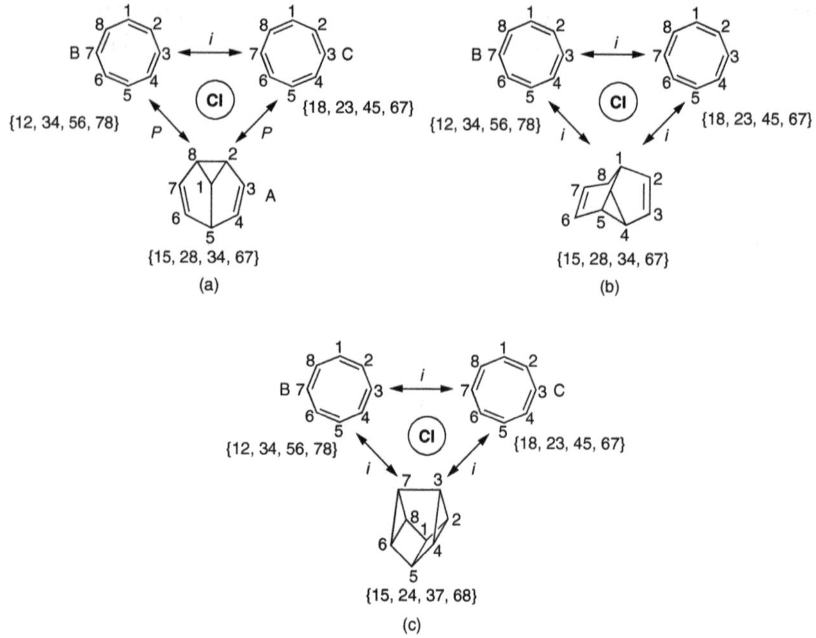

Figure 6.5 Phase-rule loops for photoreactions of COT. (a) The COT ↔ semibullvalen conversion. All reactions in the loop constructed from the two Kekulé forms of COT and semibullvalene are of Hückel type; from the number of pairs of electrons that exchange spin, it is seen that the loop is ip² and encloses a conical intersection. (b) Loop for the COT ↔ tricyclo[3.3.0.0²,⁶]octa-3,7-diene conversion and (c) for the COT ↔ cuneane conversion.

loop that encloses a conical intersection, as shown in Figure 6.5a. Similarly, the loops including tricyclo[3.3.0.0²,⁶]octa-3,7-diene (Figure 6.5b) and cuneane (Figure 6.5c) are i³ and enclose a conical intersection. The formation of 3,6-dihydropentalene involves two C–H bonds, and a total of six electron pairs have to be included in the analysis. As the hydrogens from carbons 1 and 5 migrate to other carbons, they leave unpaired electrons in orbitals on carbons 1 and 5 other than π orbitals; electrons in these orbitals will be denoted by 1' and 5'. Thus, the spin pairing in 3,6-dihydropentalene is {12,39,45,610,78, 1'5'} and differs from the spin pairing

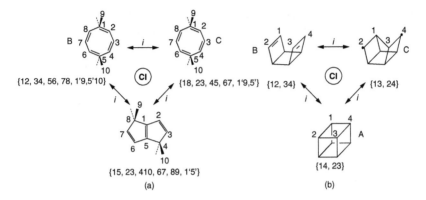

Figure 6.6 Phase-rule loops (a) for the COT ↔ 3,6-dihydopental-ene conversion and (b) for the *syn*-tricyclo[4.2.0.0^{2,5}]octa-3,7-diene ↔ cubane conversion. The former involves more than one electron per atom; the carbon orbitals involved in C$_1$–H and C$_5$–H bonds are denoted by 1' and 5', respectively.

{12,34,56,78,1'9,5'10} in COT (with 1'-9 and 5'-10 for the C–H bonds) in four electron pairs (Figure 6.6). Thus, an ip^2 loop results, which again encloses a conical intersection. Finally, the loop for the formation of cubane from *syn*-tricyclo[4.2.0.0^{2,5}] octa-3,7-diene displayed in Figure 6.6b is equivalent to the situation shown in Figure 6.4b.

In many cases the information derived from these loops may be sufficient to estimate a starting geometry for a computational location of the conical intersection. However, more detailed information is available. If phase inversion between B and C (antiaromatic transition state with wave function |B-C>) is assumed, and if the position vectors of the three structures A, B, and C with wave functions |A>, |B>, and |C>, respectively, are denoted by r_A, r_B, and r_C, the phase-preserving coordinate $Q_I = 2r_A - r_B - r_C$ (I for in-phase) connecting A with B – C and the phase-inverting coordinate $Q_O = r_B - r_C$ (O for out-of-phase) connecting B with C are appropriate coordinates to describe motion in the region enclosed by the loop. Thus, the conical intersection lying within the region encircled by the three anchors may by found by moving first along the phase-inverting reaction coordinate from the reactant B to the antiaromatic transition state B-C and then along the phase-preserving coordinate to the third anchor A. A practical method for locating the conical intersection based on these ideas has been proposed by Zilberg and Haas[25] and applied to determine all conical intersections involved in complex photochemistry of 1,4-cyclohexadiene.

6.5 SEMIEMPIRICAL STRUCTURE DETERMINATIONS OF CONICAL INTERSECTIONS

On the basis of the phase-change rule discussed in the previous section, some examples of structure determination of conical intersections using the semiempirical MNDOC-CI method will now be given. First, $2H$-azirine (**11**) is considered, which on irradiation yields nitrile ylides (**12**).[26] According to earlier work,[27] triplet intermediates are involved, whereas more recent experiments indicate that **12** is formed in the singlet ground state via a fast radiationless process.[28] To decide between these two alternatives one has to look for an S_1–S_0 conical intersection, as there is ample evidence[29] that fast deactivation of singlet states generally occurs via a conical intersection.

To apply the phase-change rule discussed in the previous section, we choose the reactant A with wave function |A> and the two chemically equivalent products B and C with wave functions |B> and |C>, which differ by a rotation of the methylene group through 180°, as anchors as indicated in Figure 6.7. The transition from B to C involves a HOMO-LUMO crossing and hence is phase-inverting, whereas the transitions from A to B or C are equivalent; the total number of phase changes on the closed-loop A-B-C-A is thus odd, and a conical intersection must be contained within the loop. The

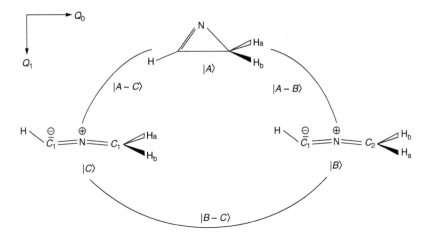

Figure 6.7   The 2*H*-azirine ↔ nitrile ylide conversion. The loop constructed from the azirine and the two isomeric nitrile ylides, which differ by a rotation of the methylene group by 180°, is either ip^2 or i^3 and encloses a conical intersection. The coordinates Q_I and Q_O are indicated.

two coordinates Q_I and Q_O along which the conical intersection is formed are seen from the diagram to correspond to the CC stretch vibration or the opening of the C–N–C angle γ and to the torsional angle θ of the CH$_2$ group, respectively. That is, an S_1–S_0 conical intersection is expected in the range of 60° < γ < 180° and –90° < θ < 90°, where $\theta = 0°$ corresponds to the planar structure.

Starting with $\gamma = 120°$ and $\theta = 0°$, the MNDOC-CI method yields a conical intersection, the structure of which is shown in Figure 6.8. The values $\theta = 0°$ for the rotational angle and $\gamma = 143,7°$ for the bond angle are as expected; the bond distances are between those of azirine and nitrile ylide, especially because the distance between nitrogen and the methylene carbon $r_{CN} = 1.34$ Å has decreased and is equal to the double-bond value $r_{CN} = 1.34$ Å in the nitrile ylide. To completely establish the mechanism, it remains to be shown that the conical intersection is accessible without a barrier from the Franck–Condon geometry and that the nitrile ylide can be reached from the conical intersection. This is easily

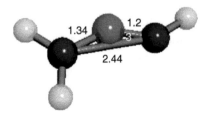

Figure 6.8 The 2*H*-azirine ↔ nitrile ylide conversion; structure at the S_1–S_0 conical intersection as determined by semiempirical MNDOC-CI calculations (AS = 2 – 2', reference configurations Φ_0 and $\Phi_{1\to1'}$).

done by calculating an appropriate cross-section through the S_1 surface and by geometry optimization on the S_0 surface, starting at the conical intersection. Next we will determine the structures of the conical intersections that are to be expected from the loop diagrams in Figure 6.5 and Figure 6.6 for the photochemistry of COT.

To obtain a suitable starting geometry for optimizing a conical intersection, the vectors $\boldsymbol{Q}_I$ and $\boldsymbol{Q}_O$ are considered; $\boldsymbol{Q}_I$ points from one of the COT structures to the other one, whereas $\boldsymbol{Q}_O$ corresponds to a movement from the middle between B and C, which corresponds to the barrier for this reaction, to the product A. Therefore, the conical intersection is expected to be found on a path from this barrier toward the product, and a suitable starting geometry is obtained by a deformation of the regular octagon toward the product geometry. By this procedure, the structures of the S_1–S_0 conical intersections for the formation of semibullvalene (**5**), tricyclo[3.3.0.0^{2,6}]octa-3,7-diene (**6**), cuneane (**7**), and 3,6-dihydropentalene (**8**) shown in Figure 6.9 were obtained; their energies relative to COT in a boat-like conformation are schematically plotted together with the calculated excitation energies of COT in Figure 6.10a, whereas in Figure 6.10b the corresponding data for *syn*-tricyclo[4.2.0.0^{2,5}] octa-3,7-diene (**9**) and cubane (**10**) are given. From these data it can be seen that the conical intersections for the formation of **7** and **8** are higher in energy than the Franck–Condon geometry of the S_1 state of COT; these photoreactions are

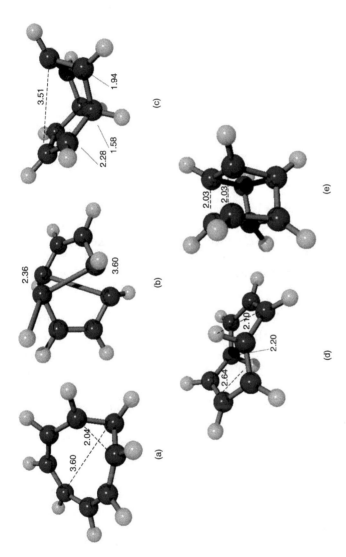

Figure 6.9 Conical intersections involved in the photochemistry of COT. Structures at the S_1–S_0 conical intersection as determined by semiempirical MNDOC-CI calculations (AS = 2 – 2', reference configurations Φ_0 and $\Phi_{1\rightarrow1'}$) (a) for the COT $\leftrightarrow$ semibullvalen conversion, (b) for the COT $\leftrightarrow$ tricyclo[3.3.0.0^{2,6}]octa-3,7-diene conversion, (c) for the COT $\leftrightarrow$ cuneane conversion, (d) for the COT $\leftrightarrow$ 3,6-dihydopentalene conversion, and (e) for the *syn*-tricyclo[4.2.0.0^{2,5}]octa-3,7-diene $\leftrightarrow$ cubane conversion.

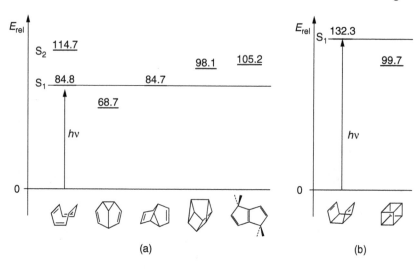

Figure 6.10   (a) Conical intersections involved in the photochemistry of COT. MNDOC-CI energies in kcal/mol relative to the ground-state energy of COT, as well as calculated excitation energies of COT. (b) The corresponding data for the hypothetical formation of cubane from *syn*-tricyclo[4.2.0.0^{2,5}]octa-3,7-diene.

therefore rather unlikely to proceed via S_1 excitation. From Figure 6.10b, however, the conical intersection for the cubane formation is seen to be much lower in energy than the Franck–Condon geometry of the S_1 state of **9**, even though this reaction cannot be observed.

In view of the many conical intersections on the S_1 surface of COT, it is important to verify that the calculated conical intersections really belong to the reactions under consideration. Therefore, geometry optimizations were performed at the geometries of each of the conical intersections as well as very slightly distorted geometries, which in agreement with the diagrams in Figure 6.5 and Figure 6.6 all led to the desired products and to COT, verifying that the ground-state reactions are barrierless in all cases considered. A more detailed discussion of the reaction mechanisms would require the search for barriers on the excited-state surface in order to prove the accessibility of the conical intersections. Thus, from the existence of a HOMO-LUMO crossing in the reaction of *syn*-tricylo[4,2,0,0^{2,5}]octa-3,7-diene to cubane because of the inversion of the natural orbital order,[30] the existence of an excited-state barrier on the reaction path from the Franck–Condon geometry toward the conical intersection can be inferred.

6.6 *AB INITIO* VERIFICATION OF SEMIEMPIRICAL RESULTS

Section 6.3 mentioned that, in spite of the qualitative agreement of the semiempirical results for the [1,7]-H shift reaction of CHT with all experimental observations, the high value of 23 kcal mol^{-1} calculated for the excited-state barrier suggested a redetermination on the basis of more reliable *ab initio* methods. Starting from MNDOC-CI structures, CASSCF-optimized structures are obtained that are, in general, in perfect agreement with the semiempirical results, the only exception being the S$_2$ minimum **2** because of a different ordering of states at this geometry. The energies of the various stationary points are compared with the MNDOC-CI values in Table 6.1. On the CASSCF level, the barrier is reduced to approximately 13 kcal mol^{-1}, whereas CASPT2 calculations at the CASSCF-optimized geometry yield a value of 3.8 kcal mol^{-1} for the excited-state barrier, in fair agreement with the photophysics observed for CHD and CHT (for details, see Reference 18).

An important verification of this mechanism comes from comparing the nonequivalent barriers in 1-substituted CHT. If CN-CHT and Me-CHT are chosen as examples with electron-accepting and electron-donating substituents, respectively, CASSCF optimization of the excited-state minima and of the transition structures for sigmatropic shifts away from the substituents and toward the substituent yields the results shown in Figure 6.11. In agreement with experimental observations, the smaller barrier is found for migration away from the acceptor and toward the donor substituent. Moreover, the differences in barrier height, which amount to 11.0 kcal mol^{-1} for CN and 4.0 kcal mol^{-1} for CH$_3$, are in excellent agreement with the experimentally observed product distribution. According to the Arrhenius equation, these values correspond to a distribution of 100:0 in the case of CN and 0.2:99.8 in the case of CH$_3$ for

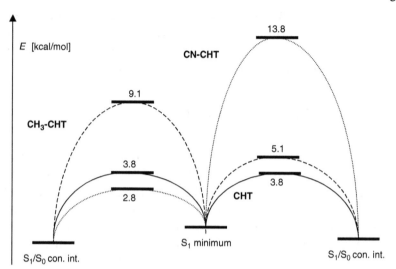

Figure 6.11 The [1,7]-H shift reaction in substituted CHTs. Schematic plot of the barrier heights E in kcal/mol from CASPT2D calculations for the hydrogen shift toward (right side) and away from (left sides) the substituent in Me-CHT (broken line) and CN-CHT (dotted lines). For CHT (solid lines) the two barriers are equivalent.

products of a sigmatropic shift away from and toward the substituent, as compared with the experimental values of 100:0 and 2:98.[17] This very good agreement confirms the importance of the excited-state barrier and confirms the mechanism for the [1,7]-sigmatropic shift derived from the calculated PESs shown in Figure 6.3.

As a second example of the *ab initio* verification of semiempirical results, we will consider the 2*H*-azirine ring cleavage to form nitrile ylide. The CASSCF results for the structure at the conical intersection shown in Figure 6.12a agree very well with those of the semiempirical MNDOC-CI calculation. As for Q_I and Q_O, the motion along the gradient difference vector x_1 (Figure 6.12b) and the nonadiabatic coupling vector x_2 (Figure 6.12c) correspond to an opening of the

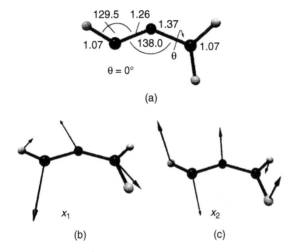

(a)

(b)   (c)

Figure 6.12 The S_1–S_0 conical intersection of 2*H*-azirine. (a) Geometry (distances in Å, angles in degrees), (b) gradient difference vector x_1, and (c) nonadiabatic coupling vector x_2.

C–N–C angle and a rotation of the CH_2 group, respectively; this is to be expected, because both sets of coordinates refer to the same physical entities. Using a 6-31G* basis set, the energy of the conical intersection relative to the ground-state energy of 2*H*-azirine is calculated to be 54 kcal/mol at the CASSCF level and 47 kcal/mol at the CASPT2 level, as compared with the value of 25.6 kcal/mol obtained from the semiempirical MR-CI calculation. This is to be compared with the vertical n → π^* excitation energies, which are calculated by the MNDOC-CI, CASSCF, and CASPT2 methods to be ΔE = 102.5, 138.7, and 118.4 kcal/mol, respectively.

In Figure 6.13, the S_0 and S_1 energies in the region of the conical intersection are plotted against x_1 and x_2, yielding a double cone. To establish possible reaction paths on the S_0 surface, geometry optimizations were carried out starting at geometries lying on a circle with radius 0.01 √amu·bohr around the conical intersection. The arrows in Figure 6.13 start at these geometries and point in the direction of steepest

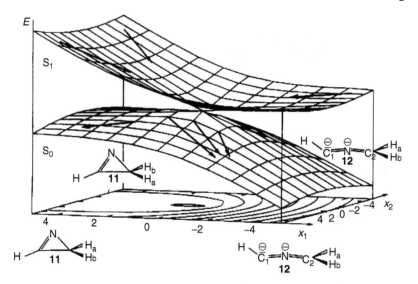

Figure 6.13   The S_1–S_0 conical intersection of 2*H*-azirine. S_1 and S_0 PESs calculated by the CASSCF method using the VDZ basis implemented in MOLPRO96[36] as a function of x_1 and x_2 in units of 10^{-2} √amu·bohr and contour diagram for S_0. The arrows point toward the reaction valleys that develop on the S_0 surface.

decent. They lead toward the two chemically equivalent nitrile ylides **12a** and **12a'**, which differ by a rotation of the CH$_2$ group by 180°. In addition, two reaction valleys leading to the reactant **11a** and, due to the symmetry of the problem, to its equivalent **11a'** with the CH$_2$ group rotated 180° develop at a distance of 0.02 √amu·bohr from the conical intersection.

The energetics of the S_1 photoreaction of 2*H*-azirine as well as the thermal ground-state reaction as obtained from CASPT2 calculations are summarized in Figure 6.14. From this diagram it can be concluded that the photolysis of 2*H*-azirine to form nitrile ylide occurs from the nπ*-excited S_1 state by way of an S_1–S_0 conical intersection. Because the reaction paths on the S_1 surface from the Franck–Condon geometry to the conical intersection as well as on the S_0 surface from the conical intersection to the nitrile ylide are

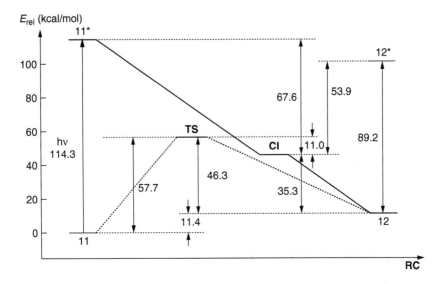

Figure 6.14 The C–C bond cleavage reaction of 2*H*-azirine. CASPT2 energies E_{rel} are plotted relative to the energy of the ground-state minimum to yield schematic reaction paths for the ground-state and the excited-state reaction.

barrierless, a fast deactivation is to be expected in agreement with experimental observation. The photochemical back reaction involves a barrier of 11 kcal/mol, which should be easily overcome in view of the high translational energy gained by relaxation of the excited nitrile ylide to the S_1–S_0 conical intersection, whereas the ground-state reaction is forbidden in both directions.

These examples show that semiempirical MR-CI calculations are very well suited to determine characteristic points on ground- as well as excited-state surfaces including conical intersections. They can therefore form the basis for a discussion of photochemical reaction mechanisms of organic molecules as well as the starting point for more sophisticated theoretical investigations. *Ab initio* calculations on the CASSCF and CASPT2 level are definitely necessary if reliable results for the energetics of the reaction under investigation are required.

6.7 TRIPLET PHOTOREACTIONS

Triplet biradicals or biradicaloids are frequently intermediates in organic photoreactions, particularly in reactions initiated from the $n\pi^*$ state of photoexcited carbonyl compounds.[29] In the course of the reaction the triplet species undergoes intersystem crossing (ISC) to the singlet state before proceeding to the final products. Because product formation on the ground-state surface S_0 is very fast and does not allow for major conformational change, the triplet-state geometry that is most favorable for ISC determines the ratio of the products in the overall reaction.

There are three mechanisms of spin flipping: solvent-induced spin relaxation (spin-lattice relaxation), SOC, and hyperfine coupling (HFC)[29] (see also Chapter 3). Although the first of these mechanisms is quite slow in the absence of paramagnetic impurities, HFC is important in biradicals in which the two radical centers are relatively far apart (1,6-biradicals and longer),[31] and SOC dominates in short biradicals, which are observed in numerous photochemical reactions. For these systems, the order of magnitude of SOC is about 0.1 to 5 cm^{-1}, which is much larger than that of the typical HFC, which is about 0.0001 cm^{-1}. We will therefore concentrate on the SOC mechanism only.

Two conditions have to be satisfied for the SOC mechanism to become efficient: the T_1–S_0 energy gap has to be small and SOC should be large.[32] Thus, estimating the overall ISC probability requires determination of the PES to locate the accessible areas of a small T_1–S_0 separation as well as evaluation of the SOC for each geometry of interest. The semiempirical MNDOC-CI method has been extended to evaluate SOC in the context of configuration interaction calculations based on Rumer spin eigenfunctions and the second quantization approach[12] such that PES and SOC surfaces may be determined at the same level of theory. Results obtained from such calculations are in good accord with available *ab initio* data, yielding. for example, a maximum value of SOC = 0.44 cm^{-1} for twisted ethylene as compared with the MCSCF value of SOC = 0.6 cm^{-1}.[33]

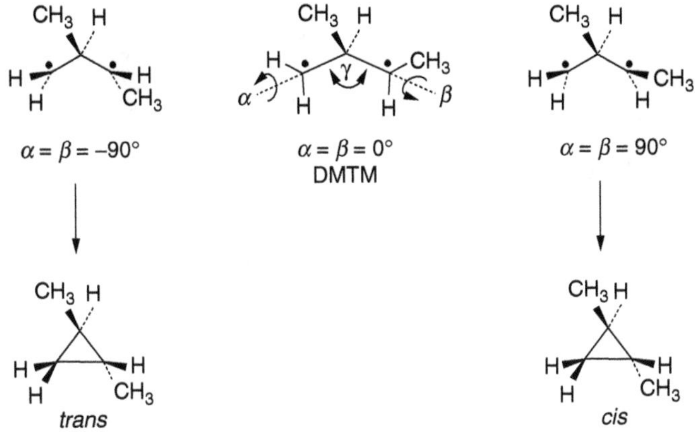

Scheme 6.3

As an example, the ring-closure reaction of 1,2-dimethyltrimethylene **15** (Scheme 6.3) will be considered. PES and SOC surfaces may be calculated for the variation of the C–C–C angle γ and the conrotatory ($\alpha = \beta$) and disrotatory ($\alpha = 180° - \beta$) motion of the terminal methylene groups. In **15** there are two modes of conrotatory motion ($\alpha = \beta$) of the radical centers, leading to stereoisomeric cyclization products: Rotation by positive ($\alpha > 0$) and negative ($\alpha < 0$) values of the rotational angle yields the *cis-* and *trans-*dimethylcyclopropane, respectively (see Scheme 6.3). S_0 and T_1 PESs together with SOC surfaces are shown for both modes of rotation in Figure 6.15. (Geometry taken from Reference 32 with idealized methyl groups; radical centers were rotated by α in steps of 15°, whereas the C–C–C angle γ was varied in steps of 5°. The CI wave functions include all CSFSs within a 3-3' active space.) For the *anti* mode of rotation, a minimum-energy valley is calculated for the triplet PES, which leads from the planar structure ($\alpha = 0°$) to the local minimum M_{trans} at $\alpha = -90°$ and corresponds to nearly free rotation of the radical centers; geometries along this valley are unfavorable for ISC because the S_0 state lies energetically above the T_1 state. However, in the region of large SOC, in which the terminal methylene groups are rotated by $\alpha > 45°$ toward a face-to-face

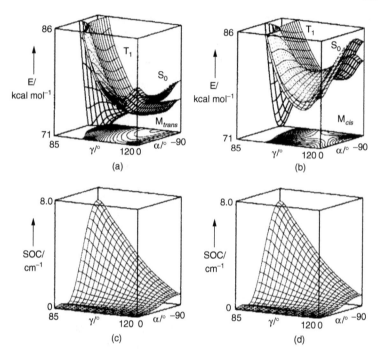

Figure 6.15 The ring closure reaction of 1,2-dimethyltrimethyl-ene. T_1 and S_0 PESs from MNDOC-CI calculations for conrotatory motion of the radical centers toward (a) the *syn* ($\alpha > 0°$) and (b) the *anti* conformation ($\alpha < 0°$), as well as SOC surfaces (c) for $\alpha > 0°$ and (d) for $\alpha < 0°$.

arrangement (Scheme 6.3), decreasing the C–C–C valence angle to a value of $\gamma = 105°$ raises the energy only slightly (1 to 2 kcal/mol) and leads to singlet-triplet degeneracy ($E_{ST} = 0$) and therefore to geometries very favorable to ISC. For the *syn* mode of rotation, a barrier of approximately 3.4 kcal/mol separates the local minimum M_{cis} at $\alpha = 90°$, which is 5.0 kcal higher in energy than M_{trans}, from the planar structure ($\alpha = 0°$) that is 1.0 kcal mol^{-1} lower in energy than the barrier. This is particularly apparent from the contour diagrams of the triplet PE and SOC surfaces shown in Figure 6.16, where the T_1–S_0 intersection E_{TS} is indicated by heavy lines. For comparison, the corresponding results for the unsubstituted trimethylene are also shown.

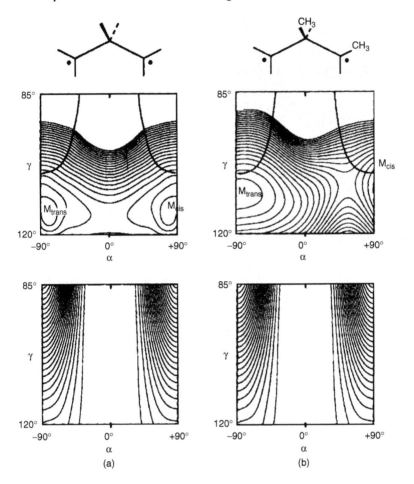

Figure 6.16 The ring-closure reaction of 1,2-dimethyltrimethylene. Contour diagrams of the T_1 surface and the SOC surface (a) of trimethylene and (b) of its 1,2-dimethyl derivative. Triplet-singlet intersection $E_{ST} = 0$ is indicated by heavy lines.

From these results it can be concluded that the reactive structure of optimal ISC is characterized by a face-to-face orientation of the radical centers and a C–C–C angle γ slightly smaller than for the triplet minima. As the singlet PES drops steeply toward the cyclopropane minimum for small values of γ, the triplet reaction yields preferably cyclic products. The conditions for optimal ISC are similar for both minima M_{cis} and M_{trans}; therefore, the stereochemical differentiation between *cis*- and *trans*-substituted products is due, in this case, to the energy difference of the two minima M_{cis} and M_{trans}. This explains the experimental observation that the triplet-sensitized photoreaction of *cis*- as well as *trans*-3,4-dimethyl-1-pyrazoline yields preferably *trans*-1,2-dimethylcyclopropane, and negligible amounts of acyclic products, in contrast to the singlet photoreaction, which occurs preferably with retention of configuration and yields appreciable amounts of acyclic products.[34]

6.8 CONCLUSIONS

Semiempirical MR-CI calculations based on analytic gradients represent a very powerful and efficient tool for the determination of excited-state PESs and in combination with the qualitative Zilberg-Haas method,[5] for locating conical intersections for larger organic molecules. One is, therefore, not restricted to discuss model systems, and systematic studies of structural variations and substituent effects on organic photoreactions are well within reach. Such studies can be very useful for exploratory work on organic photoreactions, as all essential features of the excited-state PESs are represented qualitatively correctly on the semiempirical level of theory, and relative energies, in particularly geometries at the stationary points, are reproduced quite well. To obtain absolute values for excitation energies or barrier heights, however, sophisticated *ab initio* calculations are required, for which the semiempirical results represent a good starting point. Typically, excitation energies from semiempirical MR-CI calculations are too low, whereas those from *ab initio* CASSCF calculations are too high; reliable values require CASPT2 calculations, as may be seen from the data collected in Table 6.1. Also, the reduction of the excited-state barrier for the [1,7]-H shift in cycloheptatriene from a semiempirical value of 23.3 kcal/mol to a CASSCF value of 13.6 kcal/mol and a CASPT2 value of 3.8 kcal/mol discussed in Section 6.6 appears to be quite representative.

Finally, semiempirical MR-CI calculations based on analytic gradients are sufficiently fast for "on-the-fly" determinations of excited-state trajectories, which are essential for more detailed information on the mechanisms of organic photoreactions. Because the results of such studies depend on barrier heights, semiempirical methods will not reproduce quantitative aspects correctly, but studies based on suitable model potentials as well as systematic studies of structural variations and substituent effects might be very valuable in getting deeper insight into the details of organic photoreactions.

APPENDIX

The MNDOC-CI program[11] and some representative input files are available on request from the author. The *ab initio* CASSCF calculations for 2*H*-azirine[18] reported in Section 6.6 were performed using the GAUSSIAN98 program.[14] The active space was chosen to contain the nitrogen lone-pair orbital σ_N and the bonding and antibonding π_{CN} and σ_{CC} orbitals; thus, an AS of six electrons in five orbitals results, which may be denoted in terms of the symmetry labels a' and a" as AS 2,2' (occ. 10,3; closed 7,1). CASPT2 calculations based on a state-averaged CASSCF reference function and the same AS were carried out at the CASSCF optimized stationary points using the VDZ basis and the methods implemented in MOLPRO96.[35] For the CASSCF calculations on CHT an active space of eight MOs that were primarily composed of carbon p_π AOs and the s AO of the migrating hydrogen. CASPT2 results were obtained with the MOLCAS4.0 program[36] and the ANO basis set (C: 3s2p1d, H: 2s1p). Starting geometries were always taken from the MNDOC-CI results.

REFERENCES

1. Thiel, W., *J. Mol. Struct.*, 398, 1–6, 1997.

2. Bernardi, F., Olivucci, M., and Robb, M.A., *Chem. Soc. Rev.*, 321–328, 1996.

3. Izzo, R. and Klessinger, M., *J. Comp. Chem.*, 21, 52–62, 2000.

4. Klessinger, M., *Angew. Chem. Int. Ed. Engl.,* 34, 549–551, 1995.

5. Zilberg, S. and Haas, Y., *Chem. Eur. J.,* 5, 1755–1765, 1999.

6. Foresman, J.B., Head-Gordon, M., Pople, J.A., and Frisch, M.J., *J. Phys. Chem.,* 96, 135–149, 1992.

7. Shepard, R., *Adv. Chem. Phys.,* 69, 63–200, 1987.

8. Thiel, W., *J. Am. Chem. Soc.,* 103, 1413-1420, 1981; Thiel, W., *J. Am. Chem. Soc.,* 103, 1420–1425, 1981; Schweig, A. and Thiel, W., *J. Am. Chem. Soc.,* 103, 1425–1431, 1981.

9. Klessinger, M., Pötter, T., and van Wüllen, C., *J. Theor. Chim. Acta,* 80, 1–7, 1991.

10. Bearpark, M.J., Robb, M.A., and Schlegel, H.B., *Chem. Phys. Lett.,* 223, 269–274, 1994.

11. Izzo, R. and Klessinger, M., *J. Comp. Chem.,* 21, 52–62, 2000.

12. Böckmann, M., Klessinger, M., and Zerner, M.C., *J. Phys. Chem.,* 100, 10570–10579, 1996.

13. Roos, B.O., *Adv. Chem. Phys.,* 69, 399–445, 1987.

14. Frisch, M.J., Trucks, G.W., Schlegel, H.G., Scuseria, G.E., Robb, M.A., Cheeseman, J.R., Zakrzewski, V.G., Montgomery, J.A., Jr., Stratman, R.E., Burant, J.C., Dapprich, S., Millam, J.M., Daniels, A.D., Kudin, K.N., Strain, M.C., Farkas, O., Tomasi, J., Barone, V., Cossi, M., Cammi, R., Menucci, B., Pomelli, C., Adamo, C., Clifford, S., Ochterski, J., Peterson, G.A., Ayala, P.Y., Cui, Q., Morokuma, K., Malick, D.K., Rabuvk, A.D., Ragavachari, K., Foresman, J.B., Ciolowski, J., Ortiz, J.V., Stefanov, B.B., Liu, G., Liashenko, A., Piskorz, B., Komaromi, I., Gomperts, R., Martin, R.L., Fox, D.J., Keith, T., Al-Laham, M.A., Peng, C.Y., Nanayakkara, A., Gonzales, C., Challacombe, M., Gill, P.M.W., Johnson, B., Chen, W., Wong, M.W., Andres, J.L., Head-Gordon, M., Replogle, E.S., and Pople, J.A., *GAUSSIAN98 (Revision A.7),* Gaussian, Inc., Pittsburgh, 1998.

15. Andersson, K., Malmquist, P.-A., and Roos, B.O., *J. Phys. Chem.,* 94, 5483–5488, 1990; Andersson, K., Malmquist, P.-A., and Roos, B.O., *J. Phys. Chem.,* 96, 1218–1226, 1992.

16. Dauben, W.G. and Cargill, B., *Tetrahedron,* 12, 186–189, 1961.

17. TerBorg, A.P., Razenberg, E., and Kloosterziel, H., *J. Chem. Soc. Commun.,* 23, 1210, 1967.

18. Steuhl, H.-M., Bornemann, C., and Klessinger, M., *Chem. Eur. J.*, 5, 2404–2412, 1999.

19. Reid, P.J., Wickham, S.D., and Mathies, R.A., *J. Phys. Chem.*, 96, 5720–5724, 1992.

20. Trushin, S.A., Diemer, S., Fu, W., Kompa, K.L., and Schmid, W.E., *Phys. Chem. Chem. Phys.*, 1, 1431–1440, 1999.

21. Reid, P.J., Shreve, A.P., and Mathies, R.A., *J. Phys. Chem.*, 97, 12691–12699, 1993.

22. Longuet-Higgins, H.C., *Proc. R. Soc. London A*, 344,147–156, 1975.

23. Zilberg, S. and Haas, Y., *Chem. Phys.*, 259, 249–261, 2000.

24. Haas, Y. and Zilberg, S., *J. Photochem. Photobiol. A*, 144, 221–228, 2001.

25. Zilberg, S. and Haas, Y., *Phys. Chem. Chem. Phys. 4, 34–42,* 2002.

26. Heimgartner, H., *Angew. Chem. Int. Ed. Engl.*, 30, 238–265, 1991.

27. Bigot, B., Sevin, A., and Devaquet, A., *J. Am. Chem. Soc.*, 6924–6929, 1978.

28. Albrecht, E., PhD dissertation, Westfäl, Wilhelms-Uiversität, Münster, 1995.

29. Klessinger, M. and Michl, J., *Excited States and Photochemistry of Organic Molecules,* VCH, New York, 1995.

30. Gleiter. R., *Angew. Chem. Int. Ed. Engl.*, 27, 27–44, 1992.

31. Closs, G.L., Forbes, M.D.E., and Piotrowak, P., *J. Am. Chem., Soc.,* 114, 3285–3294, 1992.

32. McGlynn, S.P., Azumi, T., and Kinoshita, M., *Molecular Spectroscopy of the Triplet State,* Prentice-Hall, Englewood Cliffs, NJ, 1969.

33. Caldwell, R.A., Carlacci, L., Doubleday, C.E., Furlani, T.R., King, H.F., and McIver, J.W., *J. Am. Chem. Soc.*, 110, 6901–6903, 1988.

34. Moore, R., Mishra, A., and Crawford, R.J., *Can. J. Chem.*, 46, 3305–3313, 1968.

35. Werner, H.-J., Knowles, P.J., Almlöf, J., Amos, R.D., Berning, A., Deegan, M.J.O., Eckert, F., Elbert, S.T., Hampel, C., Lindh, R., Meyer, W., Nicklas, A., Peterson, K., Piker, R., Sonte, A.J., Taylor, P.R., Mura, M.A., Pulay, P., Schuetz, M., Stoll, H., Thornsteinsson, T., and Cooper, D.L., *MOLPRO (Version 96.4)*, Universität Stuttgart, 1996.

36. Andersson, K., Blomberg, M.R.A., Fülscher, M.P., Karlström, G., Lindh, R., Malmqvist, P.-A., Neogrády, P., Olsen, J., Roos, B.O., Sadley, A.J., Schütz, M., Seijo, L., Serrano-Andres, L., Siegbahn, P.E.M., and Widmark, P.-O., *MOLCAS (Version 4.0)*, Department of Theoretical Chemistry, University of Lund, 1997.

7

Natural Bond Orbital Analysis of Photochemical Excitation, With Illustrative Applications to Vinoxy Radical

F. WEINHOLD

Theoretical Chemistry Institute and
Department of Chemistry,
University of Wisconsin,
Madison, WI 53706

CONTENTS

7.1 INTRODUCTION

The natural bond obital (NBO) method[1] was developed as a bridge between the theorist's abstract mathematical wave function ψ and the localized bonding concepts of experimental chemists. Such a bridge rests on two anchor points: one in the deeply arcane *ab initio* domain of Schrödinger's equation, the other in the equally arcane domain of structural and mechanistic chemistry. These domains naturally evolved with disparate methodologies and languages. Growing awareness that ψ encompasses "the whole of chemistry"[2] stimulated the search for improved lines of connection and communication to exploit this deeply mysterious commonality. NBO analysis is a general method for optimally expressing ψ in the language of structural chemists.

Although NBO analysis has proven quite successful for usual closed-shell species, one may question whether its utility and generality extend to species beyond the limits of ordinary chemical stability. Among the most challenging in this respect are the open-shell radical and photochemical excited species, whose instability and structural floppiness often appear to defy characterization in terms of ground-state valence concepts. The goal of this paper is to survey current *ab initio* calculational and NBO analysis techniques for photochemical species, with particular reference to low-lying states of the benchmark vinoxy radical (CH_2CHO).

Throughout this paper we seek to illustrate the many similarities between ground-state and excited-state NBO descriptors, as well as their characteristic differences. We believe these illustrations justify considerable optimism that key features of excited-state potential-energy surfaces can be fruitfully described in terms of NBO donor-acceptor concepts that strongly parallel those for ground states. In this manner we hope to expand awareness of available NBO-based techniques for excited states, encouraging intrepid spectroscopic investigators to explore and extend applications of NBO donor-acceptor concepts to frontier areas of molecular and supramolecular photochemistry.

7.2 OVERVIEW OF NBO LEWIS STRUCTURE CONCEPTS FOR CLOSED SHELLS

7.2.1 Lewis-Type NBOs of Carbonyl Chromophores

The NBO paradigm (as currently implemented[3]) can be described as an optimal *ab initio* realization of the chemist's Lewis structure concept. For stable closed-shell species, the familiar Lewis structure diagram depicts an idealized pattern of valence electron pair assignments to one-center (1-c) non-bonding and two-center (2-c) bonding regions of the molecule. Such localized electronic patterns can be reexpressed in the language of quantum mechanics by writing the total N-electron wave function ψ as the antisymmetrized product of doubly occupied valence bond–like orbitals (determined by the NBO variational procedure) that are localized in atomic or diatomic regions. For example, the Lewis structure diagram of the prototype carbonyl chromophore compound, formaldehyde (**1**),

corresponds to an approximate Lewis-type wave function ψ_L that can be represented schematically as

$$\begin{array}{c} \text{H} \\ \diagdown \\ \quad\quad \text{C}=\ddot{\text{O}}: \\ \diagup \\ \text{H} \end{array}$$

1

$$\psi_L = \hat{A}\left\{\left(c_C\right)^2 \left(c_O\right)^2 \left(\sigma_{CH}\right)^2 \left(\sigma_{CH'}\right)^2 \left(\sigma_{CO}\right)^2 \left(\pi_{CO}\right)^2 \left(n_O'\right)^2 \left(n_O''\right)^2\right\} \quad (7.1)$$

with familiar electron configuration symbols (c = core, n = valence nonbonding, σ_{AB} = sigma bond, π_{AB} = pi bond) and $\hat{A}$ the antisymmetrizer. Expressed somewhat heuristically, the NBO procedure corresponds to choosing *optimal* 1-c and 2-c orbitals in Equation 7.1, such that the deviation between ψ_L and the true ψ (or its best available approximation) is the smallest possible. Intrinsically, such natural bond orbitals are variationally chosen by ψ itself (effectively as ψ's local *eigen*orbitals) to give the best possible Lewis-like description.

NBOs are closely tied to underlying natural atomic orbitals (NAOs) which, in turn, form the basis of natural population analysis (NPA),[4] a widely used method for assigning atomic charges and orbital populations. The NAO-based natural atomic charges exhibit many conceptual and numerical advantages compared with older Mulliken charge and orbital population assignments and are now widely recommended for general usage.[5] The forms and occupancies of NBOs (as well as natural resonance theory weightings to be described in Section 7.2.3) are perforce fully consistent with NPA atomic charge assignments. By construction, the complete orthonormal set of NAOs are variationally optimal effective atomic orbitals in the molecular environment, ordered in occupancy to ensure the most rapid possible expansion of the electron density. NBOs inherit the optimal character of constituent NAOs and extend the compact intrinsic description of electron density from the atomic to the molecular Lewis structure level. Further details of the NAO and NBO variational algorithms are described in Reference 3a and elsewhere,[6] but for practical purposes it is necessary only to know that these algorithms are incorporated in leading electronic structure systems (Gaussian 9X, Jaguar, NWChem, Q-Chem, GAMESS, PQS, ADF, and others)[7] that are routinely used by organic chemists.

Figure 7.1 presents contour diagrams for the Lewis-type NBOs of formaldehyde (panels a, c, e, g, h), calculated at the B3LYP/6-311++G** level[8] of hybrid density functional theory (DFT). As is immediately apparent, the bonding σ_{CH}, σ_{CO}, and π_{CO} NBOs exhibit the familiar sigma and pi shapes of idealized textbook depictions. The resemblance to expected textbook forms is further emphasized when each A–B bond is written as a linear combination (LC) of its constituent natural hybrid orbitals (NHOs) h_A and h_B with polarization coefficients c_A and c_B, namely,

$$\sigma_{AB} = c_A h_A + c_B h_B . \qquad (7.2)$$

The upper entries in Table 7.1 summarize details of the LC-NHO expansions for valence Lewis-type NBOs of formaldehyde at the B3LYP/6-311++G** level. The carbon and oxygen hybrids of σ_{CO} and σ_{CH} have the expected sp^2-like composition, and the polarization coefficients vary in the expected way with elec-

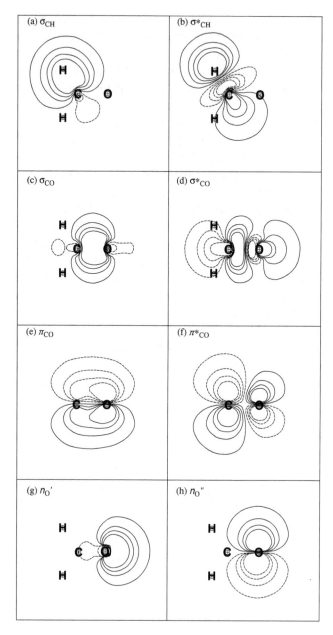

Figure 7.1 NBOView two-dimensional contour images of formaldehyde valence-shell NBOs of Lewis type (a,c,e,g,h) and non-Lewis type (b,d,f).

TABLE 7.1 Occupancies and Compositions of Lewis-Type NBOs of Formaldehyde at Various Theoretical Levels[a]

NBO	Method	Occ. (e)	$c_A h_A + c_B h_B$
σ_{CH}	B3LYP	1.9931	$0.7535(sp^{2.01})_C + 0.6574(s)_H$
	HF	1.9933	$0.7466(sp^{2.02})_C + 0.6653(s)_H$
	CISD	1.9667	$0.7473(sp^{2.02})_C + 0.6645(s)_H$
	MP2	1.9674	$0.7485(sp^{2.01})_C + 0.6631(s)_H$
	CCD	1.9585	$0.7476(sp^{2.02})_C + 0.6642(s)_H$
σ_{CO}	B3LYP	1.9997	$0.5865(sp^{1.91})_C + 0.8099(sp^{1.45})_O$
	HF	1.9998	$0.5843(sp^{1.88})_C + 0.8155(sp^{1.39})_O$
	CISD	1.9784	$0.5837(sp^{1.89})_C + 0.8120(sp^{1.45})_O$
	MP2	1.9730	$0.5829(sp^{1.90})_C + 0.8125(sp^{1.49})_O$
	CCD	1.9741	$0.5829(sp^{1.89})_C + 0.8125(sp^{1.45})_O$
π_{CO}	B3LYP	1.9999	$0.5871(p)_C + 0.8095(p)_O$
	HF	2.0000	$0.5655(p)_C + 0.8247(p)_O$
	CISD	1.9474	$0.5845(p)_C + 0.8114(p)_O$
	MP2	1.9381	$0.5890(p)_C + 0.8082(p)_O$
	CCD	1.9365	$0.5922(p)_C + 0.8058(p)_O$
n_O'	B3LYP	1.9891	$(sp^{0.68})_O$
	HF	1.9891	$(sp^{0.71})_O$
	CISD	1.9692	$(sp^{0.68})_O$
	MP2	1.9648	$(sp^{0.66})_O$
	CCD	1.9651	$(sp^{0.68})_O$
n_O''	B3LYP	1.8781	$(p)_O$
	HF	1.9051	$(p)_O$
	CISD	1.8705	$(p)_O$
	MP2	1.8510	$(p)_O$
	CCD	1.8656	$(p)_O$

[a] The calculated total energy (a.u.) and percentage accuracy of Natural Lewis Structure description (% ρ_L, in parentheses) for each method are B3LYP:–114.541848 (99.08%); HF:–113.901670 (99.25%); CISD:–114.227808 (98.11%); MP2:–114.241580 (97.88%); CCD:–114.241392 (97.86). All calculations were performed with the 6-311++G** basis set at optimized B3LYP geometry.

tronegativity. (The polarity of the σ_{CO} bond is more apparent in the polarization coefficients of Table 7.1 than in the graphical contours of Figure 7.1c.) The oxygen lone pairs correspond to the expected low-energy sigma-type (end-on, s-rich) n_O' and high-energy pi-type (p-rich) n_O'', with the latter being the active p_y orbital of spectroscopic $n \rightarrow \pi^*$ carbonyl excitation.[9]

The optimized NHOs reflect subtle variations of geometry and electronegativity differences, in the manner anticipated by Bent's Rule.[10] Each directed hybrid $h_A = (sp^\lambda)_A$ is oriented with respect to other hybrids $h_A' = (sp^\lambda)_A$ on the same center to give interhybrid angles θ determined by[11]

$$\cos \theta = - (\lambda \lambda')^{-1/2}. \qquad (7.3)$$

In acyclic molecules the bonded atoms are expected to align accurately with the directional hybrids, but appreciable "bond-bending" deviations $\Delta\theta$ between the NHO direction and the line of nuclear centers become apparent in the case of steric or topological constraints (e.g., ring closing) or vicinal hyperconjugative influences (see Section 7.2.3). The NBO program routinely tabulates such bond-bending deviations for each NHO. Although fairly negligible in acyclic equilibrium species, the bond-bending deviations become important in the vertical geometry of photoexcited species. The tabulated NHO $\Delta\theta$ values then provide clues to the expected angular deformations as the nascent photospecies relaxes toward a new equilibrium geometry. These expected geometry changes lead to Franck–Condon factors that govern the intensity of electronic excitation.

It is important to recognize that the freshman chemistry-like NBOs of Figure 7.1 and Table 7.1 are obtained with no special assumptions or restrictions on the form of ψ. In fact, nearly identical NBOs are obtained for virtually any type of variational Hartree–Fock (HF), configuration interaction (CI), Møller–Plesset (MP) perturbation theory, or coupled cluster (CC) treatment, as illustrated in the remaining entries of Table 7.1. The table footnote displays the high percentage (%-ρ_L) of the total electron density that is described by the idealized Lewis-like wave function at all levels. From the correlated CISD, MP2, and CCD results, one can see that electron correlation (post-HF) effects make perceptible corrections to the forms and occupancies of NBOs, but these are insignificant compared to the broad pattern of agreement among all approximation methods. Essential features of the NBO Lewis structure description are established even at rather low levels of theory, and the convergence toward limiting true NBOs is usually insensitive to details of the numerical approximation methods, including choice of basis functions.

NBOs also exhibit the expected *transferability*[12] of localized chemical bonds from one molecular environment to

TABLE 7.2 Occupancies and Compositions of Acyl Group Lewis-Type NBOs of Acetaldehyde **2** at the B3LYP/6-311++G** Level (cf. formaldehyde NBOs of Table 7.1)

NBO	Occ. (e)	$c_A h_A + c_B h_B$
σ_{CH}	1.9882	$0.7524(sp^{2.34})_C + 0.6587(s)_H$
σ_{CO}	1.9981	$0.5873(sp^{2.04})_C + 0.8093(sp^{1.40})_O$
π_{CO}	1.9914	$0.5775(p)_C + 0.8164(p)_O$
n'_O	1.9838	$(sp^{0.71})_O$
n''_O	1.8813	$(p)_O$

another. Table 7.2 illustrates this transferability of acyl group NBOs to acetaldehyde (**2**).

Comparison with the corresponding B3LYP entries of Table 7.1 shows that the NBOs of the acyl moiety are highly conserved under methyl substitution, leading to orbital images such as Figure 7.1 that would be scarcely distinguishable to the eye. This high recognizability and transferability of NBOs is in strong contrast to canonical molecular orbitals (MOs), whose forms often vary bewilderingly, even in closely related systems.

Although the primary emphasis of this chapter is on NBO analysis of existing wave functions, it is important to realize that these orbitals may also be useful in constructing new wave functions for photochemical phenomena. An important example is the CASNBO method,[13] which uses NBOs as starting orbitals in multiconfigurational CAS (complete active space) wave functions. The CASNBO method allows the active space to be defined in the most specific and physically relevant way for a given chemical application (such as, e.g., the two-dimensional σ_{AB} and σ^*_{AB} space for a localized A–B bond-break-

ing process). Moreover, in certain cases the CASNBO procedure leads to new self-consistent solutions of the CAS-SCF
(self-consistent field) equations, inaccessible to MO-based procedures. The CASNBO procedure also seems to exhibit consistently improved numerical convergence characteristics,
compared with conventional MO-based methods. Some illustrative CASNBO applications to photochemical excited states
will be presented in the following sections.

7.2.2 Non-Lewis Delocalization Corrections

Delocalization effects are represented in NBO theory by partial occupancy of non–Lewis-type NBOs, corresponding to
departures from a perfectly localized Lewis structure description. Non-Lewis NBOs include the valence *antibonds*, that is,
the out-of-phase combinations σ^*_{AB} that accompany each in-
phase σ_{AB} combination,

$$\sigma^*_{AB} = c_B h_A - c_A h_B \qquad (7.4)$$

as well as extra-valence *Rydberg* orbitals r^* that complete the
span of the NBO basis. Quantitative effects of electronic delocalization between Lewis and non-Lewis NBOs can be estimated by perturbative or variational means.

Simple second-order perturbation theory is well adapted
to describe the energy lowering and occupancy shifts associated
with delocalizing interactions between specific Lewis and non-
Lewis NBOs. The unperturbed wave function ψ_L corresponds
to the perfectly localized Lewis structure limit, with all Lewis-
type (electron donor) NBOs fully occupied and all non-Lewis-
type (electron acceptor) NBOs completely vacant. The perturbative interaction of donor NBO σ_{AB} with acceptor NBO
σ^*_{CD} leads to the approximate second-order energy lowering

$$\Delta E^{(2)}_{\sigma \to \sigma^*} = -n_{occ} \frac{\left\langle \sigma_{AB} \left| \hat{\mathcal{F}} \right| \sigma^*_{CD} \right\rangle^2}{\epsilon_{\sigma^*} - \epsilon_\sigma} \qquad (7.5)$$

where n_{occ} is the unperturbed donor occupancy (two for closed-
shell), $\epsilon_\sigma = \left\langle \sigma_{AB} \left| \hat{\mathcal{F}} \right| \sigma_{AB} \right\rangle$ and $\epsilon_{\sigma^*} = \left\langle \sigma^*_{CD} \left| \hat{\mathcal{F}} \right| \sigma^*_{CD} \right\rangle$ are the respective

donor and acceptor orbital energies, and $\hat{\mathcal{F}}$ is the effective one-

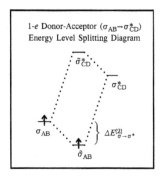

Figure 7.2 Schematic perturbation diagram for one-electron donor-acceptor interaction between an occupied (σ_{AB}) and vacant (σ_{CD}^*) orbital, leading to energy lowering $\Delta E_{\sigma \to \sigma^*}^{(2)}$; (cf. Equation 7.5 and Equation 7.6).

electron Hamiltonian (e.g., Fock or Kohn–Sham operator). The unperturbed donor σ_{AB} and acceptor σ_{CD}^* are correspondingly mixed to give the semi-localized perturbed orbitals that can be written as

$$\tilde{\sigma}_{AB} = \left(1 + \lambda^2\right)^{-1/2} \left[\sigma_{AB} + \lambda \sigma_{CD}^*\right] \tag{7.6a}$$

$$\tilde{\sigma}_{CD}^* = \left(1 + \lambda^2\right)^{-1/2} \left[\sigma_{CD}^* - \lambda \sigma_{AB}\right] \tag{7.6b}$$

where

$$\lambda = \frac{\left\langle \sigma_{AB} \left| \hat{\mathscr{F}} \right| \sigma_{CD}^* \right\rangle}{\epsilon_{\sigma^*} - \epsilon_{\sigma}} \tag{7.7}$$

is the first-order perturbative mixing coefficient. Figure 7.2 summarizes these relationships in the form of a familiar second-order perturbation theory interaction diagram (for $n_{occ} = 1$). The perturbative mixing in Equation 7.6a,b corresponds to an occupancy shift

$$\Delta q_{\sigma \to \sigma^*} = n_{occ} \left(\frac{\lambda^2}{1 + \lambda^2}\right) \tag{7.8}$$

from σ_{AB} to σ^*_{CD}. For $\lambda \to 0$ (the weak delocalization limit), these equations allow the energy lowering $\Delta E^{(2)}_{\sigma \to \sigma^*}$ and occupancy shift $\Delta q_{\sigma \to \sigma^*}$ to be approximately related by

$$\Delta E^{(2)}_{\sigma \to \sigma^*} = \left(\epsilon_{\sigma^*} - \epsilon_\sigma \right) \Delta q_{\sigma \to \sigma^*} \qquad (7.9)$$

The NBO program routinely tabulates the estimated second-order interaction energies $\Delta E^{(2)}_{\sigma \to \sigma^*}$ for all possible NBO donor-acceptor delocalizations exceeding a chosen energetic threshold.

The NBO program also permits an alternative quasi-variational deletions ($DEL keylist) method to evaluate the effects of specific NBO donor-acceptor interactions. In effect, one can choose to delete any particular interaction (or a particular non-Lewis orbital, etc.) and recalculate the variational energy and other properties as though these interactions or orbitals were absent in nature. In this approach, removal of a particular $\sigma_{AB} \to \sigma^*_{CD}$ delocalizing interaction raises the variational energy and rearranges the electronic distribution (and, therefore, the molecular geometry) to more localized form. Reoptimization of molecular geometry in the absence of a given $\sigma_{AB} \to \sigma^*_{CD}$ delocalization often allows its unique structural and energetic consequences to be clearly identified. In this manner one can isolate the NBO interaction that is the "smoking gun" responsible for a particular structural anomaly of interest. An example of this technique will be presented in Section 7.4.4.

Still another measure of electron delocalization can be seen in the forms of natural localized molecular orbitals (NLMOs).[14] By construction, each NLMO remains as close as possible to a parent NBO but includes the weak delocalization "tail" necessary to preserve exact double occupancy. A determinant of doubly occupied NLMOs is therefore unitarily equivalent to the standard MO determinant, and the Lewis-type NLMOs are able to describe all observable properties of the system just as well as canonical MOs. The perturbative expression (Equation 7.6a) approximates the form of the Lewis-type NLMO $\tilde{\sigma}_{AB}$ with parent NBO σ_{AB} and weak delocalization into σ^*_{CD}, and Equation 7.6b is the corresponding approximation for the non-Lewis NLMO $\tilde{\sigma}^*_{CD}$.

In ground-state acetaldehyde (*2*), the principal delocalizations involve vicinal hyperconjugation from the *p*-type oxygen lone pair n_O'' into the adjacent σ^*_{CH} and σ^*_{CC} antibonds

$$\Delta E^{(2)}\left(n_O'' \to \sigma^*_{CH}\right) = -23.43 \text{ kcal / mol}$$

$$\Delta E^{(2)}\left(n_O'' \to \sigma^*_{CC}\right) = -18.32 \text{ kcal / mol}$$

as illustrated in Figures 7.3a,b. Such hyperconjugation with the in-plane n_O'' has the well-known effect of stereoelectronically activating the acyl C–H and C–C bonds (the latter slightly less than the former) compared with analogous alkenes. The presence of such hyperconjugation is also shown in the form of the $\tilde{n}_O''$ NLMO

$$\tilde{n}_O'' \simeq 0.9696 n_O'' + 0.1778 \sigma^*_{CH} - 0.1475 \sigma^*_{CC} + \cdots$$

with relative delocalization tails in accord with the second-order estimates noted above. Weaker hyperconjugation is also detected between the two out-of-plane $\sigma_{CH''}$ bonds (donors) and the low-lying π^*_{CO} (acceptor), as well as a weaker delocalization in the opposite direction (Figures 3c,d).

$$\Delta E^{(2)}\left(\sigma_{CH''} \to \pi^*_{CO}\right) = -5.05 \text{ kcal / mol}$$

$$\Delta E^{(2)}\left(\pi_{CO} \to \sigma^*_{CH''}\right) = -1.58 \text{ kcal / mol}$$

The latter interactions are also evident in the slightly asymmetric form of the methyl group, reflecting slight activation of the two C–H bonds (0.006Å elongation, 1.6 smaller C–C–H bond angle) compared with the in-plane C–H bond. (Deletion of the $\sigma_{CH''} \to \pi^*_{CO}$ and $\pi_{CO} \to \sigma^*_{CH''}$ interactions and reoptimization of the CH_3CHO geometry would also demonstrate the strong hyperconjugative effect on the methyl asymmetry.) Such hyperconjugative details are easily recognized in the localized NBO or NLMO basis sets but become rather

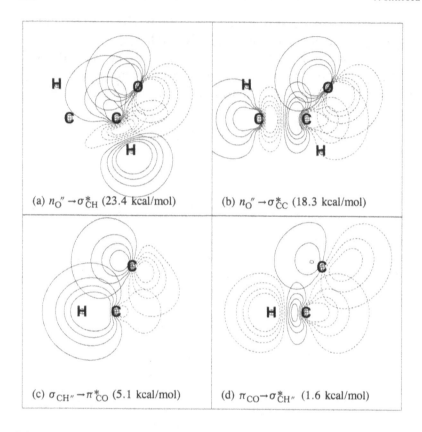

(a) $n_O'' \to \sigma^*_{CH}$ (23.4 kcal/mol)

(b) $n_O'' \to \sigma^*_{CC}$ (18.3 kcal/mol)

(c) $\sigma_{CH''} \to \pi^*_{CO}$ (5.1 kcal/mol)

(d) $\pi_{CO} \to \sigma^*_{CH''}$ (1.6 kcal/mol)

Figure 7.3 Leading NBO interactions of acetaldehyde: (a) $n_O'' \to \sigma^*_{CH}$, (b) $n_O'' \to \sigma^*_{CC}$, (c) $\sigma_{CH''} \to \pi^*_{CO}$, (d) $\pi_{CO} \to \sigma^*_{CH''}$. (c) and (d) are plotted in the contour plane of the out-of-plane C–H bond and carbonyl C, so only the nearside edge of the π_{CO} or π^*_{CO} NBO appears in the diagram.

obscured in a delocalized MO basis because of superfluous unitary transformations (symmetry adaptation and near-degeneracy mixing) that have no physical effect on the wave function or observable properties of the molecule.

7.2.3 Natural Resonance Theory and Bond Orders[15]

In cases of strong resonance delocalization, there is no longer a single dominant Lewis structure. Instead, the properties $\langle p \rangle$ of the system may be pictured as a weighted average (resonance hybrid) of two or more contributing resonance structures r, with respective properties $\langle p(r) \rangle$,

$$\langle p \rangle = \sum_r w_r \langle p^{(r)} \rangle \qquad (7.10)$$

where the $\{w_r\}$ are nonnegative weightings summing to unity. For example, an amide group cannot be described simply in terms of **3a** alone, but instead appears as a resonance hybrid of **3a,b**,

with roughly $w_{3a} \simeq 0.6$, and $w_{3b} \simeq 0.4$ relative weightings. The AB bond orders b_{AB} of the species are then evaluated as the resonance-weighted average of bond orders $b_{AB}^{(r)}$ in the contributing resonance structures

$$b_{AB} = \sum_r w_r b_{AB}^{(r)} \qquad (7.11)$$

(for example, $b_{CN} \simeq 1.4$ and $b_{CO} \simeq 1.6$ in the amide resonance hybrid **3a** $\leftrightarrow$ **3b**), and other resonance-weighted properties are obtained analogously. As is well known, the resonance concept allows a far-reaching extension[16] of structure-property relationships to species that are not well described by a single Lewis structure.

Natural resonance theory (NRT) is an optimal *ab initio* realization of the resonance weighting concepts expressed in Equation 7.10 and Equation 7.11. The necessary and sufficient condition that Equation 7.10 be satisfied for all possible density-related (one-electron) properties p is that the first-order reduced density operator[17] be expressible as such a weighted average of localized density operators $\hat{\Gamma}^{(r)}$

$$\hat{\Gamma} = \sum_r w_r \hat{\Gamma}^{(r)} \qquad (7.12)$$

For any desired bonding pattern r, one can use the NBO method to determine the optimal Lewis-like wave function $\psi^{(r)}$ and density operator $\hat{\Gamma}^{(r)}$.

Expressed somewhat heuristically, the NRT procedure corresponds to choosing optimal numerical weightings w_r such that the deviation between the right-hand side of Equation 7.12 and the true $\hat{\Gamma}$ (or its best available approximation) is as small as possible in the operator-norm sense.[18] The NRT variational algorithm is numerically implemented in all recent versions of the NBO program (version 4.0 and beyond) and can readily be used to obtain *ab initio* NRT weightings and bond orders for practically any delocalized species.

As might be anticipated, there is an intimate connection between the description of electron delocalization in terms of NBO interactions and resonance structures. Consider the vicinal $\sigma_{AB} \rightarrow \sigma^{*}_{CD}$ delocalization, starting from a *trans*-like A–B–C–D bonding skeleton as shown in **4a**.

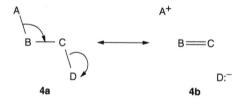

4a   **4b**

The 2e-stabilizing interaction from bond σ_{AB} into adjacent antibond σ^{*}_{CD} (with annealing of adjacent B^{+}–C^{-} formal charges to give the double-bonded B=C form) corresponds to an admixture of the charge-separated resonance structure (**4b**). Vicinal $\sigma_{AB} \rightarrow \sigma^{*}_{CD}$ interaction thus corresponds to a well-known "arrow pushing" mnemonic for deriving an alternative resonance structure. Similar examples show the general complementarity of NBO donor-acceptor and resonance concepts for describing electronic delocalization. In the acetaldehyde ground state, for example, the NRT expansion contains leading contributions corresponding to the leading $n_o^{\rightarrow}{}_{CH}$ and $n_o^{\rightarrow}\sigma^*{}_{CC}$ NBO delocalizations noted above. The NRT weightings lead to the final natural bond orders b_{AB} shown below:

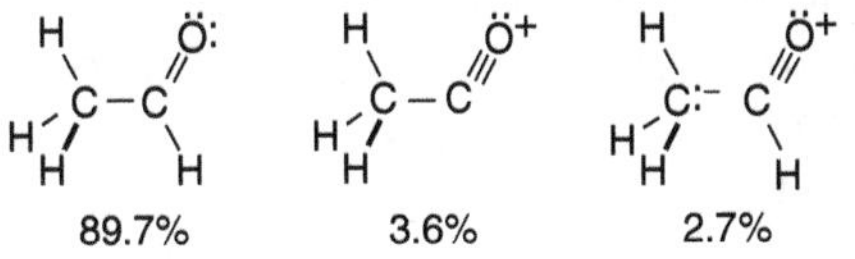

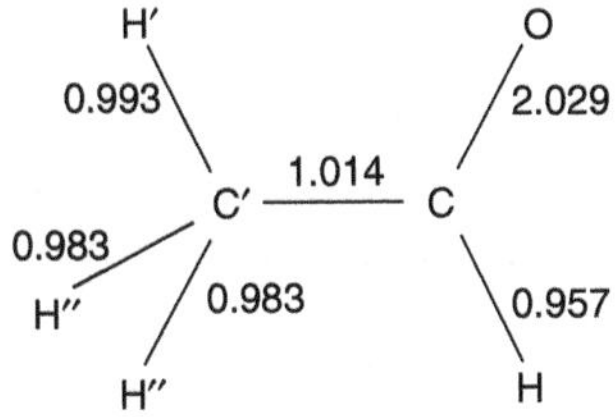

The slight variations in C–H bond orders reflect the hyperconjugative interactions noted above. However, the NRT bond orders are generally within 1 to 2% of idealized integer values, showing that this molecule is well described by the conventional single Lewis structure (**2**).

7.2.4 Alternative NBO-Based Bonding and Reactivity Indices

For completeness we briefly mention a variety of alternative bonding and reactivity indices for describing individual states or changes of state. All are based on NAO/NHO/NBO procedures, often by combining older semi-empirical bonding descriptors with modern *ab initio* NBO-based analogs. Such descriptors may be useful in characterizing the bonding in a single state, the responses to external perturbations, or the relationships between distinct states and geometries.

At the simplest level are methods for characterizing the bonding pattern of a single state and specified geometry. The BNDIDX keyword leads to printing of a variety of familiar MO-type bond-order indices as reformulated in terms of NAOs. These include the NAO-Wiberg Bond Index[19] (expressing the sums of squares of density matrix elements in the NAO basis), the "overlap-weighted NAO bond order" (expressing sums of off-diagonal NAO density matrix elements between atoms, each weighted by corresponding PNAO overlap integrals to emulate the familiar Coulson MO bond order in a nonorthogonal basis), and the Reed-Schleyer NLMO/NPA bond order[20] (expressing the "shared" covalent occupancies and overlaps contributing to NLMO composition). These indices are provided for reverse compatibility with older formulations and are considered to be superceded by the more general and accurate NRT bond order.[21]

At another level we may seek to describe the change in bonding pattern that results from a specific perturbation, perhaps as a result of chemical attack or laser-induced mode-specific vibrational excitation.[22] Such predictors are exemplified by the natural bond-bond polarizability (NBBP),[23] which expresses how a change in bond r-s (formed from bonding hybrids h_r and h_s) affects another bond t-u. The NBBP merely modernizes the original bond-bond polarizability formula of Coulson and Longuet-Higgins[24] by using elements of the NHO Fock (or Kohn–Sham) matrix in place of Hückel-type matrix elements. Regarding the NHO Fock matrix as the *ab initio* counterpart of the empirical Hückel matrix allows many earlier Hückel-type relationships to be accurately reformulated for use in the modern computational framework.

At a still more challenging level we may wish to compare NBO bonding patterns in two distinct electronic states (see, e.g., Badenhoop and Scheiner[25]), perhaps differing also in geometry. In this case we confront the problem of finding a common AO or hybrid basis for the comparison. Although NAOs and NHOs have high state-to-state transferability (see Section 7.4), such transferability cannot be strictly assumed for quantitative numerical purposes. In such a case one may choose to fix, for example, the NAOs of the ground-state species in frozen form to express the excited-state bonding pattern. The NBO program provides keywords of the form AONAO=Rnn (read from file nn) and AONAO=Wnn (write to file nn) to force usage of the AO $\rightarrow$ NAO transformation matrix taken from an earlier calculation. With this usage, frozen NAOs are imported from another state (rather than reoptimized for the current state), so that exact state-to-state comparisons of NAO-based Fock, density, and other matrices can be obtained directly. Such frozen-orbital techniques underlie the recently introduced Delta Overlap-Density index of Zimmerman and Alabugin.[26] The Delta Overlap-Density index ΔD_{rt} between bonding hybrids h_r and h_t of the ground (o) and excited (*) state is expressed as

$$\Delta D_{rt} = D^*_{rt} S^*_{rt} - D^o_{rt} S^o_{rt}$$

in terms of NHO density matrix (D) and pre-NHO overlap matrix (S) elements of the two states in the frozen-hybrid basis. Numerous examples in Reference 26 illustrate the utility of this index in predicting photochemical reactivity pathways.

7.3 EXTENSION OF NBO CONCEPTS TO OPEN-SHELL SPECIES

7.3.1 Theoretical and Practical Difficulties of Open-Shell NBO Description

The Lewis structure concept may superficially appear to lose its usefulness for open-shell species. The electrons of radical or excited-state species cannot be strictly paired as in a conventional closed-shell Lewis structure diagram. Moreover, the concept of "structure" itself seems to lose its validity in many radical species, which tend to be characterized by floppiness, large-amplitude vibrations, and general lack of structural rigidity compared with closed-shell species. The reactivity and instability of radical species might seem to preclude a useful role for the Lewis structural concepts.

However, we can recognize that the essential feature of a Lewis structure is the localized 1-c, 2-c bonding pattern rather than electron pairing *per se*. Moreover, the α and β electrons of open-shell species necessarily experience different Coulomb and exchange forces and may hence lead to different spatial distributions, spin orbitals, and localization patterns. The familiar open-shell concept of "different orbitals for different spins" can therefore be extended to a "different Lewis structures for different spins" (DLSDS) picture, where as usual we associate a Lewis structure with a specified pattern of one-center and two-center (spin-)NBOs.[27] Because the NBO procedure can be carried out separately for α and β spins, with spin orbitals of maximum allowed occupancy $n_{occ} = 1$ rather than 2, the entire NBO/NRT formalism sketched in Section 7.1 generalizes readily to open-shell species.

The physical interpretation of the DLSDS descriptors is also straightforward. In effect, electrons of α and β spin may compete for different localized connectivity patterns and hence for opposing geometrical tendencies. In cases where the α and β Lewis structure patterns coincide (i.e., electrons are paired), the structural rigidity will be similar to that of closed-shell species. However, in regions where α and β connectivity patterns differ, the molecular geometry is subject to competing influences of the two spin sets. This competition may be represented as a hybrid of distinct α and β Lewis structures, with resulting floppiness and loss of structural rigidity. The DLSDS descriptors thereby pinpoint regions of α and β structural incongruity and suggest how substituent effects may alter the weighting between α-preferred vs. β-preferred structures to drive the geometry toward one or the other connectivity pattern.

Although the formal extension of NBO/NRT concepts to open shells is straightforward, numerous practical difficulties must be faced in applying these concepts to radical and excited-state species using popular programs such as Gaussian 9X.[28] The difficulties arise both from the strongly delocalized character of open-shell species as well as the paucity of approximation methods for excited states that provide the spin-density information needed to obtain spin-NBOs:

1. The formation of an open-shell radical commonly results in a low-lying electronic hole that greatly enhances acceptor character, or a weakly bound surplus electron with correspondingly enhanced donor character. As a result, low-order perturbative approximations such as 7.5–7.8 must be replaced by multireference NRT to describe the donor–acceptor interactions.

2. Popular approximation methods are often based on UHF-like dominance of a single electronic configuration, leading to spin contamination in open-shell species with significant multiconfigurational character.

Post-HF corrections to spin-density matrices (if available at all) may then be subject to unphysical negative or Pauli-violating occupancies.

3. Only HF and DFT methods provide the effective one-electron Hamiltonian (Fock or Kohn–Sham) operator $\hat{\mathcal{F}}$ required for NBO second-order perturbation analysis or Fock matrix deletions, and these methods are generally inapplicable to excited states.

4. Among widely available methods, only CASSCF and CI techniques give rigorous bounds for excited states, allowing direct exploration of an excited-state surface. However, CASSCF calculations are generally performed by the spin-free graphical unitary group approach (GUGA)[29] that bypasses the spin densities needed to obtain spin-NBOs. Instead, the open-shell CAS density corresponds to an average of α and β spin densities, similar to the starting point for the maximum spin-paired MSPNBO description.[30] Although spin-averaged-type spatial NBOs can be obtained from this average density, such orbitals cannot represent the physical effect of spin polarization within an electron pair. The accuracy and detail of NBO analysis of CAS wave functions is therefore diminished compared with methods that retain the spin densities. In addition, NRT analysis is inapplicable to CAS spin-averaged density.

5. Among the remaining CI methods, only the CIS (CI with single excitations) method has been extensively developed for excited states, with inclusion of post-HF spin-density corrections. (As implemented in G9X, the related DFT-based "TD" method resembles the CIS method in output format but currently provides no spin-density information and is not considered further.) The CIS wave function is often of low quantitative accuracy and should be tested against more reliable methods before details of its NBO/NRT analysis are presumed useful.

6. Similar to the case of MP-based methods, the CIS spin-density corrections are evaluated in only an approximate perturbative manner and may therefore

lead to unphysical negative or Pauli-violating occupancies that result in fatal NBO run-time errors.

To remedy (or at least camouflage) the unphysical spin
densities of MP or CIS approximations, the NBO 5.0 program
provides the FIXDM keyword to replace the spin-density
matrices from the host program, with the "closest" matrices
restricted to physically allowable occupancies.[31] This replacement allows completion of many NBO tasks that previously
aborted with fatal error conditions. Although future advances
in excited-state methods will doubtless remove some of or all
the difficulties listed above, the present NBO/NRT applications to photochemical excited species are effectively limited
to CIS-level approximations with inclusion of the FIXDM
keyword, as primarily used in this chapter.

7.3.2 Necessary NBO Algorithm Modifications for Excited States

For excited states, other difficulties arise from assumptions
embedded in the standard NBO program. Because the NBO
algorithm was developed primarily in the framework of near-
equilibrium ground-state applications, certain of its conceptual and numerical details are implicitly based on the
assumed stability, uniqueness, and robustness of Lewis structures for ground states. Such assumptions are increasingly
untenable in higher excited states, and certain details of the
NBO algorithm must be modified.

1. The standard NBO algorithm attempts to maximize
   the total occupancy ρ_{Lewis} of Lewis-type (unstarred)
   NBOs, under the assumption that Lewis-type orbitals
   are formally occupied and non–Lewis-type (starred)
   orbitals are formally vacant in the best single NBO
   configuration. This assumption is obviously untenable
   for excited states, where promotions from Lewis-type
   to non–Lewis-type NBOs should be a common feature.
   We therefore define ρ_{occ} to be the total occupancy of
   the N most highly occupied spin-NBOs (whether of
   Lewis- or non–Lewis-type) and $\rho_{\text{vac}} = N- \rho_{\text{occ}}$, the
   remaining occupancy of formally vacant NBOs, and

we modify the NBO search to maximize ρ_{occ} (or minimize ρ_{vac}) when the EXCITE keyword is set. Of course, ρ_{occ} is usually the same as ρ_{Lewis} for ground-state species, but the best EXCITE configuration will often contain non-Lewis NBOs, labeled LP* (valence 1-c nonbonded), BD* (valence 2-c antibond), or RY* (Rydberg-type 1-c) in NBO program output.

2. The standard search for high-occupancy NBOs is repeated sequentially, adjusting the occupancy threshold downward by $0.1e$ intervals from $1.90e$ to $1.50e$ (or from $0.90e$ to $0.50e$ in the open-shell case) and selecting the best structure from any occupancy interval. This relatively coarse-grained search is inadequate for excited species, which may exhibit multiple alternative forms separated by only small occupancy differences. The EXCITE search therefore reduces this interval to $0.02e$ (or $0.01e$ for open shells) and searches all occupancy ranges from $1.98e$ to $1.50e$ (or $0.99e$ to $0.75e$ for open shells). This fine-grained EXCITE search restores consistent occupancy criteria for closed- and open-shell cases and makes it far less likely that distinct NBO configurations of high ρ_{occ} will be overlooked on account of too-small occupancy differences. It also ensures that the NBO criterion for distinguishing nonbonding from bonding character (i.e., at least 95% of the former is on one center only) is applied consistently in closed- and open-shell cases.

3. The standard directed-search ($CHOOSE) option formerly accepted only the most highly occupied candidate orbitals of 1-c or 2-c type to satisfy a specified Lewis diagram, without regard to whether alternative selections (perhaps differing only infinitesimally in occupancy) might give higher overall ρ_{occ}. The EXCITE search now carries out a more exhaustive examination of alternative possibilities that may lead to improved final ρ_{occ}.

Modifications 1 to 3 have been incorporated in a revised version NBO 5.X (for "eXcited states") that alters NBO variational criteria, thresholds, and search details as described above if the EXCITE keyword is set. For ground-state species, the results of NBO 5.0 and NBO 5.X, with or without the EXCITE keyword, are generally identical. However, for excited species the EXCITE search frequently returns a Lewis-like structure of higher overall ρ_{occ}.

Modification 1 also corrects a formal problem that is sometimes manifested in ground states. If nonbonded orbitals n_A and n_B are both occupied, their contribution to the wave function or electron density can be equivalently expressed in terms of bonding and antibonding σ_{AB} and σ^*_{AB} unitary transformations, namely, for two doubly occupied orbitals,

$$\hat{\mathcal{A}} \left| \cdots (n_A)^2 (n_B)^2 \cdots \right| = \hat{\mathcal{A}} \left| \cdots (\sigma_{AB})^2 (\sigma^*_{AB})^2 \cdots \right| \tag{7.13}$$

The occupied orbitals on the left in Equation 7.13 are both Lewis-type (unstarred), but σ^*_{AB} on the right is non–Lewis-type, and its occupancy is not accounted to ρ_{Lewis}. In rare cases where the standard NBO search leads to such a high-occupancy σ_{AB} and σ^*_{AB} pair, the error associated with σ^*_{AB} occupancy is not properly taken into account, and the search may return a structure that is not the best possible in the maximum-occupancy sense (as can be checked with a $CHOOSE keylist). This type of anomaly cannot occur when ρ_{occ} (rather than ρ_{Lewis}) is chosen as the variational objective function. For this reason we recommend that the generalized EXCITE search in NBO 5.X be used in all cases, including ground states, even though it requires additional computation time and usually gives the same result as NBO 5.0.

7.3.3 Lewis-Type NBOs of Vinoxy Radical

The vinoxy radical (CH_2CHO) is an important species in combustion and atmospheric chemistry, and its rich photochemistry has been the subject of benchmark studies throughout the modern photochemical era.[32] Early theoretical studies focused on the expected competition between tautomeric ethenyloxy ($CH_2=CH\dot{O}$) and formylmethyl ($\dot{C}H_2-CH=O$) radical structures,[33] and subsequent theoretical[34] and experimental[35] studies have repeatedly stressed the importance of obtaining accurate structural bonding characterizations of ground and excited states of vinoxy radical and its derivatives. Qualitative

TABLE 7.3 Theoretical and Experimental Geometrical Parameters for Vinoxy $\tilde{X}$ $^2A'$, Compared with the B3LYP/6-311++G** Model Used Here

	CAS[a]	QCISD[b]	B3LYP[c]	Exp.[d]
		Bond Lengths (Å)		
C–H	1.089	1.100	1.105	1.088
C–O	1.224	1.241	1.235	1.261
C–C'	1.436	1.434	1.423	1.408
C'–H'	1.073	1.080	1.084	1.081
C'–H''	1.073	1.081	1.084	1.080
		Bond Angles (degrees)		
C'–C–H	117.7	117.0	116.9	
C'–C–O	122.6	122.6	123.1	122.4
C–C'–H'	119.4	119.2	119.5	
C–C'–H''	121.0	121.2	121.1	
O–C–H	119.7	120.4	120.0	

[a]CAS(3,3)/6-31G**[35c]
[b]QCISD/6-31G**[35c]
[c]B3LYP/6-311++G**, author's present work.
[d]Compiled in Ref.[35c]

bonding characterizations have been inferred from symmetry properties of MOs,[34a] natural orbital occupancy patterns,[34b] structural proxies,[35d] or chemical intuition (based, e.g., on perceived parallels to allylic or other ground-state species). In this treatment we describe how to use NBO/NRT methods to obtain more detailed and quantitative bonding characterizations of vinoxy ground and excited states.

Let us first examine the ground state ($\tilde{X}$ $^2A'$) of the vinoxy radical, where a variety of familiar HF, MP, and DFT electronic structure methods are applicable. Specifically, we employ the B3LYP/6-311++G** level of theory for direct comparison with the parent closed-shell acetaldehyde molecule **2** discussed in Section 7.1. Table 7.3 displays optimized geometrical parameters calculated at this level (column 3), showing the satisfactory agreement with previous theoretical and experimental studies.

Table 7.4 exhibits some details of α and β Lewis-type NBOs for the ground-state vinoxy radical. These can be depicted in Lewis-like diagrams for the individual spin sets, where each bond-stroke (2-c) or dot (1-c) represents a single electron of α or β spin in the indicated spatial region. In this depiction, the α NBO structure corresponds to the expected vinoxy-like bonding pattern **5a**, but the β NBO structure corresponds to the formylmethyl-like pattern **5b**, leading to a type of enol-keto distinction between the two spin sets:

    5a (a)   **5b** (b)

In the depiction of unshared (1-c) electron dots around each O atom, we adopt a conventional placement of dots

$$\left(\begin{array}{c} (\sigma) \\ (\pi)\ \text{O}\ (p) \end{array} \right)$$

to denote whether the corresponding n_O NBO is of out-of-plane π-type ($n^{(\pi)}$, left), end-on σ-type ($n^{(\sigma)} = n_O'$, above), or in-plane p-type ($n^{(p)} = n_O''$, right). Thus, the symbol $\dot{\text{O}}\cdot$ in **5a** denotes that the three unshared electrons occupy all three types of NBOs, whereas the symbol $\dot{\text{O}}\cdot$ in **5b** denotes that the unshared electrons are of σ-type n_O' and in-plane n_O'' type (as in formaldehyde [1]). Where applicable, we denote β-NBOs with a bar $\bar{\sigma}_{CH}, \bar{\sigma}_{CH}^*$, etc. to distinguish them from corresponding α-NBOs.

As depicted in **5a** and **5b**, the sigma skeletal NBOs (three σ_{CH}, one σ_{CC}, and one σ_{CO}) are all formally doubly occupied, whereas the π NBOs of the C–C–O moiety exhibit a characteristic keto-enol or allylic-like pattern. Table 7.4 shows that in-plane α and β spin NBOs are nearly identical (hence, perfectly paired), and comparison of Table 7.1 and Table 7.2 shows the high transferability of skeletal sigma bonds from parent closed-shell species.

However, remarkable changes are seen in the forms of singly-occupied π_{CC} and $\bar{\pi}_{CO}$ NBOs, both of which are signifi-

TABLE 7.4 Occupancies and Compositions of Acyl Group Lewis-Type α and β Spin-NBOs of Vinoxy Radical 5 at the B3LYP/6-311++G** Level[a]

NBO	α Spin		β Spin	
	Occ.	$c_A h_A + c_B h_B$	Occ.	$c_A h_A + c_B h_B$
σ_{CH}	0.9943	$0.7576(sp^{2.26})_C + 0.6528(s)_H$	0.9944	$0.7595(sp^{2.32})_C + 0.6505(s)_H$
σ_{CO}	0.9988	$0.5905(sp^{2.11})_C + 0.8071(sp^{1.59})_O$	0.9988	$0.6001(sp^{2.11})_C + 0.8000(sp^{1.57})_O$
π_{CO}			0.9274	$0.7005(p)_C + 0.7136(p)_O$
$n_O{'}$	0.9929	$(sp^{0.62})_O$	0.9925	$(sp^{0.63})_O$
$n_O{''}$	0.9488	$(p)_O$	0.9472	$(p)_O$
n_O^{π}	0.8298	$(p_z)_O$		
σ_{CC}	0.9978	$0.7023(sp^{1.65})_C + 0.7118(sp^{1.71})_{C'}$	0.9977	$0.7165(sp^{1.61})_C + 0.6976(sp^{1.72})_{C'}$
π_{CC}	0.9989	$0.4909(p)_C + 0.8712(p)_{C'}$		
$\sigma_{CH}{'}$	0.9945	$0.7769(sp^{2.14})_{C'} + 0.6297(s)_{H'}$	0.9948	$0.7630(sp^{2.13})_{C'} + 0.6464(s)_{H'}$
$\sigma_{CH}{''}$	0.9951	$0.7736(sp^{2.19})_{C'} + 0.6336(s)_{H''}$	0.9953	$0.7594(sp^{2.18})_{C'} + 0.6507(s)_{H''}$

[a] cf. formaldehyde (Table 7.1) and acetaldehyde (Table 7.2) NBOs. C and C' denote the acyl and methylenic carbon, respectively. H is the acyl hydrogen, and H' (*cis* to O), H'' (*trans* to O) are the methylenic hydrogens.

cantly repolarized compared with closed-shell analogs. The π_{CC} bond of a closed-shell vinyl group, for example, is expected to be relatively apolar, but the actual vinoxy π_{CC} NBO is more than 75% polarized toward the methylenic C′. (Note, by contrast, that the σ_{CC} and $\bar{\sigma}_{CC}$ NBOs are quite ordinary in this respect.) As will be discussed below, this severe repolarization of π_{CC} and $\bar{\pi}_{CO}$ serves to enhance donor-acceptor stabilization and is thus a consequence of non-Lewis delocalizing interactions to be described in the following section. Figure 7.4 shows orbital contour diagrams for singly occupied π_{CC}, n_O^π, and $\bar{\pi}_{CO}$ NBOs as well as the formally unoccupied $\bar{n}_C^*$, NBO that dominate valence-shell delocalizations.

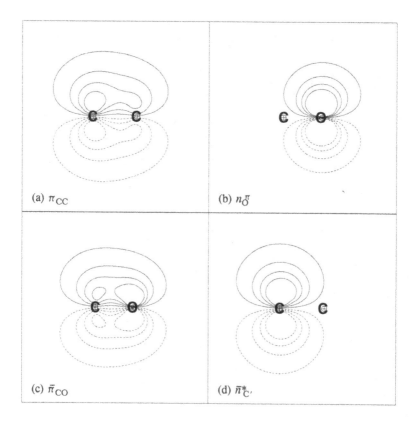

Figure 7.4 NBOView contour diagrams of partially occupied valence-shell NBOs of vinoxy radical: (a) π_{CC}; (b) n_O^π; (c) $\bar{\pi}_{CO}$; (d) $\bar{n}_{C'}^*$.

From the nearly perfect pairing of σ_{CH} and $\bar{\sigma}_{CH}$ NBOs seen in Table 7.4 and the high transferability from corresponding closed-shell NBOs in Table 7.1 and Table 7.2, one can anticipate that the vinoxy C–H bonds exhibit relatively normal bond lengths and stretching frequencies. However, the keto-enol–like bonding pattern and altered π_{CC} and $\bar{\pi}_{CO}$ polarizations suggest strong changes in the geometry and vibrational properties of the C–C–O moiety compared with parent closed-shell species. These expectations are confirmed by comparisons of optimized geometries and vibrational frequencies of parent acetaldehyde vs. vinoxy radical species.[36]

7.3.4 Vinoxy Radical Hyperconjugation

The increased role of electron delocalization in vinoxy radical is apparent from the increased occupancy of non-Lewis NBOs ($0.250e$ in the α manifold, $0.153e$ in β), more than 70% greater than in the parent acetaldehyde molecule. It is convenient to discuss the non-Lewis NBOs and donor-acceptor interactions of α (majority spin) and β (minority spin) manifolds separately, beginning with the latter.

The non-Lewis β-NBOs of vinoxy consist of the 1-c $\bar{n}_C^*$, hole (labeled "LP*" in NBO output) as well as the usual valence antibonds $\left(\bar{\sigma}_{CH}^*, \bar{\sigma}_{CC}^*, \bar{\sigma}_{CO}^*, \bar{\pi}_{CO}^*\right)$ and Rydberg orbitals. Figure 7.4d depicts the form of the most important β-acceptor $\bar{n}_C^*$, whose orbital energy (0.0874 a.u.) lies significantly below that of other non-Lewis NBOs (e.g., 0.0046 a.u. for the next-lowest $\bar{\pi}_{CO}^*$). According to the second-order estimates (5) (with $n_{occ} = 1$), the leading β-interactions are

$$\Delta E^{(2)}\left(\bar{\pi}_{CO} \to \bar{n}_{C'}^*\right) = -15.94 \text{ kcal} / \text{mol}$$

$$\Delta E^{(2)}\left(\bar{n}_O'' \to \bar{\sigma}_{CH}^*\right) = -10.65 \text{ kcal} / \text{mol}$$

$$\Delta E^{(2)}\left(\bar{n}_O'' \to \bar{\sigma}_{CC}^*\right) = -7.36 \text{ kcal} / \text{mol}$$

The latter two interactions are only slightly weaker on a per-electron basis than the corresponding interactions previously noted in acetaldehyde (Figures 7.3a,b). Figure 7.5a shows

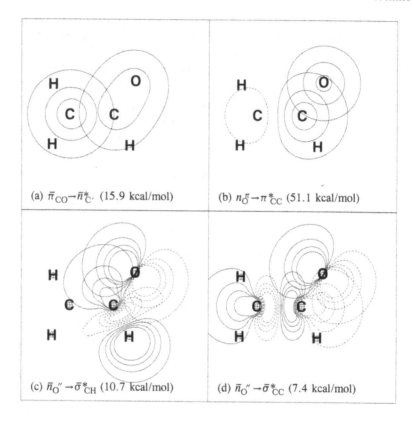

Figure 7.5 NBOView contour diagrams of leading NBO interactions in vinoxy radical: (a) $\bar{\pi}_{CO} \to \bar{n}^*_{C'}$; (b) $n^{\pi}_{O} \to \pi^*_{CC}$; (c) $\bar{n}_{O}'' \to \bar{\sigma}^*_{CH}$; (d) $\bar{n}_{O}'' \to \bar{\sigma}^*_{CC}$. The contour plane in (a) and (b) is 1 Å above the molecular plane to show slice through π-type orbitals.

the $\bar{\pi}_{CO} \to \bar{n}^*_{C'}$ interaction, which leads to a prominent delocalization tail in the $\tilde{\bar{n}}_{O}''$ NLMO

$$\tilde{\bar{\pi}}_{CO} \simeq 0.9630\bar{\pi}_{CO} + 0.2694\bar{n}^*_{C'}$$

and to significantly higher occupancy ($0.0726e$) than other β-acceptors (i.e., $0.0334e$ for $\bar{\sigma}^*_{CH}$ or $0.0170e$ for $\bar{\sigma}^*_{CC}$). The importance of $\bar{n}_{O}'' \to \bar{\sigma}^*_{CH}$ and $\bar{n}_{O}'' \to \bar{\sigma}^*_{CC}$ delocalizations is apparent in the form of the $\tilde{\bar{n}}_{O}''$ NLMO

$$\tilde{\bar{n}}_{O}'' \simeq 0.9733\bar{n}_{O}'' + 0.1718\bar{\sigma}^*_{CH} - 0.1255\bar{\sigma}^*_{CC} + \cdots$$

(analogous to acetaldehyde). However, the $\bar{\pi}_{CO} \to \bar{n}_{C}^{*}{}'$ interaction is clearly the unique feature of β-delocalization that distinguishes it from the parent closed-shell species. Repolarization of donor $\bar{\pi}_{CO}$ toward C (as observed) further strengthens the interaction with π_{CC}^{*}. Even though π_{CC}^{*} becomes partially occupied through such delocalization, the methylenic C' retains most of the cationic spin-charge and thus remains significantly more electronegative than the acyl C (of the total +0.500 formal β spin charge, +0.276 is on methylenic C' vs. +0.064 on acyl C). This effective electronegativity difference can in turn promote complementary repolarization and delocalization in the α manifold.

The delocalization effects in the α manifold are even larger than those of the β manifold. Consistent with the favorable β spin-charge distribution, the π_{CC} bond becomes strongly polarized toward the acyl carbon, thus allowing the π_{CC}^{*} antibond

$$\pi_{CC}^{*} = 0.8712\left(p\right)_{C} - 0.4909\left(p\right)_{C}{}'$$

to become a stronger acceptor for the adjacent n_{O}^{π} donor, as shown in Figure 7.5b. The powerful effect of $n_{O}^{\pi} \to \pi_{CC}^{*}$ delocalization is evident in the second-order perturbation estimate

$$\Delta E^{(2)}\left(n_{O}^{\pi} \to \pi_{CC}^{*}\right) = -51.12 \text{ kcal/mol}$$

far stronger than any interaction in closed-shell acetaldehyde. The strength of interaction is also apparent in the high occupancy of π_{CC}^{*} (0.1669e) and its large contribution to the $\bar{n}_{O}^{\pi}$ NLMO

$$\tilde{n}_{O}^{\pi} \simeq 0.9109 n_{O}^{\pi} + 0.4085 \pi_{CC}^{*} + \cdots$$

Weaker delocalizations are seen in the interaction of the in-plane p-type n_{O}'' with adjacent σ_{CH}^{*} and σ_{CC}^{*} orbitals

$$\Delta E^{(2)}\left(n_{O}'' \to \sigma_{CH}^{*}\right) = -10.18 \text{ kcal/mol}$$

$$\Delta E^{(2)}\left(n_{O}'' \to \sigma_{CC}^{*}\right) = -7.75 \text{ kcal/mol}$$

similar in magnitude to the corresponding β-interaction (and to analogous closed-shell acetaldehyde interactions). As before, these interactions contribute significant delocalization tails to the $\tilde{n}_O''$ NLMO

$$\tilde{n}_O'' \simeq 0.9740 n_O'' + 0.1653\sigma^*_{CH} + 0.1294\sigma^*_{CC} + \cdots$$

Figures 7.5c,d show the α-type $n_O'' \to \sigma^*_{CH}$ and $n_O'' \to \sigma^*_{CC}$ - interactions, which may be compared with the corresponding closed-shell interactions in Figures 7.3a,b. NBO interaction diagrams such as Figures 5a–d make it easy to visualize the much stronger delocalizations that are present in the radical compared with its closed-shell precursor.

The dominant role of non-Lewis delocalizations on vinoxy geometry and vibrational properties could also be assessed by deleting these interactions (using the $DEL keylist, a standard NBO program option[37]) and recalculating the optimized radical geometry in the absence of these interactions. An example of such deletions will be presented in Section 7.4.4 for the vinoxy-water complex.

7.3.5   NRT Description of Vinoxy Resonance

The strong non-Lewis contributions (particularly in the α manifold) suggest that no single Lewis structure representation is adequate to describe the vinoxy electronic distribution. A more accurate resonance-theoretic description will therefore be required to describe the delocalization patterns of each spin manifold.

NRT analysis of the vinoxy α spin manifold gives the leading contributions

6a (50.7%)   **6b** (43.9%)   **6c** (1.7%)

The leading NBO enol-like structure **6a** has only marginally higher NRT weighting than the alternative keto-like structure **6b** (which arises from $n_O^\pi \to \pi_{CC}^*$ delocalization of **6a**). It is therefore evident that small changes of geometry or level of description may be sufficient to discontinuously switch the preferred NBO structure from **6a** to **6b**. In this sense the NRT description is clearly preferable, for it associates continuous physical changes with continuous changes of NRT weightings and bond orders, rather than an abrupt switch from one structure to another. Remaining NRT structures such as **6c** (arising from $n_O'' \to \sigma_{CH}^*$ delocalization of **6a**) are relatively negligible compared with the leading enol/keto-like pair **6a** and **6b**. The relative NRT weightings obviously parallel the strengths of NBO donor-acceptor interactions and could be rationalized in terms of NBO orbital diagrams as shown in the previous section.

In the β manifold, the NRT expansion has leading contributions

7a (86.0%)   **7b** (7.4%)   **7c** (3.4%)

In this case the NBO structure **7a** is clearly dominant, and the secondary structures **7b** $\left(\bar\pi_{CO} \to \bar n_{C'}^*\right)$ and **7c** $\left(\bar n_O'' \to \bar\sigma_{CH}^*\right)$ provide relatively minor corrections. (Yet note that even this relatively well-localized spin set exhibits stronger resonance corrections than the closed-shell parent acetaldehyde.)

The best overall structural representation of vinoxy radical is in terms of (total) NRT bond orders, as shown:

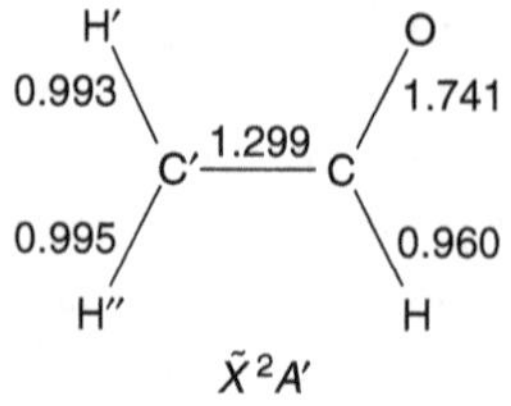

H'
0.993
C' —1.299— C
0.995
H''
O
1.741
0.960
H

$\tilde{X}\,^2A'$

The vinoxy C–H bonds are rather ordinary single bonds (b_{CH} = 0.96 to 0.99), but the skeletal C–C (b_{CC} = 1.30) and C–O (b_{CO} = 1.74) bonds deviate markedly from simple single- or double-bond character. The bond-order changes correctly reflect the qualitative geometry changes that distinguish vinoxy radical from the parent acetaldehyde molecule. From this "bottom line" structural representation, each NRT bond order can be traced back to specific resonance weightings and associated NBO donor–acceptor interactions to give an increasingly detailed picture of the radical wave function.

7.4 NBO/NRT DESCRIPTION OF PHOTOCHEMICAL EXCITED STATES

7.4.1 Low-Lying Vertical Excitations of Vinoxy Radical

Let us first consider the lowest few vertical excitations of vinoxy radical as obtained from simple CIS/6-311++G** level calculations at the $\tilde{X}\,^2A''$ equilibrium geometry. By default, the CIS calculation leads to the lowest three states of ground-state (doublet) multiplicity, which may be identified with the conventional spectroscopic labels $\tilde{A}\,^2A'$ (root = 1), $\tilde{B}\,^2A''$ (root = 2), and $\tilde{C}\,^2A'$ (root = 3) in order of ascending energy. The calculated vertical excitation energies are 2.033, 6.728, and 6.787 eV, respectively. A sample G9X input deck to perform this calculation and carry out NBO/NRT analysis on the lowest-lying $\tilde{A}$ state is shown in Example 7.1.

EXAMPLE 7.1

Sample Gaussian 9X input deck to perform NBO/NRT analysis on the lowest-lying (root = 1) $\tilde{A}^2A'$ state of vinoxy radical. The $NBO keylist also specifies use of the FIXDM keyword, production of PLOT files, and printing of the NBO $\rightarrow$ NLMO transformation matrix.

```
%chk=vinoxy.chk
#NCIS(root=1)/6-311++g** density=current POP=NBORead

CH2CHO "vinoxy radical" [1st excited (A) state]

  0    2
     6        0.132521      0.406355     0.0000
     1        0.189840      1.510183     0.0000
     8        1.168837     -0.264864     0.0000
     6       -1.167975     -0.171515     0.0000
     1       -1.271777     -1.250273     0.0000
     1       -2.056039      0.449969     0.0000

$NBO file=vin_1_cis nrt fixdm plot nbonlmo $END
```

As noted above, adoption of the CIS method necessitates use of the FIXDM keyword and rules out $\hat{\mathscr{F}}$-dependent NBO options such as second-order perturbative analysis or Fock matrix deletions. However, numerous NBO descriptors remain useful for analyzing the CIS wave function, including NBO Lewis structure, orbital shapes and occupancies, and NRT weightings and bond orders. In the following, we illustrate the use of these descriptors for the three lowest CIS excited states. Our emphasis is not on the absolute accuracy of the CIS wave functions themselves but on how the NBO/NRT descriptors provide a chemist's picture of the wave functions, useful for predicting structural rearrangements and reactivity patterns on the excited-state potential-energy surfaces. Similar analysis techniques are applicable to any improved wave functions up to and including exact solutions of the fixed-nucleus Schrödinger equation.

The vertical excitations represent only an isolated point of the full multidimensional potential-energy diagram. As a simple example, Figure 7.6 displays a cross-section of the potential-energy diagram for rigid twisting of the radical around the C–C bond. Whereas the energy levels at torsion angle $\phi = 0$ correspond to the vertical excitations to be discussed in this section, the variations at other angles allow one to qualitatively envision how the torsional barrier for vinyl twisting varies from state to state. The pronounced variations in torsional dependence give clues to the electronic nature of each state that may be compared with the NBO description of the state at $\phi = 0$. (More quantitative treatment of adiabatic torsional dependence is given in Reference 34c.)

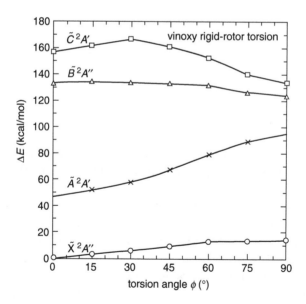

Figure 7.6 Qualitative CIS/6-311++G** potential-energy diagram (relative to the B3LYP/6-311++G** ground state) for rigid-rotor twisting of vinoxy radical from planar ($\phi = 0°$) to perpendicular ($\phi = 90°$).

7.4.1.1 First Excited $\tilde{A}\ ^2A'$ State

Table 7.5 exhibits details of α and β Lewis-type NBOs for vinoxy $\tilde{A}\ ^2A'$, corresponding to the Lewis structure representations **8a** and **8b**.

8a (α) **8b** (β)

By comparing these NBO Lewis structure diagrams with those for the $\tilde{X}\ ^2A''$ state, one can recognize that the localized NBO representation of the $\tilde{X}\ ^2A'' \rightarrow \tilde{A}\ ^2A'$ transition is

$$\left(\bar{n}_{C'}^*\right)^0 \left(\bar{\pi}_{CO}\right)^1 \rightarrow \left(\bar{n}_{O}''^*\right)^0 \left(\bar{\pi}_{CC}\right)^1$$

a formal keto-enol π-bond switch in the β manifold, expected to be of low energy. From the evident spin pairings in **8a** and **8b**, one can also recognize that the $\tilde{A}$ state is aptly identified as the "n_O-hole state" (or $\bar{n}_O''^*$state), related to the well-known $n_O'' \rightarrow \pi_{CC}^*$ state of a parent closed-shell carbonyl.

Table 7.5 exhibits the extreme (and opposite) polarization of π_{CC} and $\bar{\pi}_{CC}$ NBOs in the $\tilde{A}$-state, similar to the polarization of π_{CC} in the $\tilde{X}$ state (and presumably having a similar donor-acceptor origin). The formal $(\pi_{CC})^2$ double occupancy therefore corresponds to considerable spin polarization (singlet diradical character) in the C–C region. Overall, the $\tilde{A}$-state NBOs are similar to those of the ground-state radical and (with the π_{CC} and $\bar{\pi}_{CC}$ exception noted above) to those of parent closed-shell species. Such comparisons indicate that the high transferability of NBOs extends from state to state as well as from molecule to molecule.

Perhaps the most noteworthy feature of the NHOs is the slight rehybridization of the acyl carbon hybrid in the σ_{CC} bond (i.e., from $sp^{1.65}$ to $sp^{1.48}$ in the α hybrid, or $sp^{1.61}$ to $sp^{1.39}$

TABLE 7.5 Similar to Table 7.4, for Vertical $\tilde{A}\,^2A'$ Excited State (CIS/6-311++G**)[a] $\%-\rho_L = 98.97\%$ (α), 98.49% (β)

NBO	α Spin		β Spin	
	Occ.	$c_A h_A + c_B h_B$	Occ.	$c_A h_A + c_B h_B$
σ_{CH}	0.9957	$0.7350(sp^{2.22})_C + 0.6780(s)_H$	0.9525	$0.8112(sp^{2.27})_C + 0.5847(s)_H$
σ_{CO}	0.9991	$0.5538(sp^{2.44})_C + 0.8327(sp^{1.44})_O$	0.9990	$0.6010(sp^{2.56})_C + 0.7992(sp^{1.37})_O$
n_O'	0.9944	$(sp^{0.69})_O$	0.9939	$(sp^{0.73})_O$
n_O''	0.9782	$(p)_O$	[0.0964]	$[(p)_O]^*$
n_O^{π}	0.9227	$(p_z)_O$	0.9391	$(p_z)_O$
σ_{CC}	0.9970	$0.6894(sp^{1.48})_C + 0.7243(sp^{1.69})_{C'}$	0.9675	$0.7553(sp^{1.39})_C + 0.6553(sp^{1.87})_{C'}$
π_{CC}	0.9992	$0.3997(p)_C + 0.9166(p)_{C'}$	0.9953	$0.8698(p)_C + 0.4935(p)_{C'}$
$\sigma_{CH'}$	0.9949	$0.7774(sp^{2.23})_{C'} + 0.6290(s)_{H'}$	0.9950	$0.7539(sp^{2.12})_{C'} + 0.6570(s)_{H'}$
$\sigma_{CH''}$	0.9954	$0.7795(sp^{2.13})_{C'} + 0.6264(s)_{H''}$	0.9924	$0.7591(sp^{2.01})_{C'} + 0.6509(s)_{H''}$

in β) and the resulting misalignments (bond bending) in acyl bond angles. For the hybridizations shown in Table 7.5, the interhybrid angle θ_{CCH} is calculated from Equation **3** to be ~124° in both spin sets, whereas the actual $\sphericalangle_{CCH}$ bond angle is ~117°. One can therefore predict that $\tilde{A}$-state adiabatic relaxation should lead to significant opening of the C–C–H bond angle compared with the vertical $\tilde{X}$-state geometry.

Why does the $\tilde{X}$-state $\left(\bar{n}_{C'}^{*}\right)^{0}\left(\bar{\pi}_{CO}\right)^{1}$ lie lower than $\tilde{A}$-state $\left(\bar{n}_{O''}^{*}\right)^{0}\left(\bar{\pi}_{CC}\right)^{1}$? A prominent reason can be seen in Figure 7.7, which depicts the distinguished pair of NBOs in each state. The filled $\bar{\pi}_{CO}$ and vacant $\bar{n}_{C'}^{*}$ are co-aligned for favorable $\bar{\pi}_{CO} \to \bar{n}_{C'}^{*}$ donor-acceptor interaction in the $\tilde{X}$ state (Figure 7.7, left), but the filled $\bar{\pi}_{CC}$ and vacant $\bar{n}_{O''}^{*}$ of the $\tilde{A}$ state are in orthogonal planes (Figure 7.7, right), allowing the $\bar{n}_{C'}^{*}$ acceptor to interact only with weaker $\bar{\sigma}_{CH}$ and $\bar{\sigma}_{CC}$ donors in the latter case. This difference clearly favors the $\left(\bar{n}_{C'}^{*}\right)^{0}\left(\bar{\pi}_{CO}\right)^{1}$ configuration.

Why is the torsion barrier of the $\tilde{A}$ state so much higher than that of the $\tilde{X}$ state? Compared with the predominant C–C single-bond character in the $\tilde{X}$ state ($b_{CC} = 1.30$), the $\tilde{A}$-

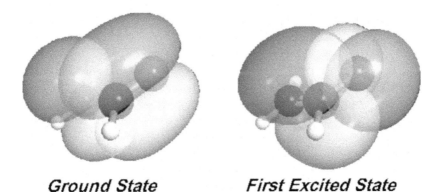

Ground State   **First Excited State**

Figure 7.7 Leading NBO interactions $\bar{\pi}_{CO} \to \bar{n}_{C'}^{*}$ of the $\tilde{X}$ state vs. $\bar{\pi}_{CC} \to n_{O''}^{*}$ of the $\tilde{A}$ state.

state structure has a predominant C–C double-bond character ($b_{CC}= 1.79$), as shown in **8a** and **8b**. The about threefold barrier increase is therefore in accord with this strongly increased double-bond character.

A more detailed description of the $\tilde{A}$ electronic distribution is given by the NRT expansion. For α spin the leading contributions are

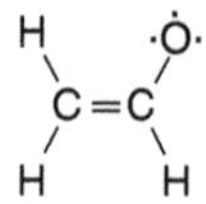

9a (63.6%) 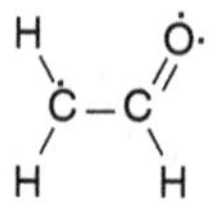   **9b** (34.0%) 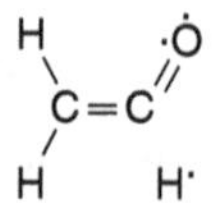   **9c** (0.8%)

and for β spin

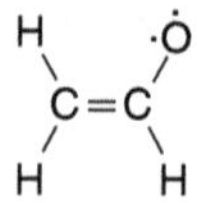

10a (86.4%) 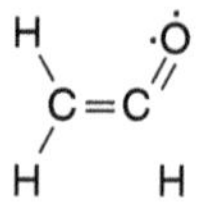   **10b** (5.1%) 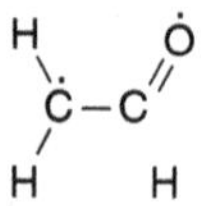   **10c** (3.9%) 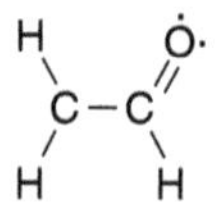   **10d** (3.6%)

The total NRT bond orders therefore lead to the composite representation

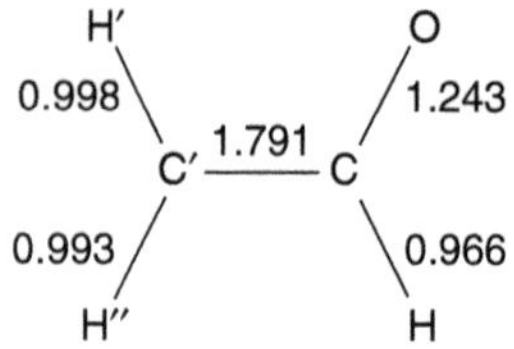

$\tilde{A}\,^2A'$

emphasizing the pronounced "vinoxy" like (C=C double-bond and C–O single-bond) character in both spin sets.

Comparison of the NRT bond orders in excited $\tilde{A}\,^2A'$ vs. ground-state $\tilde{X}^2A''$ suggests that the nascent photoproduct has high vibrational excitation in C–C and C–O bond stretches but little in C–H stretching. The calculated bond orders predict adiabatic evolution toward longer C–O and shorter C–C bonds, with attendant torsional stiffening. Other aspects of $\tilde{A}$-state equilibrium geometry are discussed below.

7.4.1.2 Second Excited $\tilde{B}\,^2A''$ State

The $\tilde{B}$ state leads to the NBOs shown in Table 7.6, corresponding to leading NBO Lewis structures.

11a (α)   **11b** (β)

Comparison of **11b** with **5b** shows that the $\tilde{X}^2A'' \to \tilde{B}^2A''$ excitation can be primarily associated with the localized β excitation

$$\bar{\pi}_{CO} \to \bar{n}_O^\pi$$

which promotes an electron from a low-energy bonding NBO to a higher-energy nonbonding NBO. However, there is also an evident shift from an enol-like to a keto-like structure in the α manifold. The radical hole character is therefore distributed over all three skeletal C–C–O atoms but most prominently in $\bar{n}_{C'}$. The spin distribution can also be aptly described in terms of doublet-coupled "*tri*radical character" in the π system.[35d]

TABLE 7.6 Similar to Table 7.4, for vertical $\tilde{B}\ ^2A''$ Excited State[a]

NBO	α Spin		β Spin	
	Occ.	$c_A h_A + c_B h_B$	Occ.	$c_A h_A + c_B h_B$
σ_{CH}	0.9942	$0.7657(sp^{2.38})_C + 0.6432(s)_H$	0.9671	$0.7332(sp^{2.58})_C + 0.6800(s)_H$
σ_{CO}	0.9986	$0.6243(sp^{1.95})_C + 0.7812(sp^{1.95})_O$	0.9657	$0.5825(sp^{1.86})_C + 0.8128(sp^{2.07})_O$
π_{CO}	0.9974	$0.6435(p)_C + 0.7655(p)_O$		
n_O'	0.9931	$(sp^{0.51})_O$	0.9898	$(sp^{0.48})_O$
n_O''	0.9501	$(p)_O$	0.9061	$(p)_O$
n_O^π			0.9264	$(p_z)_O$
σ_{CC}	0.9975	$0.7139(sp^{1.68})_C + 0.7002[sp^{1.88})_{C'}$	0.9584	$0.7082(sp^{1.66})_C + 0.7060(sp^{1.84})_{C'}$
σ_{CH}'	0.9958	$0.7799(sp^{2.02})_{C'} + 0.6259(s)_{H'}$	0.9960	$0.7604(sp^{2.03})_{C'} + 0.6495(s)_{H'}$
σ_{CH}''	0.9964	$0.7736(sp^{2.08})_{C'} + 0.6337(s)_{H''}$	0.9947	$0.7538(sp^{2.13})_{C'} + 0.6571(s)_{H''}$
n_C	0.8219	$(p)_C$	[0.0128]	$[(p)_C]$
$n_{C'}$			[0.2095]	$[(p)_{C'}]$

[a]Percent $\rho_L = 97.87\%(\alpha),\ 97.30\%(\beta)$.

The β Lewis structure **11b** is quite unusual compared with ground-state species. The formal β spin-charge is +1/2 on each C atom but 1/2 on the oxygen, corresponding to anionic character (and hence, enhanced donor character) at O. The $\tilde{B}$-state β-structure therefore manifests a type of amphiphilicity (cationic character at each C, anionic character at O) that is expected to lead to enhanced donor–acceptor interactions in this spin set.

The greater delocalization in the β manifold is manifested in the lower overall percent ρ_L values (97.87% [α] vs. 97.30% [β]) as well the individual NBO occupancies. In the α manifold, the only significantly depopulated Lewis-type NBOs are n_C (due to the usual $n_{C'} \to \pi^*_{CO}$ delocalization) and n_O'' (due to $n_O'' \to \sigma^*_{CH}, n_O'' \to \sigma^*_{CC}$). However, in the β manifold, only the two methylenic C–H bonds are well localized (occupancies > 0.99e), whereas other Lewis NBOs exhibit losses up to 10%. The forms of the $\tilde{B}$-state NBOs closely resemble their counterparts in other states, and the interhybrid angles of the planar bonding skeleton show negligible bond bending in this state.

Let us now consider the expected hyperconjugative delocalizations in the unique β structure **11b**. This electronic pattern immediately suggests that a planar structure may be unfavorable, because the formal $\bar{n}^*_{C'}$ hole then has no possible donor–acceptor interactions with adjacent donors because of strict σ-π separation in this geometry. Twisting about the C–C bond (to break σ-π symmetry) is expected to turn on vicinal $\bar{\sigma}_{CH} \to \bar{n}^*_{C'}$ and $\bar{\sigma}_{CO} \to \bar{n}^*_{C'}$ interactions that stabilize the β manifold. Because such nonplanarity simultaneously weakens the $\bar{n}_{C'} \to \pi^*_{CO}$ delocalization in the α manifold, the net tendency toward nonplanarity is likely to be weak (as shown in Figure 7.6) as a result of the conflicting geometrical tendencies of the two spin sets. Torsional relaxation in the $\tilde{B}$ state is therefore expected to differ qualitatively from that in the lower states.

Quantitative NRT weightings provide additional details of the delocalizations. The calculated weightings of the α spin set show a pronounced shift toward the keto-like structure (thereby diminishing the enolate-like character that would put anionic spin charge at O in both spin sets):

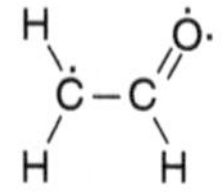

12a (63.4%) 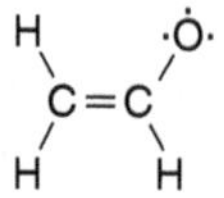　　**12b** (31.2%) 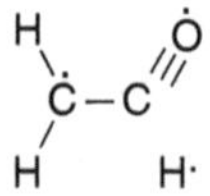　　**12c** (1.8%)

The β-spin weightings are[38]

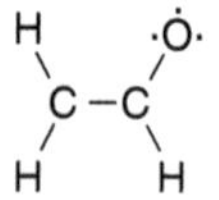

13a (75.7%) 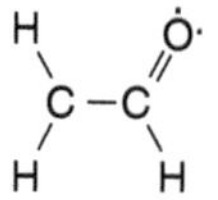　　**13b** (14.7%) 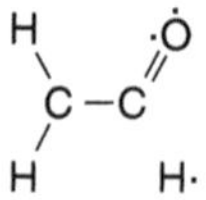　　**13c** (4.4%) 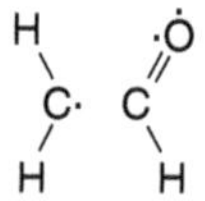　　**13d** (3.2%)

showing the expected enhancement of the $\bar{n}_O'' \rightarrow \bar{\sigma}_{CH}^*$ (**13c**) and $\bar{n}_O'' \rightarrow \bar{\sigma}_{CC}^*$ (**13d**) delocalizations as well as the delocalization of $\bar{n}_O^\pi$ into the adjacent $\bar{n}_C^*$ (**13b**). These weightings lead to the total NRT bond orders shown below:

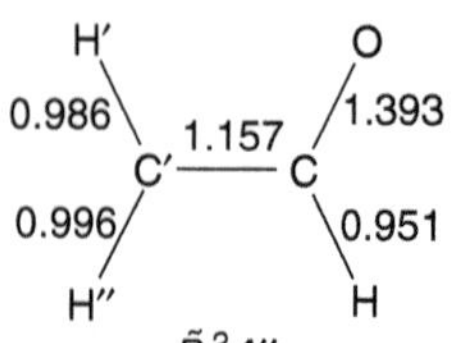

The composite bond orders show the significantly reduced valency at the acyl carbon ($V_C \simeq 3.5$) compared with either an idealized keto-like or enol-like structure.

The NRT bond orders again suggest rather strong adiabatic relaxations in the C–C–O skeleton, but in this case toward elongation of both C–C and C–O bonds. Nascent vibrational excitation patterns for the $\tilde{B}$ state (with $\Delta b_{CC} = -0.14$ and $\Delta b_{CO} = -0.35$) are therefore expected to differ significantly from those in the $\tilde{A}$ state (with $\Delta b_{CC} = +0.47$ and $\Delta b_{CO} = -0.50$). The low C–C bond order also suggests a greatly reduced $\tilde{B}$-state torsional barrier compared with other low-lying states.

7.4.1.3 Third Excited $\tilde{C}\,^2A'$ State

The $\tilde{C}$ state leads to the NBOs shown in Table 7.7, corresponding to NBO Lewis structures.

14a (α)   **14b** (β)

Comparison of **14b** with **5b** shows that the $\tilde{X}^2A'' \to \tilde{C}^2A'$ excitation corresponds primarily to NBO excitation

$$\left(\bar{n}_{O'}\right)^1\left(\bar{\pi}_{CO}{}^1\right) \to \left(\bar{n}_O^\pi\right)^1\left(\bar{\pi}_{CC}\right)^1$$

to create the localized $\bar{n}_{O'}^*$-hole state. In the ground state, $n_{O'}$ is the s-rich hybrid that is directed opposite to the C–O bond, chemically inert because of its low energy and unfavorable orientation for donor-acceptor interactions. Promotion of an electron from $\bar{n}_{O'}$ to a p-rich $\bar{n}_O^\pi$ orbital raises the state energy, but this increase is partially offset by the strong $\bar{n}_O^\pi \to \bar{\pi}_{CC}^*$ delocalization that is thereby opened up. As a

TABLE 7.7 Similar to Table 7.4, for Vertical $\tilde{C}$ $^2A'$ Excited State[a]

NBO	α Spin		β Spin	
	Occ.	$c_A h_A + c_B h_B$	Occ.	$c_A h_A + c_B h_B$
σ_{CH}	0.9930	$0.7494(sp^{2.21})_C + 0.6621(s)_H$	0.9644	$0.7803(sp^{1.83})_C + 0.6254(s)_H$
σ_{CO}	0.9982	$0.5762(sp^{2.38})_C + 0.8173(sp^{1.15})_O$	0.9993	$0.4787(sp^{4.12})_C + 0.8780(sp^{0.04})_O$
n_O'	0.9911	$(sp^{0.88})_O$	[0.1931]	$[(sp^{22.0})_O]^*$
n_O''	0.9655	$(p)_O$	0.9611	$(p)_O$
n_O^π			0.8525	$(p_z)_O$
σ_{CC}	0.9971	$0.6956(sp^{1.52})_C + 0.7184(sp^{1.66})_{C'}$	0.9629	$0.7301(sp^{1.20})_C + 0.6834(sp^{1.74})_{C'}$
π_{CC}			0.9936	$0.8489(p)_C + 0.5286(p)_{C'}$
σ_{CH}'	0.9955	$0.7766(sp^{2.19})_{C'} + 0.6301(s)_{H'}$	0.9657	$0.7660(sp^{2.24})_{C'} + 0.6429(s)_{H'}$
σ_{CH}''	0.9958	$0.7718(sp^{2.19})_{C'} + 0.6301(s)_{H''}$	0.8854	$0.7786(sp^{2.05})_{C'} + 0.6275(s)_{H''}$
$n_{C'}$	0.8418	$(p_z)_C$		

[a] Percent ρ_L = 98.13%(α), 96.21%(β).

result, both α and β manifolds are expected to exhibit the strong keto-enol resonance, with β favoring the enol form (to permit the $\bar{n}_O^\pi \to \pi_{CC}^*$ delocalization) and the α complementary keto form. The β manifold of the $\tilde{C}$ state is therefore significantly more delocalized than lower-lying states (percent $\rho_L = 96.2\%$).

The NBOs are generally similar to analogs in lower states, but the vacated $\bar{n}_O^*{}'$ rehybridizes to nearly pure-p form (as Bent's Rule anticipates). Simultaneously, the O hybrid of the $\bar{\sigma}_{CO}$ bond becomes nearly pure-s, increasing the effective electronegativity and shortening the effective covalent radius of the oxygen atom. The O atom of the $\tilde{C}$ state should therefore manifest distinctive NMR shielding, quadrupole coupling, and other properties related to its s-rich bonding. The rehybridization also leads to significant bond bendings in the β manifold for acyl C hybrids to H (7.1°) and C (5.3°). Adiabatic relaxation of the $\tilde{C}$ state should therefore be accompanied by significant angular deformation around the acyl carbon.

The NRT weightings describe the α manifold as

15a (57.7%) **15b** (38.6%) **15c** (1.2%)

and the β manifold as

16a (59.6%) **16b** (14.0%) **16c** (7.1%) **16d** (5.0%)

Note that the NBO α-structure **14a** (**15b**) receives somewhat less weighting than the enol form **15a** in the final NRT expansion. Note also the significant 1,3-interaction that leads to partial epoxide-like ring character in **16c**.

The total NRT bond orders lead to the composite $\tilde{C}$ - state representation

$$\tilde{C}\,^2 A'$$

The similarity to the $\tilde{A}$ $^2A'$ state bonding representation is conspicuous (the only difference lying in the half-occupancy of n_O'' in the $\tilde{A}$ state vs. n_O' in the $\tilde{C}$ state), and the adiabatic relaxation patterns might be expected to be similar. The similarity of initial $\tilde{A}$- vs. $\tilde{C}$-state relaxation is also suggested by the small-angle torsional dependence in Figure 7.6, but the $\tilde{C}$ state appears to go through an avoided crossing with some higher-lying state near $\phi \approx 30°$. Thus, the details of $\tilde{C}$-state adiabatic relaxation depend on fuller investigation of the excited-state potential energy surfaces. These surfaces can be expected to include barriers associated with avoided crossings or conical intersections, corresponding to near-degeneracies of distinct NBO configurational "diabatic precursor" states.

7.4.2 Adiabatic Relaxation in Vinoxy Excited States

Reoptimization of the geometry of each excited state from the initial ground-state geometry leads to the results summarized in columns 2 to 3 of Table 7.8. The $\tilde{A}$ state optimizes smoothly

TABLE 7.8 Equilibrium Bond Lengths A–B (Å) and Angles A–B–C (degrees) for Optimized Geometry of Low-Lying $\tilde{X}$, $\tilde{A}$, $\tilde{B}$ Vinoxy Radical States and $\tilde{X}-\tilde{A}_{cx}$, $\tilde{B}-\tilde{C}_{cx}$ Conical Intersections at B3LYP, CIS, or CASNBO(7,5) Levels (all 6-311++G** basis)

| | B3LYP | | CIS | | | CASNBO | | |
	$\tilde{X}$	$\tilde{A}$	$\tilde{B}$	$\tilde{X}$	$\tilde{A}$	$\tilde{B}$	$\tilde{X}-\tilde{A}_{cx}$	$\tilde{B}-\tilde{C}_{cx}$
				Bond Lengths (Å)				
C–H	1.1053	1.0766	1.0966	1.0896	1.0784	1.0749	1.0792	1.0816
C–O	1.2347	1.3718	1.2482	1.2170	1.3331	1.3591	1.3392	1.3718
C–C'	1.4231	1.3178	1.4101	1.4393	1.3358	1.4494	1.3197	1.4794
C'–H'	1.0837	1.0745	1.0780	1.0737	1.0741	1.0721	1.0737	1.0759
C'–H''	1.0839	1.0729	1.0764	1.0735	1.0723	1.0719	1.0716	1.0763
				Bond Angles (degrees)				
C'–C–H	116.9	125.1	114.2	117.5	123.8	121.8	124.1	117.3
C'–C–O	123.1	123.6	127.4	122.9	125.1	121.6	128.4	123.0
C–C'–H'	119.5	122.2	119.4	119.6	122.0	120.2	122.6	119.4
C–C'–H''	121.1	119.3	120.8	120.8	119.2	119.7	118.4	118.7
H–C–C'–H'	180.0	180.0	210.7	180.0	180.0	188.3	180.0	224.8
H'–C'–C–O	0.0	0.0	25.7	0.0	0.0	2.8	0.0	86.7
H''–C'–C–O	180.0	180.0	205.3	180.0	180.0	185.6	180.0	240.4

to planar geometry and the $\tilde{B}$ state to a rotameric twisted form with torsional angle ~30°. However, the CIS $\tilde{C}$- state optimization from vertical planar geometry is found to be unstable, leading eventually to the same result as $\tilde{B}$-state optimization. As mentioned in Section 7.3.1 (cf. Figure 7.7), the $\tilde{C}$-state surface is expected to exhibit multiple minima that correspond to strong torsional distortions and altered NBO configurations.

Table 7.9 shows available comparison $\tilde{X}$, $\tilde{A}$, and $\tilde{B}$ geometrical parameters from experimental or previous theoretical determinations. The agreement with CIS-optimized geometry is satisfactory for the $\tilde{A}$ state, but that for the $\tilde{B}$ state reveals gross discrepancies (i.e., ~0.1 Å differences in C–O and C–C bond lengths). The twisted CIS $\tilde{B}$-state geometry contrasts with the planarity reported[35b,c] in previous theoretical and experimental studies. The soft $\tilde{B}$-state torsional dependence (Figure 7.7) and strong torsional coupling to C–C–O skeletal modes evidently amplify the weaknesses of the CIS approximation to give a qualitatively incorrect equilibrium geometry in this case.

We therefore reoptimized the ground- and excited-state geometries using a CASNBO(7,5) multiconfigurational SCF treatment, as shown in columns 4 to 6 of Table 7.8. The 40-configuration active space was defined by initial NBOs (from a starting B3LYP ground-state calculation) to include the π_{CC} and π_{CC}^* (or π_{CO} and π_{CO}^*) orbitals at C and all remaining non-bonding orbitals (n_O', n_O'', π_{CO}^*) at O, corresponding formally to three doubly occupied, one singly occupied, and one unoccupied active space orbital. A sample G9X input deck for a single-point CASNBO calculation at the optimized $\tilde{B}$-state geometry is shown in Example 7.2. Accurate CASNBO convergence was obtained to the full G9X threshold in all cases, whereas corresponding MO-based CAS treatments gave ill-behaved convergence that prevented successful optimization. (However, partially or fully converged CASMO energies were obtained at final CASNBO-optimized geometries that agreed satisfactorily with CASNBO values.) Frequency analysis

TABLE **7.9** Comparison Best-Available (Experimental, Where Available, or Highest-Level Theoretical) Geometrical Parameters

	$\tilde{X}^{35c}$	$\tilde{A}^{35c}$	$\tilde{B}^{35d}$
	Bond Lengths (Å)		
C–H	1.088	1.098	1.09
C–O	1.261	1.348	1.38
C–C'	1.408	1.333	1.43
C'–H'	1.081	1.074	1.09
C'–H''	1.088	1.072	1.09
	Bond Angles (degrees)		
C'–C–H	119.2	123.9	120.0
C'–C–O	122.4	124.7	123.1
C–C'–H'	119.2	121.8	120.0
C–C'–H'	121.0	119.6	120.0

Source: Data from Osborn, D.L., Choi, H., Mordaunt, D.H., Bise, R.T., Neumark, D.M., and Rohlfing, C.M., *J. Chem. Phys.,* 106, 3049–3066, 1997; Williams, S., Harding, L.B., Stanton, J.F., and Weisshaar, J.C., *J. Phys. Chem. A,* 104, 9906–9913, 10131–10138, 2000.

shows that the optimized $\tilde{A}$ and $\tilde{B}$ structures are both true minima.[39]

EXAMPLE 7.2

Sample Gaussian 9X input deck to evaluate CASNBO(7,5)/6-311++G** wave function for the optimized $\tilde{B}$ state of vinoxy radical. The first job evaluates and checkpoints the NBOs for a nearby DFT ground-state calculation. The second job reads these NBOs from the checkpoint file, permutes NBO 3 with 9 and 84 with 13 (in each spin set) to bring the desired orbitals into the active space, and finds the self-consistent CAS-NBO(7,5) solution for the CI third root (second excited state).

```
%mem=12000000
%chk=vinoxy.chk
#N b3lyp/6-311++g** POP=NBORead
```

CHOCH2 "vinoxy radical" (g.s.) in approximate B-state geometry

```
    0   2
    6            0.088939      0.449508     -0.005353
    1            0.195197      1.516046      0.075691
    8            1.239196     -0.273987     -0.006877
    6           -1.210464     -0.193272      0.005947
    1           -1.281127     -1.261736      0.059446
    1           -2.098482      0.400171     -0.083684
```

$NBO file=v2_caso aonbo=c $END

```
—Link1—
%mem=12000000
%chk=vinoxy.chk
#pcas(7,5,nroot=3)/6-311++g** POP=NBORead guess=(alter,read)
```

CHOCH2 "vinoxy radical" (B-state), CAS/NBO-optimized, E=-152.2381928

```
    0   2
    6            0.088793      0.448918     -0.005296
    1            0.193434      1.515758      0.074413
    8            1.239846     -0.273643     -0.006569
    6           -1.210532     -0.193325      0.005708
    1           -1.283947     -1.261646      0.058389
    1           -2.097825      0.401469     -0.082725

    3      9
   84     13

    3      9
   84     13
```

$NBO file=v2_casnbo $END

Both the active space and basis set of our CAS-NBO(7,5)/6-311++G** treatment are somewhat larger than those of previous CAS studies.[34,35] The calculated CASNBO excitation energies from the $\tilde{X}$ state (−152.368151 a.u.) are 0.88 eV for the $\tilde{A}$ state and 3.53 eV for the $\tilde{B}$ state, in good agreement with the values[35a] (0.99 and 3.56 eV, respectively) inferred from the apparent experimental band origins. The geometrical agreement with previous CAS studies, as well as CIS (Table 7.8) and experimental (Table 7.9) values, is satisfactory for the $\tilde{X}$ and $\tilde{A}$ states. For the $\tilde{B}$ state our final CASNBO geometry is in greatly improved agreement (~0.01 to 0.02 Å) with the experimentally inferred C–C–O skeletal geometry, and the skeletal twisting is greatly reduced compared with the CIS geometry. Nevertheless, residual nonplanarity (~5°) is evident in our best CASNBO structure, in contrast to previous CAS studies.[40]

Spin-averaged NBOs of the CAS equilibrium $\tilde{A}$ and $\tilde{B}$ wave functions are summarized in Table 7.10. The qualitative similarities to the corresponding CIS results for vertical excitations (Table 7.5 and Table 7.6) are apparent, with each CAS-type NBO resembling an effective average of the corresponding CIS-type α and β spin-NBOs. The n_O'' NBO is clearly recognizable as the singly occupied radical site NBO of the $\tilde{A}$ state, whereas the $\tilde{B}$ state exhibits the tri-radical character of three singly occupied NBOs: n_O^π, n_C, and $n_{C'}$. The latter orbitals correspond to the highly spin-polarized π_{CC} spin-NBOs of the CIS-type description, representing the singlet–diradical character of the π_{CC} region. The qualitative persistence of the NBO Lewis-structure patterns through large geometry changes on the excited-state surfaces parallels the familiar robustness of ground-state NBO Lewis-structure patterns.

Let us now compare the spin-averaged CAS-type NBOs with those of corresponding CIS wave functions for the same equilibrium geometry, as shown in Table 7.11. The average of α, β CIS-type NBOs closely resembles the CAS-type NBOs (Table 7.10) for the $\tilde{A}$ state, but gross discrepancies are evident for the $\tilde{B}$ state. From the β manifold one can easily recognize that the CIS $\tilde{B}$ state is the $n_{C'}$-hole configuration that correlates diabatically with the CIS $\tilde{C}$ state in vertical geometry (Table 7.8). The mismatch between CAS and CIS states could also be easily recognized from NPA atomic charges or

TABLE 7.10 NBOs of CASNBO(7,5)/6-311++G** Wave functions for Optimized Ã and B̃ Equilibrium Structures

NBO	Ã state (percent ρ_L = 98.87%)		B̃ state (percent ρ_L = 95.65%)	
	Occ.	$c_A h_A + c_B h_B$	Occ.	$c_A h_A + c_B h_B$
σ_{CH}	1.9659	$0.7655(sp^{2.32})_C + 0.6434(s)_H$	1.9920	$0.7515(sp^{2.02})_C + 0.6598(s)_H$
σ_{CO}	1.9968	$0.5846(sp^{2.82})_C + 0.8113(sp^{2.03})_O$	1.9972	$0.5881(sp^{2.54})_C + 0.8088(sp^{2.48})_O$
n_O'	1.9914	$(sp^{0.49})_O$	1.9935	$(sp^{0.40})_O$
n_O''	1.0175	$(p)_O$	1.9591	$(p)_O$
n_O^{π}	1.8923	$(p_z)_O$	1.0619	$(p_z)_O$
σ_{CC}	1.9861	$0.7170(sp^{1.27})_C + 0.6971(sp^{1.63})_{C'}$	1.9970	$0.7105(sp^{1.55})_C + 0.7037(sp^{1.87})_{C'}$
π_{CC}	1.9207	$0.6787(p)_C + 0.7344(p)_{C'}$		
σ_{CH}'	1.9868	$0.7678(sp^{2.19})_{C'} + 0.6407(s)_{H'}$	1.9918	$0.7628(sp^{2.05})_{C'} + 0.6467(s)_{H'}$
σ_{CH}''	1.9832	$0.7682(sp^{2.24})_{C'} + 0.6402(s)_{H''}$	1.9913	$0.7603(sp^{2.08})_{C'} + 0.6495(s)_{H''}$
n_C			0.9125	$(p)_C$
n_C'			1.0177	$(p)_{C'}$

TABLE 7.11 Similar to Table 7.4, for Equilibrium $\tilde{A}$ and $\tilde{B}$ Excited States [CIS/6-311++G**//CASNBO(7,5)/6-311++G** level] (cf. Table 7.10)

NBO	α Spin		β Spin	
	occ.	$c_A h_A + c_B h_B$	occ.	$c_A h_A + c_B h_B$
		$\tilde{A}_{eq}$ state: percent $\rho_L(\alpha) = 99.13\%$		percent $\rho_L(\beta) = 99.08\%$
σ_{CH}	0.9942	$0.7443(sp^{2.27})_C + 0.6678(s)_H$	0.9717	$0.7884(sp^{2.34})_C + 0.6152(s)_H$
σ_{CO}	0.9987	$0.5591(sp^{2.85})_C + 0.8291(sp^{2.06})_O$	0.9982	$0.6056(sp^{2.91})_C + 0.7958(sp^{1.94})_O$
n_O'	0.9959	$(sp^{0.49})_O$	0.9954	$(sp^{0.51})_O$
n_O''	0.9867	$(p)_O$	[0.0423]	$[(p)_O]^*$
n_O^π	0.9401	$(p_z)_O$	0.9649	$(p_z)_O$
σ_{CC}	0.9970	$0.6988(sp^{1.28})_C + 0.7154(sp^{1.57})_{C'}$	0.9891	$0.7335(sp^{1.23})_C + 0.6797(sp^{1.67})_{C'}$
π_{CC}	0.9991	$0.5061(p)_C + 0.8625(p)_{C'}$	0.9950	$0.8228(p)_C + 0.5683(p)_{C'}$
σ_{CH}'	0.9932	$0.7772(sp^{2.24})_{C'} + 0.6293(s)_{H'}$	0.9935	$0.7606(sp^{2.18})_{C'} + 0.6492(s)_{H'}$
σ_{CH}''	0.9920	$0.7775(sp^{2.28})_{C'} + 0.6288(s)_{H''}$	0.9912	$0.7624(sp^{2.20})_{C'} + 0.6471(s)_{H''}$
		$\tilde{B}_{eq}$ state: percent $\rho_L(\alpha) = 98.97\%$		percent $\rho_L(\beta) = 96.36\%$
σ_{CH}	0.9957	$0.7566(sp^{2.05})_C + 0.6538(s)_H$	0.9814	$0.7741(sp^{1.91})_C + 0.6330(s)_H$
σ_{CO}	0.9981	$0.5979(sp^{2.50})_C + 0.8016(sp^{2.02})_O$	0.9895	$0.4529(sp^{3.81})_C + 0.8916(sp^{0.22})_O$
n_O'	0.9944	$(sp^{0.50})_O$	[0.1992]	$[(sp^{16.7})_O]^*$
n_O''	0.9774	$(p)_O$	0.9740	$(p)_O$
n_O^π	0.9067	$(p_z)_O$	0.9524	$(p_z)_O$
σ_{CC}	0.9966	$0.6959(sp^{1.58})_C + 0.7181(sp^{1.86})_{C'}$	0.9903	$0.7357(sp^{1.57})_C + 0.6773(sp^{2.08})_{C'}$
π_{CC}	0.9953	$0.4615(p)_C + 0.8872(p)_{C'}$	0.7579	$0.8344(sp^{15.1})_C + 0.5511(sp^{45.7})_{C'}$
σ_{CH}'	0.9958	$0.7809(sp^{2.05})_{C'} + 0.6247(s)_{H'}$	0.9905	$0.7635(sp^{2.09})_{C'} + 0.6458(s)_{H'}$
σ_{CH}''	0.9952	$0.7741(sp^{2.15})_{C'} + 0.6331(s)_{H''}$	0.9647	$0.7648(sp^{2.02})_{C'} + 0.6443(s)_{H''}$

other valence descriptors. Hence, the CIS approximation is unreliable for determining the energetics and NBO/NRT descriptors of the $\tilde{B}$ state, even in the near-experimental CAS-optimized geometry. This example illustrates how NBO analysis can spotlight chemically significant differences between one approximate wave function and another.

To illustrate the NRT description of an equilibrium excited state, we therefore confine attention to the $\tilde{A}$ state, where the CIS wave function appears at least qualitatively reasonable compared with the more accurate CAS description (for which NRT analysis is inapplicable; Section 7.2.1). For the equilibrium $\tilde{A}$-state species at the CIS/6-311++G** level, the NRT structures and weightings obtained for the α manifold are

17a (90.7%)   **17b** (6.8%)

and those for the β manifold are

18a (91.9%)   **18b** (2.9%)   **18c** (2.4%)

The leading $\tilde{A}$-state equilibrium structures **17** and **18** are similar to the vertical structures **9** and **10** but shifted to more localized enol forms, leading to equilibrium NRT bond orders.

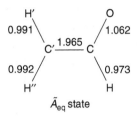

$\tilde{A}_{eq}$ state

Thus, the equilibrium $\tilde{A}$-state species corresponds closely to an idealized localized vinoxy-like Lewis bonding pattern, whereas other CH_2CHO states exhibit significant formylmethyl character. The $\tilde{A} \leftarrow \tilde{X}$ changes in CC and CO bond lengths (Table 7.8) are qualitatively consistent with the calculated NRT bond-order changes (Δb_{CC} = +0.67, Δb_{CO} = −0.68). However, too few data points are available to judge whether general state-to-state bond order–bond length correlations are of useful accuracy. This interesting question remains open to future research.

7.4.3 Conical Intersections and Related Internal Conversion Pathways

Let us now examine the special regions of near or exact degeneracy of potential-energy surfaces, leading to high probability of de-excitation (electronic surface hopping) without fluorescent emission. In a diatomic molecule such degeneracy corresponds to one-dimensional curve-crossing (real or avoided), whereas the corresponding feature for polyatomics is a multidimensional *conical intersection* or "funnel" crossing.[41] If $\mathbf{R}$ denotes the collective nuclear geometry, the conical intersection $\vec{\mathbf{R}}_c$ between, for example, the $\tilde{X}$ and $\tilde{A}$ potential energy surfaces, is identified as the minimum energy for which

$$E_{\tilde{X}}\left(\vec{\mathbf{R}}_c\right) \simeq E_{\tilde{A}}\left(\vec{\mathbf{R}}_c\right) \tag{7.14}$$

where "=" denotes a true intersection and "~" an avoided crossing. In the near-degenerate region the probability of nonadiabatic $\tilde{X} \leftarrow \tilde{A}$ transitions is greatly enhanced, in accord with the Landau–Zener model.[42] Conical intersections therefore provide highly favorable gateways for nonradiative internal conversion from one adiabatic surface to another. Eventual patterns of rearrangement or vibrational-energy disposal on the lower surface are likely to be imprinted with memory of the initial conditions imposed by the conical portal.

In the G9X framework, a conical intersection is determined[43] by variational minimization of the energy at which intersection (Equation 7.14) of adjacent states (i.e., crossing of CAS roots) occurs. State averaging is used to reduce the CAS bias toward one state or the other, so the calculated energy of the conical degeneracy need not agree exactly with individual CAS energies calculated separately at the same geometry. In each case we present NBO analysis of the state-averaged CASNBO wave function for the formal upper root at the optimized conical geometry (i.e., the $\tilde{A}$-like root for the $\tilde{X}$–$\tilde{A}$ intersection or the $\tilde{C}$-like root for the $\tilde{B} - \tilde{C}$ intersection). For the alternative lower root, we used a single-point CASNBO(7,5)/6-311++G** calculation (no state averaging) at the same geometry.

A converged vinoxy $\tilde{X}$–$\tilde{A}$ conical intersection was successfully located at 12.8 kcal/mol above the $\tilde{A}$-state equilibrium level,[44] with geometry shown in the " $\tilde{X}$–$\tilde{A}_{cx}$" column in Table 7.8. A $\tilde{B} - \tilde{C}_{cx}$ conical interaction was similarly located[45] at 12.2 kcal/mol above the $\tilde{B}$-state equilibrium, with geometry shown in the final column of Table 7.8. Figure 7.8 shows ball-and-stick diagrams of the CASNBO-optimized equilibrium and conical intersection structures for visual comparison.

Table 7.12 summarizes the total and relative energetics (E and ΔE) and NBO localizability (percent ρ_L) of the low-lying equilibrium and conical intersection species, all calculated at the CASNBO(7,5)/6-311++G** level. (Note that the sharply lower percent ρ_L value for the $\tilde{B}$ state is principally due to the inherent inaccuracy of the CAS-GUGA spin averaging for the highly spin-polarized π_{CC} bond pair; if NBOs for C–C π-bonding are ignored, the localizability of the $\tilde{B}$ state is fairly comparable with that of the lower states.) Although details of the individual Lewis-structures vary widely (see

TABLE 7.12   Total Energy E (a.u.), Relative Energy ΔE (kcal/mol), and NBO Localizability Percent ρ_L for Low-Lying Equilibrium and Conical Intersection Geometries (cf. Table 7.8) of Vinoxy Radical at the CASNBO(7,5)/6-311++G** Level

	E (a.u.)	ΔE (kcal/mol)	Percent ρ_L
$\tilde{X}$	-152.3690399	0.0	98.51
$\tilde{A}$	-152.3359533	20.8	98.87
$\tilde{B}$	-152.2381928	82.1	95.65
$\tilde{X}-\tilde{A}_{cx}$	-152.3156165	33.5	99.18
$\tilde{B}-\tilde{C}_{cx}$	-152.2186809	94.3	95.25

below), the overall accuracy of a localized NBO Lewis structure description is found to be comparable in all these cases,
showing the general usefulness of localized NBO descriptions
for a variety of photospecies.

Table 7.13 shows details of the NBOs and occupancies
for $\tilde{X}-\tilde{A}_{cx}$ and $\tilde{B}-\tilde{C}_{cx}$ conical intersection structures. Comparison of Table 7.10 and Table 7.13 shows that the forms
and occupancies of the equilibrium $\tilde{A}$-state NBOs are well
conserved at the $\tilde{X}-\tilde{A}_{cx}$ conical intersection geometry. Similarly. the equilibrium $\tilde{B}$-state NBOs are well conserved at the
$\tilde{B}-\tilde{C}_{cx}$ conical intersection geometry. These examples illustrate the transferability and recognizability of NBO struc-
tural elements across broad stretches of the excited-state
potential-energy surfaces.

Visual comparison of Figure 8b,c (or examination of the
corresponding geometrical parameters in Table 7.8) shows
that the $\tilde{X}-\tilde{A}_{cx}$ conical intersection geometry is achieved with
only modest distortions from the equilibrium $\tilde{X}$-state or $\tilde{A}$-
state geometry. However, Figure 8c,e show that the $\tilde{B}-\tilde{C}_{cx}$
conical intersection requires strong twisting from equilibrium $\tilde{B}$-state geometry. Moreover, the CH_2 and CHO groups
become significantly pyramidalized at both ends of the torsional C–C axis, and the two methylenic C–H bonds become
distinctly inequivalent.

What is the physical origin of the $\tilde{B}-\tilde{C}_{cx}$ conical intersection, and how can we understand its weirdly distorted
geometry? Qualitatively, we can associate the conical intersection with the near-degeneracy of n_O^{π}-radical vs. the $n_{O''}$-

TABLE 7.13 Similar to Table 7.10, for both States (Roots) Found at Optimized $\tilde{X}-\tilde{A}_{cx}$ and $\tilde{B}-\tilde{C}_{cx}$ Conical Intersection Structures[a]

	$\tilde{B}-\tilde{A}_{cx}$ ($\tilde{A}$-root)		$\tilde{X}-\tilde{A}_{cx}$ ($\tilde{X}$-root; $E=-152.2985534$)	
NBO	Occ.	$c_A h_A + c_B h_B$	Occ.	$c_A h_A + c_B h_B$
σ_{CH}	1.9651	$0.7654(sp^{2.41})_C + 0.6436(s)_H$	1.9834	$0.7592(sp^{2.52})_C + 0.6508(s)_H$
σ_{CO}	1.9965	$0.5885(sp^{2.80})_C + 0.8085(sp^{2.09})_O$	1.9953	$0.5881(sp^{2.99})_C + 0.8088(sp^{2.47})_O$
n_O'	1.9905	$(sp^{0.48})_O$	1.9918	$(sp^{0.48})_O$
n_O''	1.0085	$(p)_O$	1.9750	$(p)_O$
n_O^{π}	1.9046	$(p_z)_O$	1.0165	$(p_z)_O$
σ_{CC}	1.9819	$0.7154(sp^{1.23})_C + 0.6987(sp^{1.59})_{C'}$	1.9871	$0.7094(sp^{1.13})_C + 0.7048(sp^{1.52})_{C'}$
π_{CC}	1.9980	$0.6831(p)_C + 0.7303(p)_{C'}$	1.9412	$0.7060(p)_C + 0.7082(p)_{C'}$
σ_{CH}'	1.9859	$0.7716(sp^{2.21})_{C'} + 0.6361(s)_{H'}$	1.9836	$0.7709(sp^{2.25})_{C'} + 0.6370(s)_{H'}$
σ_{CH}''	1.9818	$0.7709(sp^{2.29})_{C'} + 0.6370(s)_{H''}$	1.9769	$0.7606(sp^{2.35})_{C'} + 0.6373(s)_{H''}$

	$\tilde{B}-\tilde{C}_{cx}$ ($\tilde{C}$-root)		$\tilde{B}-\tilde{C}_{cx}$ ($\tilde{B}$-root)	
NBO	Occ.	$c_A h_A + c_B h_B$	Occ.	$c_A h_A + c_B h_B$
σ_{CH}	1.9651	$0.7654(sp^{2.41})_C + 0.6436(s)_H$	1.9920	$0.7515(sp^{2.02})_C + 0.6598(s)_H$
σ_{CO}	1.9965	$0.5885(sp^{2.80})_C + 0.8085(sp^{2.09})_O$	1.9972	$0.5881(sp^{2.54})_C + 0.8088(sp^{2.48})_O$
n_O'	1.9905	$(sp^{0.48})_O$	1.9935	$(sp^{0.40})_O$
n_O''	1.0085	$(p)_O$	1.9591	$(p)_O$
n_O^{π}	1.9046	$(p_z)_O$	1.0619	$(p_z)_O$
σ_{CC}	1.9819	$0.7154(sp^{1.23})_C + 0.6987(sp^{1.59})_{C'}$	1.9970	$0.7105(sp^{1.55})_C + 0.7037(sp^{1.87})_{C'}$
π_{CC}	1.9980	$0.6831(p)_C + 0.7303(p)_{C'}$		
σ_{CH}'	1.9859	$0.7716(sp^{2.21})_{C'} + 0.6361(s)_{H'}$	1.9918	$0.7628(sp^{2.05})_{C'} + 0.6467(s)_{H'}$
σ_{CH}''	1.9818	$0.7709(sp^{2.29})_{C'} + 0.6370(s)_{H''}$	1.9913	$0.7603(sp^{2.08})_{C'} + 0.6495(s)_{H''}$

[a] In each case the nominal upper root is the wave function from the state-averaged CASNBO opt = conical run, and the nominal lower root is a single-point (non-state-averaged) CASNBO calculation at the same geometry.

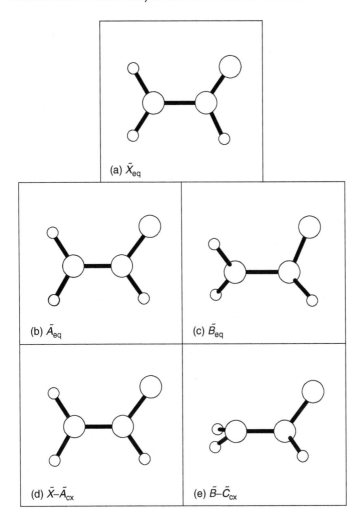

(a) $\tilde{X}_{eq}$

(b) $\tilde{A}_{eq}$

(c) $\tilde{B}_{eq}$

(d) $\tilde{X}$–$\tilde{A}_{cx}$

(e) $\tilde{B}$–$\tilde{C}_{cx}$

Figure 7.8 Optimized CASNBO(7,5)/6-311++G** structures (cf. Table 7.8) for low-lying vinoxy equilibrium and conical intersection species.

radical NBO configurations, that is, with the half-filled p_O orbital lying either in-plane ($n_{O''}$ -radical) or out-of-plane (n_O^π -radical) with respect to the skeletal C–C–O plane. In planar geometry, the n_O^π -radical can participate in favorable allylic-like resonance with the adjacent π_{CC} bond, whereas in

90°-twisted geometry only the n_O'' orbital can be stabilized by such interactions. Near-equivalence of these configurations may therefore be expected for angles intermediate between these extremes, as observed (H-C-C'-H' $\simeq$ 45°). In the twisted geometry of weakened allylic conjugation, the two methylenic C–H bonds participate asymmetrically in vicinal hyperconjugative interactions with the CHO moiety, leading to asymmetric geometry distortions that further strengthen these interactions.

We can obtain a more quantitative description of the competition between n_O^π and n_O'' configurations by examining the $\tilde{B}$-state torsional behavior in greater detail. For this purpose, we display in Figure 7.9 the calculated adiabatic torsional potential for the $\tilde{B}$ state, with all remaining variables optimized with respect to chosen values of dihedral angle $\phi =$ H'–C'–C–O in the range of $0° \leq \phi \leq 180°$. As shown in Figure 7.9, the $\tilde{B}$-state torsional potential exhibits complex overall shape, with a barrier height of about 7 kcal/mol. Pronounced pyramidalization of the acyl group is apparent at the transition state (i.e., H'–C'–C–H = 226°), leading to an asymmetrical conformational profile near ϕ = 90°. The torsional potential also reveals distinct changes in NBO electronic character, as shown by the distinct symbols for n_O^π-radical (circles) or n_O''-radical (squares) configurations at each torsion angle. One can see, as expected, that the n_O''-radical configuration tends to be favored for angles near ϕ = 90°, whereas the n_O^π-radical configuration is favored for near-planar geometries near $\phi \simeq$ 0° or 180°. Cross-overs between n_O^π- and n_O''-radical character are apparent near 65° and 130°, and the former was taken as the starting geometry for the conical intersection search that culminated in the $\tilde{B} - \tilde{C}_{cx}$ species of Figure 7.8e. The behavior in the range of 125° < ϕ < 130° was not investigated further, due to severe instability in CASSCF convergence and geometry optimization.

Figure 7.10 shows a blow-up of the indicated NBO cross-over region of the torsional potential, with fine-grained variations ($\Delta\phi$ = 1°) in the range of 60° < ϕ < 75°. The distinct trend-lines of n_O^π-radical (circles, solid line) vs. n_O''-radical (squares, dotted line) segments indicate a curve-crossing near $\phi_x \simeq$ 67° and $E(\phi_x)$ = –152.2309 a.u., about 4.8 kcal/mol above

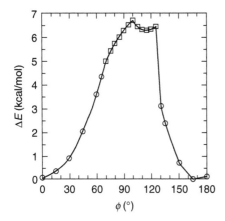

Figure 7.9 Adiabatic torsion potential $\Delta E(\phi)$ for vinoxy $\tilde{B}$ state, showing changes of n_O^{π} (circles) vs. n_O'' (squares) radical NBO character with O–C–C′–H′ dihedral angle ϕ (all other variables optimized) at the CASNBO(7,5)/6-311++G** level.

the $\tilde{B}$-state equilibrium geometry. Although a casual examination of the coarse-grained potential energy profile (Figure 7.9) might not suggest the fundamental electronic rearrangement near ϕ_x, the NBO configurational assignments make it easy to locate and characterize this crossing.

Of course, the apparent crossing shown in Figure 7.10 is not a true degeneracy in the full space of geometrical variables, because the values of adiabatically optimized nontorsional coordinates are different at ϕ_x, as summarized in Table 7.14. The most important mismatches are those involving pyramidalization at the methylenic center, as depicted in Figure 7.11. In addition, there is significant closing of the skeletal C′–C–O angle (by ~12°) and opening of the CC′H″ angle (by ~4°) in the n_O'' -radical species, but bond length mismatches are quite small (i.e., –0.018 Å and +0.006 Å for the key skeletal CO and CC bonds). Removing these mismatches requires raising the energy by 7.4 kcal/mol to reach the true $\tilde{B} - \tilde{C}_{cx}$ conical intersection shown in Figure 7.8e. Because the latter lies about 5 kcal/mol above the $\tilde{B}$-state torsional barrier (Figure 7.9), it presumably plays no significant role in low-energy $\tilde{B}$-state vibrational dynamics.

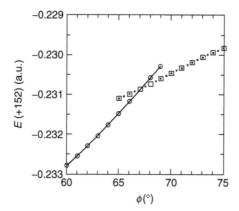

Figure 7.10 Expanded view of adiabatic $\tilde{B}$-state torsional potential in the region of the torsional crossing of n_O^π-radical (circles, solid line) and n_O''-radical (squares, dotted line) species (cf. Figure 7.9). The crossing occurs at $E_x \simeq -152.23087$ a.u. and $\phi_x \simeq 67.1$.

TABLE 7.14   Optimized Bond Lengths and Angles for Near-Crossing $\tilde{B}$-state n_O^π-Radical and n_O''-Radical Species [CASNBO(7,5)/6-311++G** level] at Constrained Torsion Angle $\phi = 67$ (cf. Figures 7.10 and 7.11)

	n_O^π-Radical	n_O''-Radical
Bond Lengths (Å)		
C–H	1.0776	1.0790
C–O	1.3850	1.3668
C–C$'$	1.4469	1.4532
C$'$–H$'$	1.0773	1.0737
C$'$–H$''$	1.0772	1.0750
Bond Angles (degrees)		
C$'$–C–H	119.96	119.74
C$'$–C–O	119.22	106.75
C–C$'$–H$'$	119.69	119.94
C–C$'$–H$''$	117.13	121.32
H–C–C$'$–H$'$	216.41	198.09
H$'$–C$'$–C–O	[67.00]	[67.00]
H$''$–C$'$–C–O	214.21	248.90

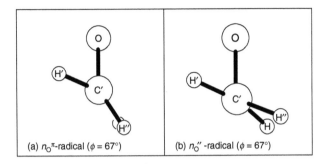

(a) n_O^π-radical ($\phi = 67°$)　　(b) n_O''-radical ($\phi = 67°$)

Figure 7.11 Newman projection views along the C–C axis of $\tilde{B}$-state vinoxy radical near the torsional curve-crossing for (a) n_O^π-radical and (b) n_O''-radical species (cf. Figure 7.10, Table 7.9), showing the greater pyramidalization at the methylenic center in the former case.

However, the torsional $n_O^\pi - n_O''$ crossing in Figure 7.10 evidently provides a favorable portal for low-energy vibrational relaxation and internal conversion. Taking account of the zero-point torsional energy of the equilibrium $\tilde{B}$ state (0.67 kcal/mol), we estimate that vibrational excitation of only 4.1 kcal/mol (~1440 cm^{-1}) is sufficient to achieve the torsional crossing at $\phi_x = 67°$ that leads to efficient radiationless crossing to the alternative surface. This calculated $n_O^\pi - n_O''$ threshold is in excellent agreement with the observed 1400 cm^{-1} fluroresence quenching above the $\tilde{B}$-state band origin,[35c] a conspicuous (and previously unexplained) feature of the photoexcitation spectrum. This picture is also consistent with the observed[35c] mode-specific effect of a torsional excitations in promoting dissociation.

Table 7.15 displays NBOs and occupancies of the n_O^π- and n_O''-radical species at the torsional crossing. Despite strong geometric deformation and electronic excitation, the resemblance of these NBOs to one another as well as to those of lower-lying species is quite apparent. Thus, one can conclude that a compact set of NBOs (or their NHO hybrid constituents) provide a useful basis set for concise, accurate, and descriptive valence-shell configurational assignments over

TABLE 7.15 Similar to Table 7.13, for the n_O^{κ}-Radical (Figure 7.11a) and n_O''-Radical (Figure 7.11b) Species at the Torsional Crossing $\phi_x = 67$.

| NBO | n_O^{κ}-Radical | | n_O''-Radical | |
	Occ.	$c_A h_A + c_B h_B$	Occ.	$c_A h_A + c_B h_B$
σ_{CH}	1.9898	$0.7500(sp^{2.05})_C + 0.6614(s)_H$	1.9868	$0.7526(sp^{2.03})_C + 0.6585(s)_H$
σ_{CO}	1.9967	$0.5851(sp^{2.66})_C + 0.8110(sp^{2.66})_O$	1.9901	$0.5761(sp^{2.73})_C + 0.8174(sp^{2.46})_O$
n_O'	1.9941	$(sp^{0.38})_O$	1.9928	$(sp^{0.41})_O$
n_O''	1.9632	$(p)_O$	1.2885	$(p)_O$
n_O^{π}	1.0204	$(p_z)_O$	1.6727	$(p_z)_O$
σ_{CC}	1.9966	$0.7081(sp^{1.56})_C + 0.7062(sp^{1.82})_C{'}$	1.9866	$0.7041(sp^{1.65})_C + 0.7101(sp^{1.84})_C{'}$
σ_{CH}'	1.9881	$0.7575(sp^{2.09})_C{'} + 0.6487(s)_H{'}$	1.9912	$0.7621(sp^{2.12})_C{'} + 0.6474(s)_H{'}$
σ_{CH}''	1.9872	$0.7575(sp^{2.23})_C{'} + 0.6528(s)_H{''}$	1.9877	$0.7613(sp^{2.05})_C{'} + 0.6484(s)_H{''}$
n_C	0.9655	$(sp^{65.1})_C$	1.0541	$(sp^{29.0})_C$
n_C'	0.9837	$(sp^{68.6})_C{'}$	0.9152	$(p)_C{'}$

broad regions of the excited-state potential-energy manifold. The departures from idealized single and double occupancies of these NBOs provide the starting point for describing delocalization and resonance effects in a manner that parallels that for ground-state species. As seen in Table 7.15, these departures tend to be greater than those in near-equilibrium ground-state species (see, e.g., the occupancy $1.2885e$ of the singly occupied n_O'' -radical orbital). Nevertheless, the delocalization patterns can be visualized and rationalized in terms of the pre-NBO overlap of available donor and acceptor orbitals (or, if available, from corresponding Fock matrix elements and second-order interaction energy estimates), paralleling the analysis in closed-shell equilibrium species.

7.4.4 Complexation and Solvation Effects

Solution-phase photochemical events differ appreciably from those in the gas phase. In the absence of radiative emission, the excess vibrational energy of a gas-phase photospecies can be removed only by dissociative processes that convert internal vibrational energy into the relative kinetic energy of dissociating photofragments. The nascent photospecies may therefore explore large regions of the potential-energy surface before migrating into a dissociative exit channel, and the eventual fragmentation pattern reflects surface features far from equilibrium. However, in solution phase the enforced interactions with nearby solvent molecules provide efficient mechanisms for intermolecular vibrational energy transfer that rapidly relax the species toward the nearest local minimum. Solution-phase photochemistry therefore tends to be dominated by near-equilibrium features of the potential-energy surface, whereas gas-phase photochemistry reflects the more exotic characteristics of far-from-equilibrium features.

The efficiency of solvent-induced vibrational quenching is strongly enhanced as solute-solvent interactions increase in strength and specificity. Particularly important are the ubiquitous *hydrogen bonding* interactions of aqueous solutions. As a first step toward bridging the gap between gaseous

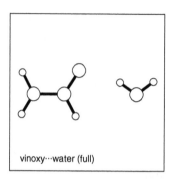

Figure 7.12 Equilibrium geometry of vinoxy···water complex (B3LYP/6-311++G** level) (cf. Table 7.16.)

and solution phases, let us therefore consider the effect of complexation by a single water molecule, i.e., a solvated vinoxy$(H_2O)_n$ cluster in the low-density limit $n = 1$.

Figure 7.12 displays the equilibrium B3LYP/6-311++G** geometry of the $CH_2CHO···H_2O$ dimer, with detailed geometrical parameters presented in Table 7.16. The calculated binding energy is $\Delta E_{H\text{-bond}} = 6.20$ kcal/mol (uncorrected for effects of zero-point energy or basis set superposition error, both expected to be relatively unimportant for current qualitative purposes). As the geometry suggests, an O···H-O hydrogen bond is formed between radical and water molecules. What is the effect of this H-bond on ground states and excited states of vinoxy radical?

Table 7.17 exhibits NBOs of the equilibrium $\tilde{X}$-state vinoxy···water complex in a form directly comparable with the isolated vinoxy radical of Table 7.14. Subtle changes of Lewis structure are seen throughout but most notably in the π_{CO} bond repolarization (from 50.9 to 54.0% on O) and occupancy increase (by 0.0068e). More significant population shifts are found in the valence antibond occupancies (not shown) and net atomic and molecular unit charges (not shown), which correspond to an overall transfer of 0.01836e from the vinoxy to the water moiety. As suggested by Equation 7.9, population shifts Δq of this magnitude are expected to correspond to stabilization energies of the order of 0.02 a.u. (~12 kcal/mol),

TABLE 7.16   Optimized Bond Lengths and Angles of Vinoxy-Water Complex in Equilibrium $\tilde{X}$-State (B3LYP/6-311++G^{**}; cf. Figure 7.12) and $\tilde{A}$-State (CIS/6-311++G^{**}; cf. Figure 7.14)[a]

	$\tilde{X}$	$\tilde{A}$
Bond Lengths (Å)		
C–H	1.1020	1.0762
C–O	1.2396	1.3791
C–C′	1.4215	1.3162
C′–H′	1.0836	1.0744
C′–H″	1.0835	1.0728
O···O′	2.8641	3.2176
O′–H_f	0.9609	0.9409
O′–H_b	0.9713	0.9426
Bond Angles (degrees)		
C′–C–H	117.80	125.70
C′–C–O	122.54	123.23
C–C′–H′	119.48	122.29
C–C′–H″	120.94	119.21
C–O···O′	100.71	140.87
O···O′–H_f	119.74	101.83
H_f–O′–H_b	105.55	106.04
H–C–C′–H′	179.98	179.78
H′–C′–C–O	359.99	0.01
H″–C′–C–O	179.98	180.09
C′–C–O···O′	179.93	85.90
C–O···O′–H_f	186.14	182.26
O···H_f–O′–H_b	1.75	1.72

[a] H_f, H_b are, respectively, the free and H-bonded protons of the H_fO′–H_b water monomer.

given that the separation of donor-acceptor orbital energies is a large number of order unity in atomic units. Further details and consequences of the intermolecular delocalization (charge transfer) of ~2% of an electron charge from solvent to radical are discussed below.

Subtle solvent-induced shifts are also seen in the NRT resonance weightings. In vinoxy-water, the leading α-spin structures corresponding to **6a** and **6b** are

TABLE 7.17 Similar to Table 7.4, for Vinoxy···Water Ground-State Complex (B3LYP/6-311++G**)[a]

NBO	α spin		β spin	
	Occ.	$c_A h_A + c_B h_B$	Occ.	$c_A h_A + c_B h_B$
σ_{CH}	0.9942	$0.7643(sp^{2.20})_C + 0.6448(s)_H$	0.9943	$0.7651(sp^{2.26})_C + 0.6439(s)_H$
σ_{CO}	0.9985	$0.5894(sp^{2.14})_C + 0.8078(sp^{1.61})_O$	0.9985	$0.5974(sp^{2.15})_C + 0.8020(sp^{1.58})_O$
π_{CO}			0.9343	$0.6783(p)_C + 0.7348(p)_O$
$n_O{'}$	0.9916	$(sp^{0.63})_O$	0.9909	$(sp^{0.64})_O$
$n_O{''}$	0.9476	$(p)_O$	0.9462	$(p)_O$
n_O^{π}	0.8338	$(p_z)_O$		
σ_{CC}	0.9975	$0.7037(sp^{1.66})_C + 0.7105(sp^{1.72})_C{'}$	0.9974	$0.7169(sp^{1.62})_C + 0.6971(sp^{1.73})_C{'}$
π_{CC}	0.9989	$0.4984(p)_C + 0.8669(p)_C{'}$		
$\sigma_{CH}{'}$	0.9946	$0.7777(sp^{2.14})_C{'} + 0.6287(s)_H{'}$	0.9948	$0.7639(sp^{2.13})_C{'} + 0.6453(s)_H{'}$
$\sigma_{CH}{''}$	0.9950	$0.7750(sp^{2.18})_C{'} + 0.6320(s)_H{''}$	0.9952	$0.7609(sp^{2.18})_C{'} + 0.6489(s)_H{''}$

[a] NBOs of the water monomer are omitted.

$$
\begin{array}{ccc}
\overset{H}{\underset{H}{\diagdown}}C=C\overset{\cdot\dot{O}\cdot\ (\text{water})}{\underset{H}{\diagup}} &
\overset{H}{\underset{H}{\diagdown}}\dot{C}-C\overset{\dot{O}\cdot\ (\text{water})}{\underset{H}{\diagup}} &
\overset{H}{\underset{H}{\diagdown}}C=C\overset{\cdot\dot{O}\ (\text{water})}{\underset{H\cdot}{\diagup}\!\!\!/}\\
\textbf{19a}\ (51.8\%) & \textbf{19b}\ (42.3\%) & \textbf{19c}\ (1.5\%)
\end{array}
$$

whereas the β-spin structures corresponding to **7a–7c** are

$$
\begin{array}{ccc}
\overset{H}{\underset{H}{\diagdown}}C-\underset{H}{\overset{\diagup}{C}}\overset{\dot{O}\cdot\ (\text{water})}{\diagup\!\!\!/} &
\overset{H}{\underset{H}{\diagdown}}C=C\overset{\dot{O}\cdot\ (\text{water})}{\underset{H}{\diagup}} &
\overset{H}{\underset{H}{\diagdown}}C-C\overset{\dot{O}\ (\text{water})}{\underset{H\cdot}{\diagup\!\!\!/}}\\
\textbf{20a}\ (86.2\%) & \textbf{20b}\ (6.7\%) & \textbf{20c}\ (2.9\%)
\end{array}
$$

The overall NRT bond orders of the complexed vinxoy moiety are

$$
\begin{array}{c}
\overset{0.007}{O\cdots\cdots(\text{water})}\\
\underset{0.992}{H'}\diagdown\ \overset{1.306}{}\ \diagup\ 1.729\\
\underset{0.995}{\diagup}C'\!-\!C\diagdown 0.965\\
H''\qquad H
\end{array}
$$

$\tilde{X}^2 A'$ complex

Comparison with the free vinoxy radical shows an ~1% shift toward the enolate-like structures **19a** and **20b** (largely at the expense of intramolecular hyperconjugation structures such as **19c** and **20c**). This resonance shift increases the anionic (Lewis base) character at oxygen, thereby enhancing the donor-acceptor H-bonding interaction with water (which leads to a weak partial bond order of 0.007 between the monomers). Many other examples of such synergy between intra- and intermolecular resonance interactions have been recognized in the phenomenon of "resonance-assisted H-bonding" (RAHB).[46]

The presence of the H-bond is shown by the characteristic intermolecular NBO delocalization from n_O'' (primarily) of the vinoxy radical to the $\sigma_{O'H}^*$ antibond of the water molecule.

The strength of the attractive $n_O'' \rightarrow \sigma_{O'H}^*$ interaction is estimated as 6.6 kcal/mol (sum of α and β interactions) by second-order perturbation theory (Equation 7.5). The importance of the $n_O'' \rightarrow \sigma_{O'H}^*$ interaction for the H-bonded geometry and energetics can be assessed by deleting the $\left\langle n_O''|\hat{\mathscr{F}}|\sigma_{O'H}^* \right\rangle$ matrix element and reoptimizing the geometry (using the \$DEL keylist option) in the absence of this interaction. The result of carrying out this procedure while holding the internal vinoxy geometry fixed as in Figure 7.12 but reoptimizing the internal and relative coordinates of the attached water monomer is shown in Figure 7.13.

Compared with the full calculation (Figure 7.12), the \$DEL-optimized structure of Figure 7.13 is dramatically

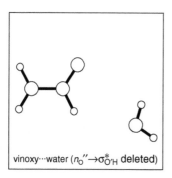

vinoxy···water ($n_O'' \rightarrow \sigma_{O'H}^*$ deleted)

Figure 7.13 \$DEL-optimized geometry of vinoxy···water complex (B3LYP/6-311++G** level), with intermolecular $\left\langle n_O''|\hat{\mathscr{F}}|\sigma_{O'H}^* \right\rangle$ delocalization interaction deleted.

altered. The net binding energy is approximately halved (to ΔE_{bind} = 3.2 kcal/mol), the intermolecular O···H separation is elongated by approximately 1 Å (to $R_{O···H}$ = 2.93 Å, beyond van der Waals contact), and the water monomer is reoriented nearly perpendicular to the normal near-linear O···HO H-bonded arrangement. In addition, the water geometry relaxes back to nearly the isolated monomer form, losing the distinctive R_{OH} bond elongation (by ~0.01 Å) that is a ubiquitous feature of H-bond geometry. The most important energetic and geometrical signatures of H-bonding, therefore, are uniquely associated with the single $n_O'' \rightarrow \sigma^*_{O'H}$ intermolecular interaction. This is consistent with the conclusion from NBO analyses of many systems that nonclassical $n \rightarrow \sigma^*$ donor-acceptor delocalization generally constitutes the single most important contribution to H-bonding.[6]

The $n_O'' \rightarrow \sigma^*_{O'H}$ interaction results in transfer of ~0.02e from the donor n_O'' of vinoxy into the acceptor $\sigma^*_{O'H}$ of water. The n_O'' donor therefore becomes somewhat less available for intramolecular hyperconjugative interactions that depend on the same orbital, weakly raising the effective vinoxy ground-state energy (and reducing hyperconjugative structural effects) relative to an isolated vinoxy radical, as noted above.

However, a more significant effect is expected on the first excited $\tilde{A}$-state complex, in which one of the n_O'' electrons is promoted into the π system. About half the H-bonding attraction (~3 kcal/mol $\simeq$ 1000 cm^{-1}) is lost by removal of the $n_O'' \rightarrow \sigma^*_{O'H}$ interaction in the β spin set, but this is partially compensated by the reverse $\bar{\sigma}_{O'H} \rightarrow \bar{n}_{o''}^*$ delocalization that is thereby opened by emptying of the n_O'' orbital. From the relative pre-NBO overlap integrals for the two interactions ($S_{n\sigma}/S_{n\sigma^*} \simeq 0.6$) and the expected Mulliken-type proportionality to energetic (Fock/Kohn-Sham) matrix elements, we might expect that the residual ~40% effect on the vertical $\tilde{A} \leftarrow \tilde{X}$ promotion energy corresponds to a blue-shift on the order of ~400 cm^{-1}. The calculated vertical $\tilde{X} \leftarrow \tilde{A}$ blue shifts $\Delta\nu_{\text{H-bond}}$ at the CIS/6-311++G** (450 cm^{-1}) or CASNBO(7,5)/6-311++G** (250 cm^{-1}) levels are consistent with this simple estimate.

Geometric relaxation of the $\tilde{A}$-state complex naturally results in reorientation of the water monomer (Lewis acid) to

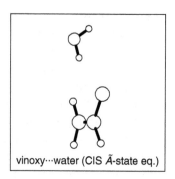

vinoxy···water (CIS Ã-state eq.)

Figure 7.14 Optimized Ã-state vinoxy···water complex (CIS/6-311++G** level), showing nonplanar equilibrium geometry.

achieve more favorable coordination with the Lewis base. As suggested by the NBO Lewis structure of free vinoxy (Table 7.5), the n_O^π lone pair provides the next-best candidate for $n_O^\pi \rightarrow \sigma_{O'H}^*$ delocalization. (The end-on $n_{O'}$ lone pair is too *s*-rich and low in energy to serve as a good Lewis base.) The water monomer therefore swings out of the plane of vinoxy into a nearly perpendicular arrangement as shown in Figure 7.14. Geometrical parameters are given in Table 7.16.

In this nonplanar geometry the intramolecular $n_O^\pi \rightarrow \pi_{CC}^*$ resonance is weakened at the expense of the inter-molecular $n_O^\pi \rightarrow \sigma_{O'H}^*$ interaction, and the radical reverts to a much more localized enol-like form. The NRT description of the relaxed Ã-state complex exhibits these effects, as shown in the leading α-spin structures **21a** to **21c**

21a (92.2%)   **21b** (5.2%)   **21c** (0.9%)

and β spin structures **22a** to **22c**.

22a (92.9%)   **22c** (2.2%)   **22a** (2.1%)

The composite NRT bond orders reflect the almost pure C=C–O vinoxy character in the equilibrium $\tilde{A}$ state of the complex

$\tilde{A}$- state complex

as shown also by the skeletal bond lengths in Table 7.16. It is surprising that even a weak H-bonding interaction is able to effectively disrupt the intramolecular keto-enol resonance, demonstrating the strong effect of solvent complexation on the excited state.

Table 7.18 shows details of the vinoxy NBOs of the relaxed $\tilde{A}$-state complex for comparison with free $\tilde{A}$-state vinoxy NBOs (Table 7.5). Numerous changes are evident, corresponding to effective loss of keto–enol resonance and return to relatively normal C=C and C–O bond lengths. Particularly noticeable is the reduced occupancy of the vacant $\bar{n}_O^\pi$ radical orbital (from $0.096e$ to $0.033e$, reflecting loss of hyperconjugative stabilization of the radical center) and

TABLE 7.18 Similar to Table 7.5, for Equilibrium Ã-State Geometry of Vinoxy···Water Complex (CIS/6-311++G**; cf. Figure 7.14)[a]

NBO	α spin		β spin	
	Occ.	$c_A h_A + c_B h_B$	Occ.	$c_A h_A + c_B h_B$
σ_{CH}	0.9937	$0.7350(sp^{2.18})_C + 0.6615(s)_H$	0.9770	$0.7858(sp^{2.24})_C + 0.6185(s)_H$
σ_{CO}	0.9983	$0.5528(sp^{3.08})_C + 0.8333(sp^{2.31})_O$	0.9977	$0.6029(sp^{3.10})_C + 0.7978(sp^{2.18})_O$
n_O'	0.9964	$(sp^{0.43})_O$	0.9955	$(sp^{0.46})_O$
n_O''	0.9895	$(p)_O$	[0.0330]	$[(p)_O]^*$
n_O^π	0.9539	$(p_z)_O$	0.9695	$(p_z)_O$
σ_{CC}	0.9968	$0.6995(sp^{1.25})_C + 0.7146(sp^{1.54})_{C'}$	0.9912	$0.7282(sp^{1.21})_C + 0.6853(sp^{1.62})_{C'}$
π_{CC}	0.9992	$0.5418(p)_C + 0.8405(p)_{C'}$	0.9950	$0.8134(p)_C + 0.5817(p)_{C'}$
$\sigma_{CH'}$	0.9924	$0.7777(sp^{2.26})_{C'} + 0.6287(s)_{H'}$	0.9928	$0.7631(sp^{2.20})_{C'} + 0.6463(s)_{H'}$
$\sigma_{CH''}$	0.9908	$0.7778(sp^{2.32})_{C'} + 0.6286(s)_{H''}$	0.9902	$0.7646(sp^{2.25})_{C'} + 0.6445(s)_{H''}$

[a]NBOs of the water monomer are omitted.

reduced spin polarization in σ_{CC} and π_{CC} (reflecting reduced diradical character of the C=C double bond). Further discussion of these differences is beyond the scope of this chapter, as is discussion of the accuracy of the CIS/6-311++G** description, used here primarily for illustrative purposes.

7.5 SUMMARY: FUTURE OF NBO-BASED METHODS IN COMPUTATIONAL PHOTOCHEMISTRY

How do we assess the current status and future prospects of NBO-based methods in computational modeling of photochemical phenomena?

The foregoing examples illustrate some current possibilities for NBO/NRT analysis combined with popular excited-state program options in Gaussian 98. Even cursory inspection of these examples suggests that theoretical description of excited states is in a rudimentary stage of development compared with near-equilibrium ground-state species. The *status quo* delineates both the obstacles and opportunities for future development.

Because NBO-based methods are intrinsically tied to the wave functions and densities they analyze, future progress on the NBO side is critically dependent on improved *ab initio* and DFT methods for excited states. Whereas DFT methods have revolutionized the economical description of leading dynamic correlation effects in ground-state phenomena, the restrictive conditions of the Hohenberg–Kohn theorem[47] have inhibited similar progress for excited states (outside the TD framework). Practical and accurate computational models for excited states must confront the daunting factorial-exponential increase of active-space dimensionality with the number of basis functions and correlated electrons. The CASNBO method illustrates how localized NBO transformations may help on this front by effectively reducing the basis dimensionality and more accurately pinpointing the relevant active spaces for localized or delocalized electron correlation treatments. Perhaps some combination of NBO-type dimensional reduction and DFT-type approximations of selected matrix elements can achieve the algorithmic breakthrough needed to bring real progress to excited-state studies.

Even in the framework of present excited-state computational technology, certain improvements would be desirable to extend the scope of NBO/NRT methods for standard CI/CAS applications. Of highest priority is the need for program packages to compute the first-order reduced spin-density matrices (including correlation corrections), so that full NBO analysis can be performed. Similarly important is the need to define a one-electron effective Hamiltonian $\hat{h}_{eff}$ (comparable to the Fock or Kohn-Sham operator for ground states) that would allow extension of second-order energetic analysis, steric analysis, and deletion techniques to excited states.[48] Of course, inclusion of NBO modules in leading specialized packages for excited states is the most elementary *desideratum* for promoting broader NBO-based understanding of excited-state bonding phenomena.

Aside from the necessary methodological developments, further theoretical and experimental efforts should be encouraged to bridge the gap between the small-molecule computational theorists and the practicing photochemists. From the gas-phase side, more experimental guidance is needed (i.e., from vibrationally mediated photodissociation or stimulated emission pumping studies) to test theoretical characterizations of excited-state potential-energy surfaces. However, of greater ultimate importance to the organic photochemist are systematic theoretical and experimental investigations of clusters, aimed at connecting isolated gas-phase species and condensed-phase limits. More quantitative understanding of specific solvation effects and the detailed mechanisms of intra- and intermolecular vibrational and electronic relaxation would seem essential for elucidating realistic photochemical pathways in solution. Increasing evidence suggests the importance of specific solvation effects and the inadequacy of simple continuum dielectric models (e.g., of the Onsager reaction-field type[49]) for describing photospecies in aqueous solution.

Finally, it is evident that confrontation with excited-state phenomena will bring enrichment of the NBO-bonding concepts themselves. For example, it is likely that a localized Lewis-like depiction of a general excited state must include provision for localized valence antibonds or extra-valent Rydberg-type excitations as intrinsic structural units. Such extensions of Lewis-structure concepts present both conceptual and algorithmic challenges for the future.

NOTE ADDED AFTER MANUSCRIPT COMPLETION

Matsika and Yarkony[50] have recently reported computational studies of low-lying vinoxy states that offer numerous points of numerical comparison with the present work.

ACKNOWLEDGMENTS

Thanks are due to Christine Morales for discussion and programming assistance and to Mohamed Ayoub, Fleming Crim, David Osborn, Jim Weisshaar, Martin Zanni, and Howard Zimmerman for comments on the manuscript. Computational resources provided to the University of Wisconsin Chemistry Computer Facility through NSF Grant CHE0091916 and gifts from the Intel Corporation are gratefully acknowledged.

REFERENCES

1. Foster, J.P. and Weinhold, F., *J. Am. Chem. Soc.*, 102, 7211–7218, 1980; Reed, A.E. and Weinhold, F., *J. Chem. Phys.* 78, 4066–4073, 1983.

2. "The underlying physical laws necessary for the mathematical theory of...the whole of chemistry are thus completely known," Dirac, P.A.M., *Proc. R. Soc.*, 123, 714, 1929; "In so far as quantum mechanics is correct, chemical questions are problems in applied mathematics," Eyring, H., Walter, J., and Kimball, G.E., *Quantum Chemistry*, Wiley, New York, 1944, p. iii.

3a. Weinhold, F., Natural bond orbital methods, in *Encyclopedia of Computational Chemistry*, Vol. 3, Schleyer, P.V.R. et al., Eds., Wiley, New York, 1998, pp. 1792–1811.

3b. Glendening, E.D., Badenhoop, J.K., Reed, A.E., Carpenter, J.E., Bohmann, J.A., Morales, C.M., and Weinhold, F., *NBO 5.0*, Theoretical Chemistry Institute, University of Wisconsin, Madison, 2001. (The NBO 5.0 manual is obligatory for informed use of the program and should be consulted for further details on options and keywords mentioned herein.)

3c. See http://www.chem.wisc.edu/~nbo5 for general introduction to the NBO 5.0 program, background references, and tutorials.

4. Reed, A.E., Weinstock, R.B., and Weinhold, F., *J. Chem. Phys.*, 83, 735–746, 1985.

5. See, e.g., Bachrach, S.M., in *Reviews in Computational Chemistry*, Vol. 5, Lipkowitz, K., and Boyd, D.B., Eds., VCH, New York, 1999, Ch. 3; Levine, I., *Quantum Chemistry*, 5th ed., Prentice-Hall, Upper Saddle River, NJ, 2000, pp. 507–508; NPA charges will replace Mulliken charges as the default charge descriptor in future versions of Gaussian 9X, Frisch, M.J., private communication).

6. Reed, A.E., Curtiss, L.A., and Weinhold, F., *Chem. Rev.*, 88, 899–926, 1988.

7. See contact links (www.chem.wisc.edu/~nbo5/ess.htm) in Ref. 3c for NBO-affiliated electronic structure packages.

8. For description of standard computational method and basis set designations used herein, see Foresman, J.B., and Frisch, A., *Exploring Chemistry with Electronic Structure Methods*, 2nd ed., Gaussian, Inc., Pittsburgh, PA, 1996.

9. See, e.g., Zimmerman, H.E., *Adv. Photochem.*, 1, 183–208, 1964.

10. Bent, H.A., *Chem. Rev.*, 61, 275–311, 1961.

11. Coulson, C.A., *Valence*, 2nd. ed., Oxford Univ. Press, London, 1952, Ch. 8.

12. Carpenter, J.E. and Weinhold, F., *J. Am. Chem. Soc.*, 110, 368–372, 1988.

13. Nemukhin, A.V. and Weinhold, F., *J. Chem. Phys.*, 97, 1095–1108, 1992.

14. Reed, A.E. and Weinhold, F., *J. Chem. Phys.*, 83, 1736–1740, 1985.

15. Glendening, E.D. and Weinhold, F., *J. Comp. Chem.*, 19, 593–609, 627, 1998; Glendening, E.D., Badenhoop, J.K., and Weinhold, F., *J. Comp. Chem.*, 19, 628–646, 1998.

16 Pauling, L., *Nature of the Chemical Bond*, 3rd ed., Cornell Univ. Press, Ithaca, NY, 1960.

17. Löwdin, P.-O., *Phys. Rev.*, 97, 1474–1489, 1955; Davidson, E.R., *Reduced Density Matrices in Quantum Chemistry*, Academic Press, New York, 1976.

18. The density matrix-based NRT criterion (**12**) differs from the wave function superposition *ansatz* that was conjectured[16] to underlie empirical resonance theory, but only the former can be generally consistent with the assumption[10] on which empirical applications are based.

19. Wiberg, K.B., *Tetrahedron Lett.*, 24, 1083–1096, 1968.

20. Reed, A.E. and Schleyer, P.V.R., *Inorg. Chem.*, 27, 3969–3987, 1988; *J. Am. Chem. Soc.*, 112, 1434–1445, 1990.

21. See Paper II of Ref. 15 for more complete discussion of bond-order concepts.

22. Crim, F.F., *Acc. Chem. Res.*, 32, 877–884, 1999.

23. Zimmerman, H.E. and Weinhold, F., *J. Am. Chem. Soc.*, 116, 1579–1580, 1994.

24. Coulson, C.A. and Longuet-Higgins, H.C., *Proc. R. Soc.*, A191, 39–60, 1947; A192, 16–32, 1947.

25. Badenhoop, J.K. and Scheiner, S., *J. Chem. Phys.*, 105, 4675–4691, 1996.

26. Zimmerman, H.E. and Alabugin, I.V., *J. Am. Chem. Soc.*, 122, 952–953, 2000; *J. Am. Chem. Soc.*, 123, 2265–2270, 2001.

27. Carpenter, J.E. and Weinhold, F., *J. Mol. Struct. (Theochem.)*, 169, 41–62, 1988; Weinhold, F. and Carpenter, J.E., The natural bond orbital lewis structure concept for molecules, radicals, and radical ions, in *Proceedings of the International Workshop on the Structure of Small Molecules and Ions*, Naaman, R. and Vager, Z., Eds., Plenum, New York, 1988, pp. 227–236.

28. Frisch, M.J., Trucks, G.W., Schlegel, H.B., Scuseria, G.E., Robb, M.A., Cheeseman, J.R., Zakrzewski, V.G., Montgomery, J.A., Jr., Stratmann, R.E., Burant, J.C., Dapprich, S., Millam, J.M., Daniels, A.D., Kudin, K.N., Strain, M.C., Farkas, O., Tomasi, J., Barone, V., Cossi, M., Cammi, R., Mennucci, B., Pomelli, C., Adamo, C., Clifford, S., Ochterski, J., Petersson, G.A., Ayala, P.Y., Cui, Q., Morokuma, K., Malick, D.K., Rabuck, A.D., Ragha-vachari, K., Foresman, J.B., Cioslowski, J., Ortiz, J.V., Baboul, A.G., Stefanov, B.B., Liu, G., Liashenko, A., Piskorz, P., Koma-romi, I., Gomperts, R., Martin, R.L., Fox, D.J., Keith, T., Al-Laham, M.A., Peng, C.Y., Nanayakkara, A., Gonzalez, C., Chal-lacombe, M., Gill, P.M.W., Johnson, B., Chen, W., Wong, M.W., Andres, J.L., Gonzalez, C., Head-Gordon, M., Replogle, E.S., and Pople, J.A., *Gaussian 98*, Gaussian, Inc., Pittsburgh, PA, 1998.

29. Paldus, J., *J. Chem. Phys.*, 61, 5321–5330, 1974.

30. McKinley, A.J., Ibrahim, P.N., Balaji, V., and Michl, J., *J. Am. Chem. Soc.*, 114, 10631–10637, 1992.

31. True spin density matrices are known[17] to have nonnegative eigenvalues bounded by the Pauli principle. The FIXDM key-word of NBO 5.0 replaces an unphysical input density matrix with the closest matrix satisfying physical positivity and Pauli constraints.

32. Cvetanovic, R.J., *Adv. Photochem.*, 1, 115–182, 1963.

33. Baird, N.C., Gupta, R.R., and Taylor, K.F., *J. Am. Chem. Soc.*, 101, 4531–4533, 1979; Baird, N.C. and Taylor, K.F., *Can. J. Chem.*, 58, 733–738, 1980.

34a. Dupuis, M., Wendoloski, J.J., and Lester, W.A., Jr., *J. Chem. Phys.*, 76, 488–492, 1982.

34b. Huyser, E.S., Feller, D., Borden, W.T., and Davidson, E.R., *J. Am. Chem. Soc.*, 104, 2956–2959, 1982.

34c. Yamaguchi, M., Momose, T., and Shida, T., *J. Chem. Phys.*, 93, 4211–4222, 1990; Yamaguchi, M., *Chem. Phys. Lett.*, 221, 531–536, 1994.

35a. Hunziker, H.E., Kneppe, H., and Wendt, H.R., *J. Photochem.*, 12, 377–387, 1981.

35b. DiMauro, L.F., Heaven, M., and Miller, T.A., *J. Chem. Phys.*, 81, 2339, 1984.

35c. Osborn, D.L., Choi, H., Mordaunt, D.H., Bise, R.T., Neumark, D.M., and Rohlfing, C.M., *J. Chem. Phys.*, 106, 3049–3066, 1997.

35d. Williams, S., Harding, L.B., Stanton, J.F., and Weisshaar, J.C., *J. Phys. Chem. A*, 104, 9906–9913, 10131–10138, 2000.

35e. Zhang, J., private communication.

36. For detailed discussion of vibrational properties of vinoxy radical in ground states and excited states, see Ref. 34c.

37. See, e.g., the $DEL tutorial on the NBO 5.0 website.[3c]

38. In this case the default NRT procedure was constrained (with $NRTSTR keylist) to ensure that the NBO Lewis structure **13a** was selected as the leading NRT structure, for consistency with other states. (The default NRT procedure gives a three-membered C–C–O ring structure with a highly polarized "$\pi_{C'O}$" bond [nearly all concentrated on O, as in **13a**] that complicates direct comparisons with other states.)

39. The calculated CASNBO(7,5)/6-311++G** frequencies (in cm^{-1}) may be identified with the descriptive mode labels commonly used (see Refs. 34c, 35c) to indicate approximate mode character (v_1 = asymmetric C'H'H" stretch, v_4 = C–O stretch, v_8 = C–C' stretch, v_{11} = torsion, etc.). For the $\tilde{A}$ state, the (unscaled) frequencies and labels are 3403 (v_1), 3314 (v_3), 3300 (v_2), 1725 (v_8), 1540 (v_5), 1377 (v_6), 1179 (v_4), 1021 (v_7), 953 (v_{11}) 733 (v_{12}), 630 (v_{10}), 481 (v_9). For the $\tilde{B}$ state, the corresponding values are 3430 (v_1), 3337 (v_3), 3317 (v_2), 1842 (v_4), 1545 (v_5), 1427 (v_6), 1203 (v_8, v_7), 941 (v_7, v_8), 488 (v_9, v_7), 451 (v_{10}), 314 (v_{12}), 270 (v_{11}). For comparison, the corresponding CASNBO(7,5)/6-311++G** $\tilde{X}$-state values are 3417 (v_1), 3301 (v_2), 3177 (v_3), 1675 (v_4), 1583 (v_5), 1509 (v_6), 1209 (v_8, v_7), 1025 (v_7, v_8), 979 (v_{10}), 952 (v_{12}), 537 (v_9), and 425 (v_{11}). Overall, the calculated values tend to be ~8% too high, reflecting a lack of dynamical correlation effects.

40. In earlier studies the gradient thresholds for CAS geometry convergence were often weaker than those used here, and accurate analytic frequencies were unavailable to detect false (transition state) stationary points. The slight effects of $\tilde{B}$-state torsional deformation (if present at lower CAS levels) may therefore have fallen within the numerical uncertainties.

41. Teller, E., *J. Phys. Chem.*, 41, 109–116, 1937; Herzberg, G., *Electronic Spectra and Electronic Structure of Polyatomic Molecules*, Van Nostrand Reinhold, New York, 1966, pp. 442–444.

42. Landau, L., *Phys. Z. Sowj.*, 2, 46–51, 1932; Zener, C., *Proc. R. Soc. A*, 137, 696–702, 1932; Child, M.S., *Molecular Collision Theory*, Academic Press, New York, 1974, Sec. 8.5.

43. Ragazos, I.N., Robb, M.A., Bernardi, F., and Olivucci, M., *Chem. Phys. Lett.*, 197, 217–223, 1992; Bernardi, F., Robb, M.A., and Olivucci, *Chem. Soc. Rev.*, 25, 321–328, 1996.

44. A minor difficulty arises for geometries where the occupancies of σ_{CC} and π_{CC} NBOs become degenerate, because the ordering of NBOs (and, hence, the necessary permutations to bring π_{CC} and π^*_{CC} into the active space) become unpredictable. Initial NBOs from a geometry on one side of this degeneracy may then lead to CAS convergence failure for a geometry on the opposite side. The solution is to re-initialize the NBOs and CASNBO search for the new geometry region.

45. After an initial failure from starting equilibrium $\tilde{B}$-state geometry, the starting geometry was twisted by 90°, in the expectation (cf. Figure 7.6) that the $\tilde{A}$–$\tilde{B}$ near-degeneracy lies closer to this limit.

46. Gilli, G.F., Belluci, V., Ferretti, V., and Bertolasi, V., *J. Am. Chem. Soc.*, 111, 1023–1028, 1989; Gilli, P., Verretti, V., Bertolasi, V., and Gilli, G., *Adv. Mol. Struct. Res.*, 2, 67–102, 1995.

47. Hohenberg, P. and Kohn, W., *Phys. Rev.*, 136, B864, 1964.

48. DFT-like approximation techniques might provide a Fock-like $\hat{h}_{eff}$ operator that can incorporate static and dynamic correlation effects of excited states in a mean-field sense.

49. See, e.g., Foresman, J.B., Keith, T.A., Wiberg, K.B., Snoonian, J., and Frisch, M.J., *J. Phys. Chem.*, 100, 16098–16104, 1996 and references therein.

50. Matsika, S. and Yarkony, D.R., *J. Chem. Phys.*, 117, 7198–7206, 2002.

Chapter 8

A Theoretical Approach to Solid-State Organic Photochemistry: Mechanistic and Exploratory Organic Photochemistry

HOWARD E. ZIMMERMAN

Chemistry Department, University of
Wisconsin, Madison, Wisconsin

CONTENTS

8.1 INTRODUCTION

It has long been known that the photochemistry of crystalline
compounds proceeds differently than upon irradiation in solution; one of the best examples is that of the cinnamic acids
and the research of Ciamician and Silber in 1902.[1] Control of
solid-state reactions was postulated by Cohen and Schmidt[2]
in 1964 to involve minimum molecular motion in the crystal.
This approach has been termed the "topochemical principle,"
perhaps otherwise known as least motion. Cohen and Schmidt
also advanced the idea of a molecule reacting in a cavity
created by surrounding neighbors.

This chapter aims to delineate the author's research. Nevertheless, there have been studies by others in the field aimed
at understanding the factors controlling the photochemistry of
crystalline material. Very good studies are those by Chang and
coworkers,[3] Thomas and coworkers,[4] Warshel and coworkers,[5]
Gavezzotti,[6] McBride,[7] by Ariel and coworkers,[8] and others.

However, the major impediment to exploration of the
factors controlling solid-state photochemistry has been the
disinclination of organic chemists to pursue x-ray, computational, and theoretical methods, and this being paralleled by
the physical chemists' disinclination to study organic reactions involving molecular rearrangements.

What had been lacking was a general and cohesive quantitative treatment of crystal lattice reactions. Most quantita-

tive studies dealt with individual examples of reactions. Bimolecular [2+2] cycloadditions were most commonly studied, and unimolecular rearrangements tended not to be considered. Most importantly, attention seemed to be focused on the properties of the reactant rather than considering both reactant and transition structure as is common in ground state solution chemistry.

8.2   THE MINI-CRYSTAL LATTICE, INSERTION OF REACTING SPECIES, AND DETERMINATION OF FIT

The remainder of this chapter will concentrate on the solid-state research carried out in our laboratories at Wisconsin. Our research in this area began in 1984[9] when it was recognized that with the x-ray coordinates available, one could use methodology not too different from that in ground-state chemistry. Somewhat later we termed a truncated portion of the x-ray data a "mini-crystal lattice."[10] At its simplest, it consisted of one central molecule surrounded by one layer of neighbors. However, the number of layers of neighbors is adjustable. This proved to be one of the more important advances permitting the quantitative studies described in this chapter.

To obtain the mini-crystal lattice, several programs were developed. "Smartpac"[10] was one. This was designed to use the x-ray crystallographic coordinates of a single molecule as a starting point. Then a cluster is built with at least one new corresponding atom from each molecule in the unit cell being properly displaced and the remaining atoms being derived from knowledge of the space group symmetry. This molecular cluster is then displaced to afford a mini-lattice of desired size. Finally the clusters are reduced in size to afford the desired crystal packing around the original, central molecule.

We then define the central molecule as the reactant. A second mini-crystal lattice is generated by replacing the central molecule with a partially reacted species selected as a reasonable model for the transition structure. This species is generated using theoretical generation, optimally by quantum mechanical computation. The reacting species is placed with its atoms as close as possible to those of the reactant molecule

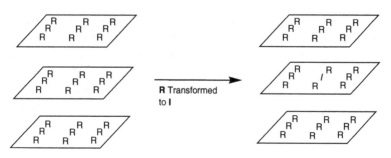

Figure 8.1 The mini-crystal lattice reacting; the central *R* is replaced by *I*.

it has replaced. Then the change in properties of the mini-crystal on conversion from reactant to transition structure is determined. One criterion is increase in energy, a second is increased overlap with neighboring molecules. In our early work we also used volume increase. Also, we used measurement of motion of atoms of the central molecule with the assumption that least motion might be operative. In this early work there was considerable reliance on molecular mechanics for assessment of energy. Transition structures were often approximated by the first intermediates encountered in reactions. In summary, the basic approach is aimed at determining the fit of a molecule as it reacts along alternative pathways. A schematic representation of the mini-crystal lattice is shown in Figure 8.1. The central reacting molecule *R* is replaced by *I*, which is a first intermediate approximating the transition structure.

8.3 EARLY EXAMPLES OF EXPERIMENTAL REACTIVITY VS. THEORETICAL PREDICTION

One of the early examples of our solid-state photochemistry is outlined in Scheme 8.1, which gives the mechanisms of the Di-π-methane[11a-c] and the Tri-π-Methane Rearrangement.[11d] After excitation and π–π bridging the cyclopropyldicarbinyl diradical can ring-open to give an allylic–benzhydryl diradical species. However, two conformations, **2-Cis** and **2-Trans**, differing in the geometry around bond *a–b* may be generated. The diradical with the transoid conformation is able to close to afford only a three-membered ring and thus the Di-π-Methane photoproduct **3**. This is the preferred pathway in solution as might be anticipated; formation of a transoid π-bond is energetically favored. In contrast, photolysis of the crystal lattice leads preferentially to the cisoid conformer and formation of the five-membered ring photoproduct **4**.[11d,e]

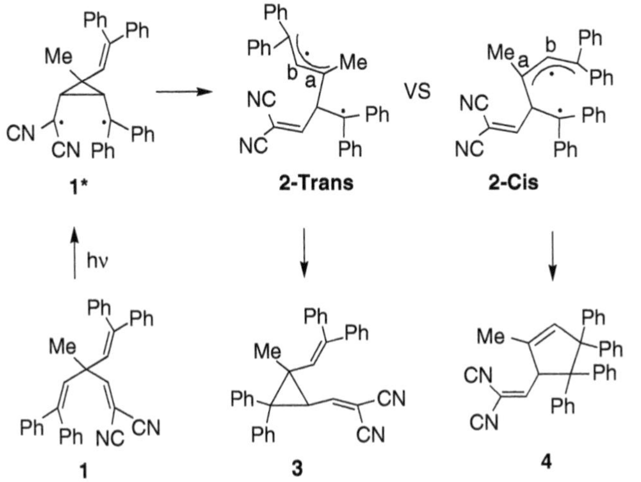

Scheme 8.1 The Di-π-Methane and Tri-π-Methane rearrangements.

In this study several theoretical measures of fit of the allylic diradical **3** in the crystal lattice were used. One was van der Waals overlap of the cisoid and transoid diradicals with the cavity of surrounding neighbors. The transoid allylic diradical required a 10% overlap increase compared with 1% overlap increase for formation of the cisoid counterpart from reactant. Thus the preference for the five-membered ring photoproduct in the crystal lattice photolysis is understood, because only that reaction pathway may occur via an intermediate that can fit in the lattice cavity.

A second example discovered in the same study was that of 1,1,5,5-tetraphenyl-3,3-dimethyl-1-penten-5-ol **5** as given in Scheme 8.2. In this case the increase in overlap with the lattice cavity computationally was 2% for the six-membered ring photoproduct **7**, 1% for the five-membered ring product **8**, and 5% for the acyclic tetraphenyl pentanone photoproduct **6**, which was the only isomer formed in solution. The reaction mechanism is described in Scheme 8.3. X-ray analysis of the

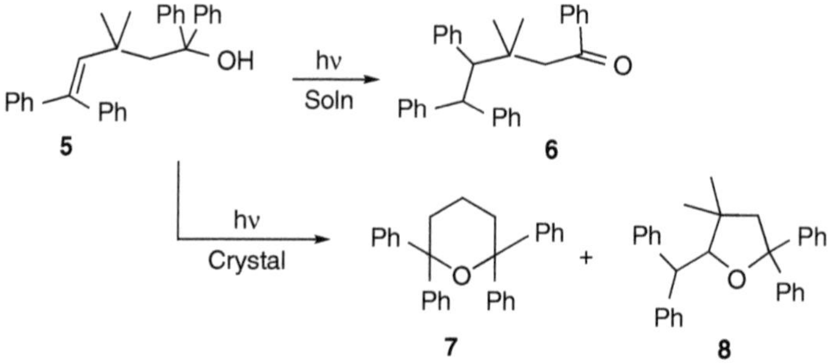

Scheme 8.2 Solid-state versus solution photochemistry of the tetraphenylpentenol **5**.

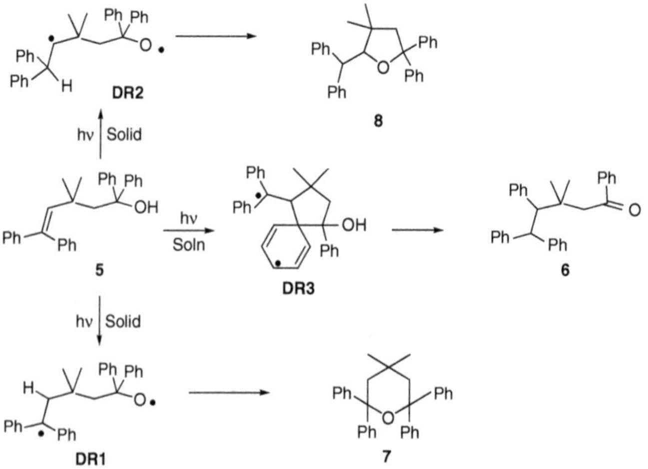

Scheme 8.3 Three reaction pathways.

reactant **5** showed it to be in a coiled conformation. Still, it would have been difficult to compare the change in molecular demands for the first steps of the tetrahydrofuran and pyran formation with that of the competing phenyl migration. Use of the mini-crystal lattice computation thus agreed with the complete change in reaction course in the solid-state process.

8.4 REACTIVITY CRITERIA IN THE TYPE B REARRANGEMENT OF SUBSTITUTED CYCLOHEXENONES

An interesting case contrasts the solid-state photochemistry of 2-methyl-4,4-diphenylcyclohexenone **9** with that observed in solution. Although only one product, **10**, is formed in solution, in the crystal lattice, a second photoproduct, **11**, is formed to an equal extent.[10] The theoretical treatment of this is considered along with some further substituted cyclohexenone photochemistry, which follows.

| **9** | **10** Formed in Solution and in Solid-State | **11** Formed Only in Solid State |

$$(8.1)$$

The photo rearrangements of a series of 5-substituted cyclohexenones both in solution and in crystal lattices were investigated with the idea of comparing experimental behavior with computational predictors.[10] The R group in Equation 8.2 ranged from methyl to ethyl to iso-propyl to t-butyl to phenyl. Photoproduct **13b** arises from migration of the phenyl group, which is cis to the substituent R, while photoproduct **13a** arises from migration of the trans phenyl group.

Interestingly, depending on the identity of the substituent at carbon 5 (i.e., the R group), different product distributions were observed experimentally, both in solution and in the solid state. The ratios of exo product **13a** to endo product **13b** ranged from 1:2.5 (for R = Me) in solution to 10:1 (for R = t-butyl). In contrast, in crystal-lattice transformations the exo product **13a** was formed exclusively for R = Et, i-Pr,-t-Bu and R = Ph. This results from migration of the C-4 trans-phenyl group.

$$\text{(8.2)}$$

In exploration of the theoretical basis of the observations, computations were run on the mini-crystal lattices. We used three criteria: (a) the fit of the first reaction intermediate (note cis and trans diradicals in Figure 8.2) in the mini-crystal lattice, (b) the amount of molecular motion in proceeding to these species by migration of the phenyl group trans to R compared with migration of the cis-phenyl group, and (c) the steric energetics. For the reaction of the enones **12** there is general agreement between the greater overlap with the surrounding cavity observed for migration of the cis-phenyl group and the experimental stereochemical preference for migration of the trans phenyl group (Figure 8.2 and Table 8.1).

The same is true for the criterion of least motion. Thus; Table 8.2 reveals that considerably less motion is involved in migration of the trans-phenyl groups compared with the cis ones. In fact, the amount of motion for cis-phenyl migration is close to twice that required for migration of the trans phenyl group. It is intuitive that the reaction process that requires the maximum molecular reorganization will tend to give a reaction intermediate that is a different shape than that of the reactant. However, we note that this correlation need not hold in all cases. One can envision a molecular rearrangement in a cavity shaped as to permit considerable reacting species

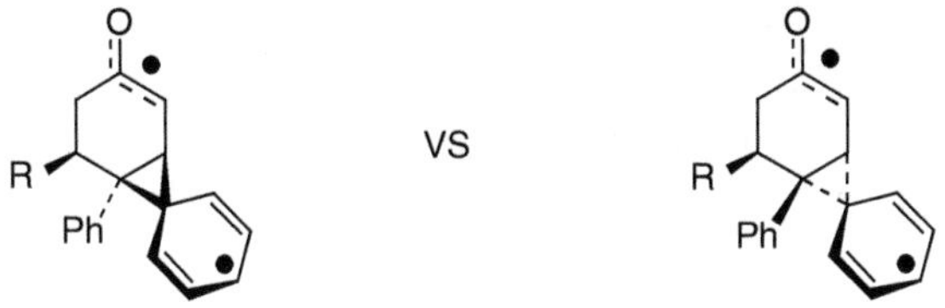

Cis Migration

Trans Migration

Figure 8.2 Two half-migrated species.

TABLE 8.1 Van der Waals Overlaps of Alternative Diradicals En Route to **13a** and **13b**[a]

	Overlap With Lattice, ΔS (Å^3)	
Enone	Cis-Phenyl Migration	Trans-Phenyl Migration
5-Methyl	20.4	13.8[b]
5-Ethyl	23.2	14.2[b]
5-i-Propyl	17.3	11.4[b]
5-t-Butyl	16.9	7.7[b]
5-Phenyl	29.4	15.1[b]

[a]Overlaps of first reaction intermediate.
[b]Leading to observed major product **13a**.

TABLE 8.2 RMS Motion in the Conversion to Phenyl-Bridged Triplet

	RMS Motion from Reactant to First Intermediate, Å	
Enone	Cis-Phenyl Migration	Trans-Phenyl Migration
5-Methyl	1.40	0.74[a]
5-Ethyl	1.48	0.71[a]
5-i-Propyl	1.46	0.70[a]
5-t-Butyl	1.05	0.46[a]
5-Phenyl	1.81	0.84[a]

[a]Reaction course leading to observed major photoproduct.

distortion without severe interference with the surrounding molecules of the cavity.

Table 8.3 summarizes the energetic results for the five cases as obtained from MM3 molecular mechanics treatment of the first intermediate in the mini-crystal lattice. Again, we note that the species formed by migration of the trans-phenyl

TABLE 8.3   Steric Energy Increases in Phenyl Bridging

	Energy Increase in Formation of the Phenyl-Bridged Species[a]	
Enone	Cis-Phenyl Migration	Trans-Phenyl Migration
5-Methyl	41.64	0.06[b]
5-Ethyl	46.41	2.89[b]
5-i-Propyl	72.96	0.51[b]
5-t-Butyl	57.84	1.22[b]
5-Phenyl	20.58	0.04[b]

[a]kcal/mole.
[b]Observed experimentally.

group is much lower in energy than that resulting from migration of the cis-phenyl. Despite the rather nice predictive power of molecular mechanics, we recognize that there is a potential fallacy in treating an open-shell species here (a triplet,) with programming not capable of assessing electronic effects and merely taking van der Waals contributions into account.

8.5   THE TYPE-B BICYCLIC REARRANGEMENT

A further test of the various criteria of solid-state reactivity was found in the type-B bicyclic rearrangement.[12,13] The mechanism is outlined in Scheme 8.4. We see that the reaction proceeds by opening of the central bond of a triplet bicyclo[3.1.0]hexenone **14T** to afford a triplet diradical **15T** that subsequently undergoes intersystem crossing with formation of a zwitterion (i.e., **15**).

One particularly interesting aspect of the chemistry is the preference in solution for the zwitterion **15** in Scheme 8.4 to undergo aryl migration from C-3 to C-2 rather than to C-4. This occurs despite computations showing that C-4 of the zwitterion is more electron deficient. However, the apparent discrepancy becomes understandable if one considers the alternative half-migrated species **exo-2,3** and **endo-2,3** vs. **exo-3,4** and **endo-3,4** as approximations to transition structures. Of course, the exo and endo specifications do not apply to the parent diphenyl example in Scheme 8.4 and are meaningful only if the geminal aryl groups of reactant **14** are differently substituted. The 3,4-bridged species has an enolate moiety that is cross-conjugated and less resonance stabilized than the 2,3-counterparts. This is confirmed by Hartree–Fock computations and even by simple Hückel calculations.

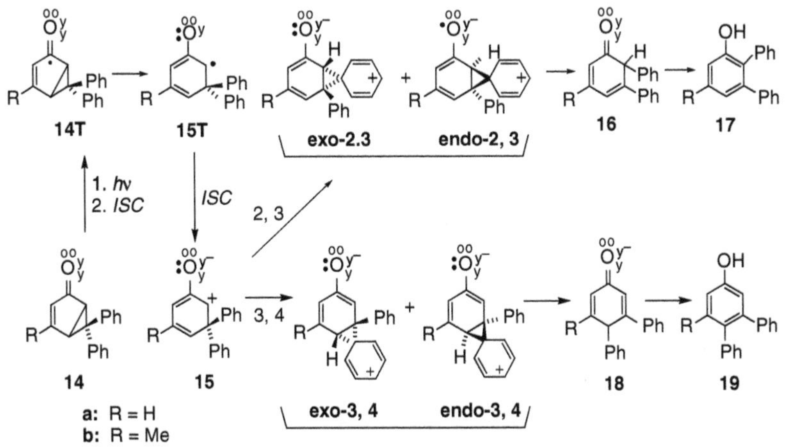

Scheme 8.4 Mechanism of the type-B bicyclic rearrangement.

Another relevant point is that when there is one phenyl group and one p-cyanophenyl group present, it is the phenyl group that migrates in solution, as might be anticipated for migration to an electron-deficient center.

These solution preferences do not hold in solid-state photolysis (see Scheme 8.5).[13] Remarkably, in the case of the endo cyanophenyl reactant **20b** there is a preferential cyanophenyl migration in total contrast to the solution behavior. In all cases much more migration to C-4 was observed than in solution.

Beyond this, in general, with such different groups (e.g., phenyl together with p-cyano or p-bromophenyl), it was always the endo group that migrated selectively (note Scheme 8.5 again). Thus, the two faces of the zwitterion no longer are equivalent but differ because of the dissymmetry of the cavity. The source of the preference of the migration of the endo group, experimentally and computationally, is understood as arising from interference of the surrounding molecules (i.e., the cavity) with the motion of a migration exo aryl moiety. Furthermore, in the case of the diphenyl bicyclic reactant,

TABLE **8.4** Theoretical Treatment of Solid-State Photo-Reactions of **22**-endo and **22**-exo

Reactant	Migrating Group	Migration Energy C-2 C-4		Overlap C-2 C-4		RMS Motion C-2 C-4	
Endo-1	*Endo*-Br-Phenyl	−30.7[a]	−8.3	17.3	31.3	1.78	1.43
	Exo-Phenyl	0.8	−14.3	42.9	37.3	1.82	2.15
Exo-1	*Exo*-Br-Phenyl	−12.7	−12.4	31.9	36.0	1.75	2.17
	Endo-Phenyl	−26.7	−26.3	16.7	22.9	1.78	1.28

[a]The experimentally observed pathways are underlined.

without isotopic labeling there is no experimental test of which group is migrating. However, computations reveal the endo migrating species (i.e., **endo-3,4**), again, to be preferred. It is not often that one is tempted to rely on computations alone without experimental verification. However the cyano- and bromo-substituted examples provide calibration points for computations. See Table 8.4 for a summary of these computations that are in excellent agreement with the experimental results described.

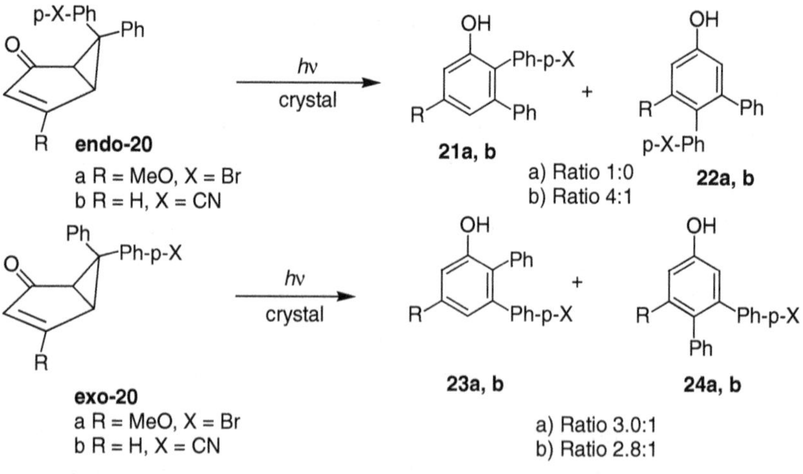

Scheme 8.5 Solid-state photochemistry of two bicyclics.

The computations used molecular mechanical treatment of the mini-crystal lattice for the preceding and various further reactions. As noted earlier, however, the weakness of molecular mechanics in dealing with such chemistry is that electronic effects are not included in the calculations. This is true for any open-shell reaction.

Additionally, computations were carried out to assess overlap with the surrounding cavity and the extent of motion of the atoms of reactant in proceeding to the ring-opened diradical and thence to the alternative aryl bridged diradicals. These results are included in Table 8.4. Both minimal energy and minimal overlap correlate with experiment, whereas least motion does not.

8.6 DIMORPHS, CRYSTAL MORPHOLOGY, AND PSEUDO CRYSTAL DISORDER

One interesting result derived from the observation that 4-p-cyanophenyl-4-phenylcyclohexenone crystallized in two separable modifications: $P2_1/c$, termed "crystal A," and $C2/c$ termed "crystal B."[14] We need to outline the reaction mechanism (see Scheme 8.6), although the Type B enone rearrangement has been discussed in another context above. We see that the reactant enone **25** is electronically excited to S_1, which rapidly intersystem-crosses to T_1. In Scheme 8.6 the aryl groups at C-4 are taken as phenyl for simplicity. However, these groups may be substituted. If one of the aryl groups is p-cyanophenyl, in solution that group migrates in preference to phenyl.[15] This is reasonable, because the bridged species (**DR1** in Scheme 8.6) has greater electron delocalization as shown in Figure 8.3.

A number of remarkable results were encountered in the solid-state photochemistry. The most exciting was that found for crystal B, which gave exclusively phenyl migration. A second exciting result was that this behavior persisted precisely to 16% conversion, at which point a 1:1 distribution of phenyl and p-cyanophenyl migration products began forming. Thus, the reaction seemed to proceed in two stages.

The occurrence of a preferential phenyl migration signifies that cavity effects in the mini-crystal lattice cavity override the electronic preference for cyanophenyl migration.

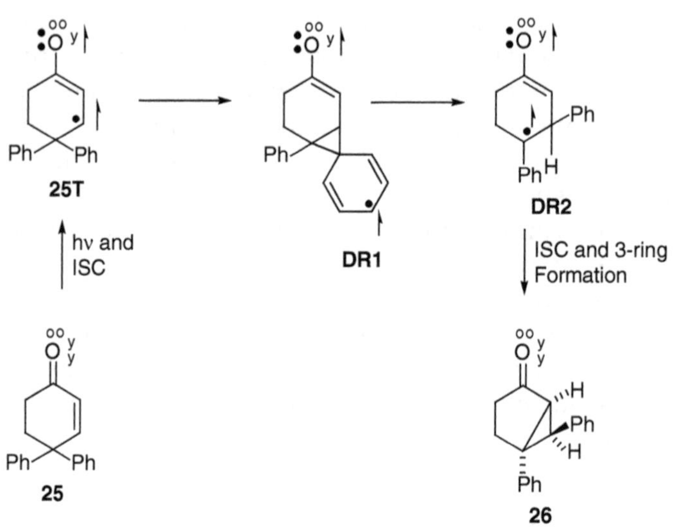

Scheme 8.6   Mechanism of the Type B enone rearrangement.

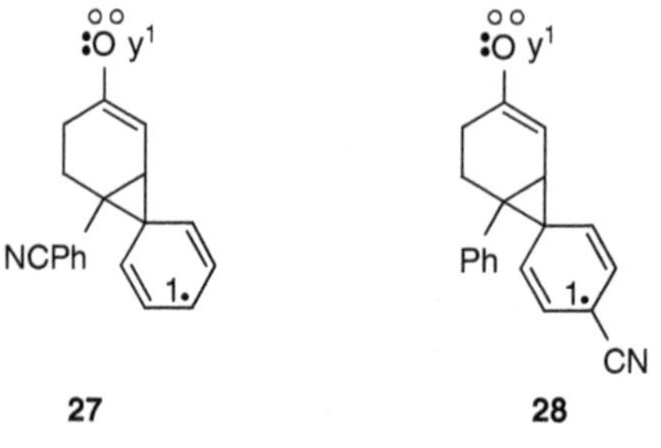

Figure 8.3   Half-migrated phenyl **27** vs. cyanophenyl **28** triplets; preference for **28**.

Thus, this is one dramatic example of how a cavity can change the course of a reaction. However, every attempt to understand this cavity effect by computationally constructing a mini-crystal lattice with diradical 1 (i.e., **DR1**) embedded and comparing this with a mini-lattice with diradical 2 (i.e., **DR2**) embedded, invariably led to a lower energy for the mini-crystal lattice with the p-cyanophenyl migration intermediate. Initially this failure was ascribed to what seemed to be crystal disorder.

However, closer computational investigation of the x-ray data revealed that the disorder was an orderly disorder. Thus, the space group is centrosymmetric, and the enone reactant molecules appear as pairs of enantiomers. Each molecule is surrounded by five neighbors. However, it turned out that one neighbor was the wrong (ie. opposite) enantiomer.

A program we termed "Pairs"[16] computes all distances between each atom of the embedded central molecule in the mini-lattice and each atom of the surrounding neighbors. The atom pairs are then sorted with increasing distances. This, then, affords a picture of the details of molecule–neighbor interactions. In the case of crystal B it was determined that when the nearest molecule was computationally replaced by its enantiomer, a mini-lattice was produced that did predict a lower energy with the p-phenyl-bridged diradical (i.e., **27**) embedded. Indeed, only with this nearest neighbor replacement was the experimental migration preference favored.

The second item of interest was the change in regioselectivity at 16% conversion. It was ascertained that at this point, each molecule now had an adjacent product molecule. In addition it was shown that the ratio of the two products (i.e., from phenyl and cyanophenyl migration) followed a simple kinetic expression, shown in Equation 8.3, where index 1 refers to stage 1 and index 2 refers to stage 2.

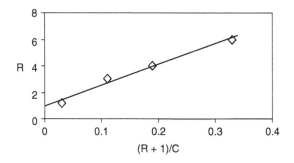

Figure 8.4 Plot of diastereomer ratio vs. ratio plus unity/conversion.

$$R = Ph1(R + 1)/C + (Ph2/CN2)$$

15   **18**   **20**

$$(8.4)$$

Reactant	Products	Ratio
15a R=H	3.0	2.0
15b R=Me	0.0	1.0

At 16% conversion there was a phase change. That the computations, predicated on the two-stage model, held experimentally, confirms the interpretation on which the kinetics was based. In any case, the observation of stages in this reaction led us to consider whether the phenomenon is general. This is considered later.

8.7 THE INERT ATOMIC SHELL MODEL

One of the weaknesses of the treatment of mini-crystal lattices with molecular mechanics is that these computations do not take into account electronic effects, and in the case of open-shell species such as triplets, electronic energy differences between reactants and intermediate and other transformation structures may be appreciable. Yet *ab initio* and even semi-empirical computations on the entirety of a mini-crystal lattice are formidable. Thus, an alternative was developed.[17] This involved generation of a cavity shell composed of inert gas atoms. The Pairs program[16] was used to identify atoms belonging to the neighboring molecules that were nearest to the reacting central molecule. The number of these nearest atoms varied. Then, all but these nearest atoms were computationally annihilated. However, at this point, the nearest atoms, mainly hydrogens but also some carbons, oxygens, and nitrogens, had dangling free valences. Each of the nearest hydrogen atoms was then replaced by helium, and the other atoms were replaced by neons. This afforded an oddly shaped inert gas shell. Next, the entirety was subjected to an *ab initio* computation in which the inert gas shell was kept fixed but the reacting molecule was geometry optimized.

The method was applied to a series of triplet type-B bicyclic (Equation 8.4) and Type B enone reactions (Equation 8.5) that we had studied earlier but with limitations

	12-Et	13-a-Et-Exo	13-b-Et-Endo	29
Solution Ratios:		*1.0*	*2.0*	*0.4*
Crystal Ratios:		*1.0*	*0.0*	*0.0*

$$(8.5)$$

TABLE 8.5A Energies[a] of Four Alternative Half-Migrated Species from Bicyclic 15a[b]

Components	Endo-2,3	Exo-2,3	Endo-3,4	Exo-3,4
Compound with shell	−816.28448	−816.78088	−816.87729	−816.81137
Shell	−55.92535	−56.42625	−56.51490	−56.45517
Compound	−760.35913	−760.35463	−760.36239	−760.35621
Experiment:	Preferred 60	—	Preferred 40	—

[a]Energies in Hartrees.
[b]ROHF/3-21G computations.

TABLE 8.5B Energies[a] of Four Alternative Half-Migrated Species from Bicyclic 15b[b]

Components	Endo-2,3	Exo-2,3	Endo-3,4	Exo-3,4
Compound with shell	−855.26722	−855.45976	−855.69397	−855.65673
Shell	−56.08549	−56.28772	−56.32353	−56.48572
Compound	−799.18173	−799.17204	−799.37044	−799.17101
Experiment:	—	—	Only Product	—

[a]Energies in Hartrees.
[b]ROHF/3-21G computations.

TABLE 8.6 Energies[a] of the Two Alternative Half-Migrated Species from Enone 12-Et[b]

Components	Cis-Endo[c]	Trans-Exo[d]
Compound with shell	−894.56091	−894.71928
Shell	−55.55763	−55.68417
Compound	−839.00328	−839.03511
Experiment	—	Only Product

[a]Energies in Hartrees
[b]ROHF/3-21G computations
[c]Precursor of **13b-Et-endo**
[d]Precursor of **13a-Et-exo** in Equation 8.5

8.8 TREATMENT OF HOST–GUEST LATTICE REACTIONS

In the examples in the previous section, the mini-crystal lattices were homogeneous and composed only of reactant molecules. However, there was that especially interesting case discussed above where two different crystal modifications of the same compound (i.e., dimorphs) were found to have different reactivity. Further examples of dimorphs proved difficult to find. Thus, the idea arose of studying inclusion compounds involving the orderly cocrystallization of reactant molecules (guests) with especially designed hosts. Beek et al.,[18a] Seebach and Beck[18b] and Toda[18c] have extensively studied complexation between taddol **30** and a variety of guest molecules. Furthermore, Toda and Tanaka found that the photochemistry of such inclusion compounds varies with the host molecule.

In our own research we felt that with each host–guest complex having different spacing around a given organic reactant, the differing reactivity would parallel the dimorph situation. The hosts we synthesized are shown in Scheme 8.7.

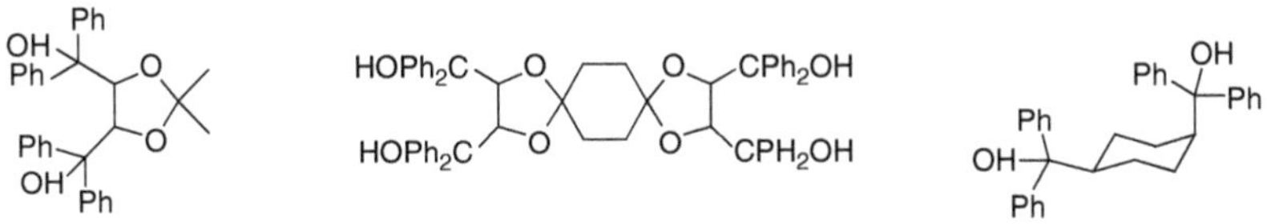

Scheme 8.7   Three hosts used for inclusion compound studies.

The host-guest photochemistry of 4-cyanophenyl-4-phenylcyclohexenone is outlined in Scheme 8.8. Experimentally, the 1:1 host–guest complex using the Seebach–Toda host **30** (see Scheme 8.7) led to exclusive formation of the bicyclic cyanophenyl migration product **34a**. Both the reactant and the photoproduct are chiral as a consequence of the use of this chiral host. Use of the chiral octaphenyl host **31** (again from Scheme 8.7) similarly afforded exclusively cyanophenyl migration; however, in this case the other enantiomer **34b** was obtained. Strikingly, with the tetraphenyl host (**32** in Scheme 8.7), the bicyclic phenyl migration product **35** resulted; with an achiral host in this case, a racemic product was obtained.

For computational and theoretical analysis of host–guest complexes, Smartpac was inadequate, because it dealt with lattices composed of the same molecules. In an extension of the capabilities of Smartpac, Icepack[16] was developed to permit construction of mini-crystal lattices for such inclusion compounds. The mini-crystal lattices constructed with Icepack and with the alternative half-migrated triplet intermediates embedded were subjected to Morokuma's ONIOM[19] combined quantum mechanical/molecular mechanical (QM/MM) computation, which permitted the central reacting species to be subjected to an *ab initio* computation using Gaussian98,[20] with the surrounding neighbors treated with molecular mechanics. The results are outlined in Table 8.7.

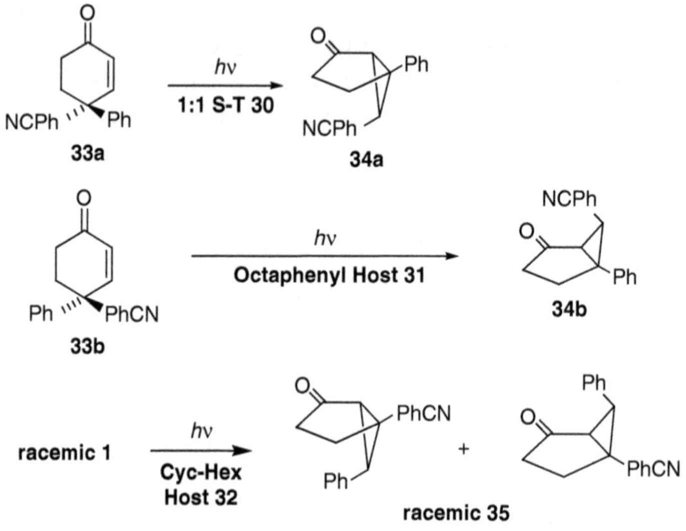

Scheme 8.8   Effect of host molecules on reaction course.

There is a remarkably good computational prediction of the reaction pathways.

TABLE 8.7   ONIOM Energies of Mini-Crystal Lattices with Alternative Migration Intermediates[a]

Inclusion Compound	Phenyl Migration Intermediate 27	CyanoPhenyl Migration Intermediate 28	Δ Energy [c]	Experiment
Seebach-Toda 1 1:1	−855.4092	−855.4249	9.9	CNPh migration
Cyclohexyldiol 2	−855.5556	−855.5523	−2.1	Phenyl migration
Benzopinacol 3	−855.2728	−855.2893	10.4	CNPh migration

[a]ONIOM(ROHF/6-31G*:MM3).
[b]Energies in Hartrees (627.5 kcal/mole).
[c]kcal/mole.

8.9 THE PHENOMENON OF REACTIONS PROCEEDING IN STAGES

In the dimorph photochemistry of 4-p-cyanophenyl-4-phenyl-cyclohexenone, there was the observation of a discontinuous change in stereoselectivity at 16% conversion. This led us to consider whether the phenomenon of reactions occurring in stages might be general. Before this, it had been common to assume that at some point, in a solid-state reaction, the crystal structure would be destroyed, and the reaction would be that of a heterogeneous solid or a liquid and thus be of no use.

We have uncovered a number of examples that parallel the dimorph case. In these, at some point in the solid-state reaction, there is a discontinuos change in reactivity as a result of a phase change. The products in stage 2 generally differ from those in stage 1. Throughout, including the dimorph example discussed above, the crystal lattices remain intact but with an orderly change in structure. One example is given in Scheme 8.9 with the reaction of 4-α-naphthyl-4-β-naphthylcyclohexenone **36**.[21]

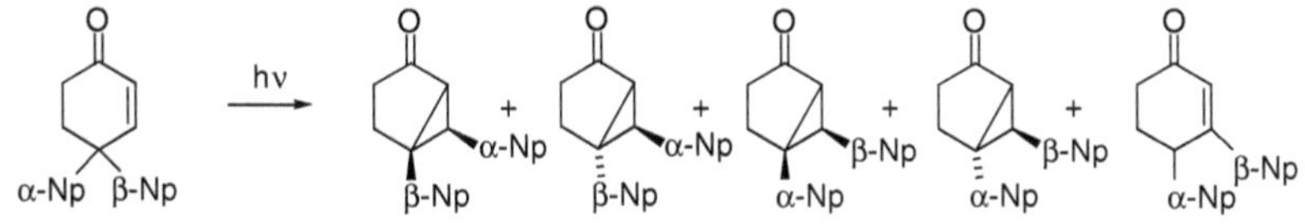

		37	38	39	40	41
Solution:		2.0	2.1	1.0	3.3	1.0
Solid-State Stage 1[a]:		1.0	0	5.1	2.3	0
Solid-State Stage 2[b]:		1.0	0	1.3	3.9	0

a. to 30 percent; b. above 30 percent

Scheme 8.9 The reaction of α-naphthyl-4-β-naphthylcyclohexenone **36** in stages.

Given a general example with two products, here termed **A** and **B**, and with a conversion labeled C, it can be shown[21,22] that Equations 8.6a and 8.6b apply where α and β are the slopes of plots of **A** and **B** versus **C**, and **R** is the ratio of **A** to B. Then, within any one stage, the incremental ratio of products, R_{inc}, remains constant and equal to α/β (note Equation 8.6b). This is confirmed experimentally for this case and others.

$$\alpha / \beta = [CR - A\,(R + 1)]/[C - B(R + 1)] \qquad (8.6a)$$

$$\alpha / \beta = R_{inc} \qquad (8.6b)$$

In the case of the solid-state photolysis of 2-methyl-4,4-diphenylcyclohex-2-enone **9**[23] outlined in Scheme 8.10, the photochemistry in stage 1 provides photoproduct **11**. Note that the ratios given in this scheme are incremental ratios and not cumulative amounts. Of two possible mechanisms for formation of product **11** — (a) direct formation of **11** from **9** by a ring contraction and (b) indirect by a fast reaction of the initially formed bicyclic ketone **10** — by labeling the skeleton with C-13, it was demonstrated that the latter alternative was occurring. In stage 2 this product is no longer formed. In stage 1, as photoproduct **10** is formed, half of it proceeds rapidly onward to afford **11**, thus keeping the observed 1:1 product ratio.

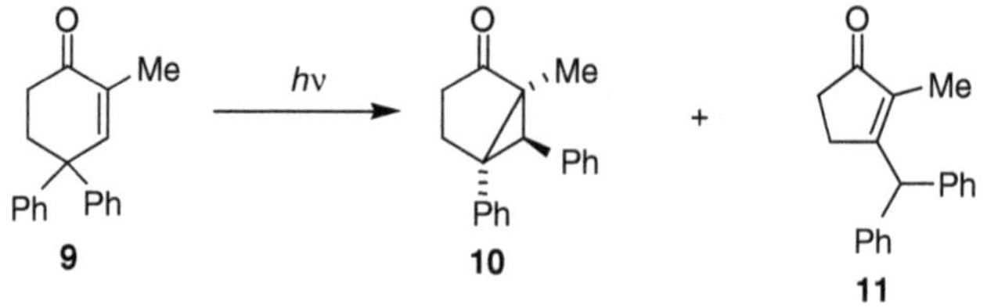

Solution:	1.0	0
Solid-state stage 1 (to10%):	1.0	1.0
Solid-state stage 2 (above 10%):	1.0	0

Scheme 8.10   Photochemistry of 2-methyl-4,4-diphenylcyclohex-2-enone **9**.

Figures 8.5a,b depict the two stages along with the sharp change in product selectivity at 10% conversion. The linearity in each stage is in accord with Equation 8.6b and confirms the constancy of the reaction processes occurring in each stage.

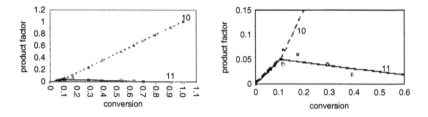

Figure 8.5a,b Photochemistry vs. conversion of the 2-methyl enone **9**.

One challenge was understanding why bicyclic photoproduct **10** in the crystal photolysis proceeds onward with facility to afford cyclopentenone **11** despite **10** being stable in solution photolyses. The problem was approached computationally by inspecting the triplet of bicyclic ketone **10** both as an isolated (e.g., gas phase) species and generated by geometry optimization in a mini-crystal lattice composed of enone **9** reactants. This was accomplished using ONIOM and Icepack but starting with an enone crystal lattice with the central molecule deformed toward bicyclic triplet and optimizing this central species. The aim here was to generate bicyclic ketone **10** in a way corresponding to its generation in the solid-state reaction and thus in a constrained form. The energies are summarized in Table 8.8. The geometries were opti-

TABLE 8.8 Variation of the Energy of Bicyclic Triplet 10T With Environment

	Ab Initio Energy, a.u.	ΔE, kcal/mol
10T$_{gas}$, optimized in gas phase	−804.7447991	0
10T$_{cr}$, optimized in crystal **9**[a]	−804.6830413	38.8
10T$_{cr'}$, optimized in crystal	−804.6909315	33.8

[a]Optimization began with an enone **9** in an enone lattice but deformed toward the geometry of **10**.

mized at a HF/6-31G* level for the gas phase and with ONIOM (HF/6-31G*:Dreiding) for the enone crystal. **10T**$_{gas}$ represents the gas phase species, whereas **10T**$_{cr}$ is the same triplet but generated in the mini-lattice. Finally, the third entry, **10T**$_{cr}$, is the same bicyclic triplet but with an adjacent cyclopentenone **11**.

Inspection of Table 8.8 reveals that **10T**$_{cr}$ is considerably higher in energy than its gas-phase counterpart (by 38.8 kcal/mole). Also, as noted, the geometry of this optimized triplet is distorted from that of the gas-phase species.

To understand the nature of the crystal distortion of the bicyclic triplet, a delta-density analysis was used. Note Equation 8.7, where the delta-density values correspond to changes in bonding on incorporation of the bicyclic triplet **10T** in the crystal lattice. D_{cr} represents the density matrix for the triplet optimized in the crystal lattice, whereas D_{gas} represents the density matrix for the isolated molecule. Then the delta density matrix ΔD reveals the bonding consequences for the molecule generated in the crystal lattice. Figure 8.6 lists the most interesting matrix elements. Bond 1–6 is dramatically weakened, and thus we understand the preferential and facile fission of this bond to afford diradical **42**, which is the precursor to cyclopentenone **11**.

$$\Delta D = D_{cr} - D_{gas} \qquad (8.7)$$

Some general commentary regarding the phenomenon of stages is required. This phenomenon takes place when there is a substantial geometric change in proceeding from reactant to product. After sufficient of the reactant molecules has been transformed, the locale surrounding any given reactant molecule have changed dramatically. This changes the reactivity dramatically as well. Furthermore, this process proceeds in a consistent manner. Thus, it is always one of a pair of neighboring molecules with a fixed geometric relationship that reacts and thus modifies the reactivity of the second member of the pair. This means that in, for example, stage 1, the process continues until all reactant molecules have had their reactivity changed in the same way. At this point, new reactivity is imposed. This requires that in the second stage the reaction is likely to be different and slower. If the second-

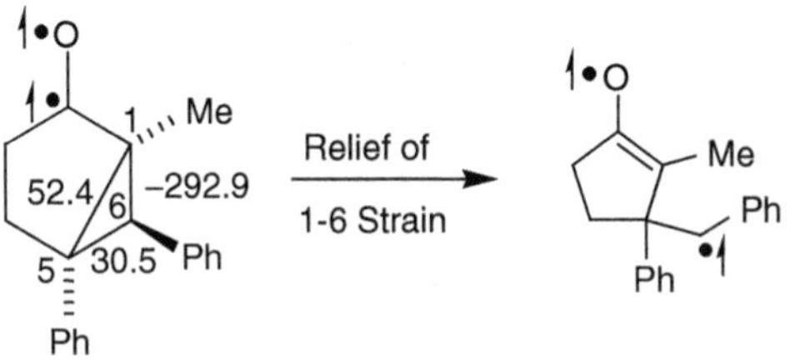

Figure 8.6 Delta-density analysis of triplet **9T** showing bond 1-6 strain. Fission forms biradical **42**. ΔD changes adjacent to bonds.

stage reaction were not slower, then that reaction course would have occurred before all molecules had a transformed neighbor. Finally, we can see that reactions involving little change in molecular shape, such as [2+2] cycloadditions, are less likely to affect the reactivity of neighboring molecules. Conversely, molecular rearrangements with large appending substituents groups are good candidates for stage behavior.

8.10 THE TYPE-C REARRANGEMENT

In early studies; the photochemistry of 4,5,5-triphenylcyclo-hexenone **43** was investigated and shown to lead to cyclobu-tanones.[24] The corresponding photochemistry of the crystalline material proved equally fascinating. Both the solu-tion photochemistry and the corresponding solid-state behav-ior are outlined in Scheme 8.11. Also included is the dramatic change in solid-state reactivity at 12% conversion. This, then, provides still another example of reactivity in stages.

   A particularly exciting result in our solid-state photo-chemistry was the formation of a new type product, bicyclic ketone **46**, in addition to the anticipated type-B product. The reaction, termed Type C,[23] involves a δ to α phenyl migration, and an overall mechanism is given in Scheme 8.12, the evi-

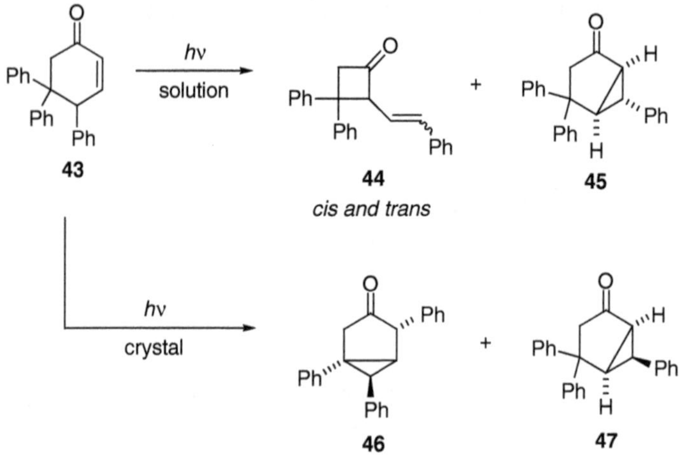

| Solid-state stage 1 (to 12%): | 1.0 | 0 |
| Solid-state stage 2 (above12%): | 1.0 | 6.0 |

Scheme 8.11 Solution and solid-state photochemistry of triphenylcyclohexenone **43**.

dence for which is outlined in the subsequent discussion. The bicyclic ketone **47** is an example of a compound whose preparation derives from only a second-stage process.

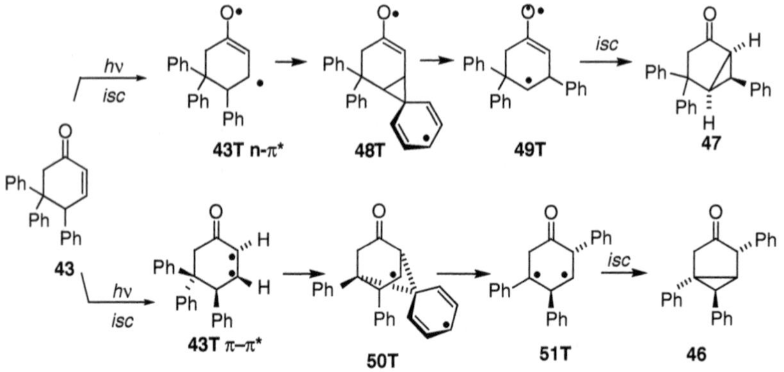

Scheme 8.12 Mechanisms of Type B and Type C enone rearrangements for enone **43**.

TABLE 8.9   Energies of Intermediate Species in the
Photochemistry of Enone **43**

Reaction species[a]	*Ab Initio* Energy, a.u.[b]	Relative Energy, kcal/mol[b]
A, S0 of enone **43**, relaxed geometry	−995.3351504	0.7
B, π-π* twisted triplet **43T** (π-π*)	−995.2492053	54.6
C, TS (**43T 50T**)	−995.2005462	85.2
D, Half-migrated triplet intermediate **50T**	−995.2359950	62.9
E, Triplet biradical **51T**	−995.2804044	35.1
F, Bicyclic product **46**, S0	−995.3361895	0
G, n-π* planar triplet **43T**	−995.2684720	42.5
H, TS **43T(n-π*) 48T**	−995.2419138	59.2
I, Half-migrated triplet intermediate **48T**	−995.2581979	49.0

[a] Geometry was optimized with the HF/6-31G* level.
[b] HF/6-31G*.

With unusual behavior in this system, we turned to
investigating the reactivity computationally. The reaction
pathways proceeding via the alternative phenyl bridged triplet species (**48T** and **50T**) are outlined in Scheme 8.12. These
were followed computationally. Two excited enone triplets —
one planar and one twisted — were found as local minima
using Hartree–Fock UHF/6-31G* computations, and the energies are given in Table 8.9 together with energies for some
other important points on the reaction hypersurface. The planar n-π* triplet was 12.1 kcal/mole lower in energy, accounting
for the solution preference of the Type B enone rearrangement
to afford photoproduct **45**, which, except for stereochemistry,
is formed by the same mechanism as its solid-state counterpart **47**. In addition, the two alternative phenyl bridged species, **48T** and **50T**, were subjected to computation both as gas-
phase and as mini-crystal lattice species. The results are given
in Table 8.10. Again, the gas-phase bridged species **48T**
derived from the n-π* triplet are about 13.9 kcal/mole lower
in energy.

The computations in the crystal lattice were done with
ONIOM, as usual. The situation is now reversed, with the

TABLE 8.10   The Two Bridged Intermediates as a Function of
Neighboring Medium

	Type B Migration (intermediate **48T**)		Type C Migration (intermediate **50T**)	
	Ab Initio Energy, a.u.[b]	Relative Energy, kcal/mol[b]	*Ab Initio* Energy, a.u.[b]	Relative Energy, kcal/mol[b]
Gas phase	−995.2581979	0	−995.2359950	0
Crystal	−995.1830517	47.2	−995.2243063	7.3

[a] Geometries were optimized with HF/6-31G* for gas phase and with ONIOM (HF/6-31G*:Dreiding) for crystal.
[b] HF/6-31G*.

Type C intermediate being lower in energy, which agrees with
the exclusive occurrence of this process in stage 1 of the solid-
state reaction.

The geometry of the π–π* triplet (**43T** π–π*) was twisted
around the enone double bond. To determine the nature of
this structural change, we resorted to our delta-density computations applied to triplets **43T(n-π*)** and **43T(π-π*)** as well
as inspection of the electronic population of the two species
(note Table 8.11 and Table 8.12 for our CASSCF-NBO analysis).

The delta-density computations showed a heavy loss of
the electron density from the n (i.e., p_y) orbital of the planar
triplet **43T(n-π*)** as well as from the carbonyl π-bond. This
is then identified as having an n-π* configuration (column 2

TABLE 8.11   Delta-Density Computation for the Two
Triphenyl Enone **43T** Triplets

Orbitals	Planar Triplet **43T(n-π*)**	Twisted Triplet **43T(π–π*)**
π C=C	−1174.9	−4483.8
π C=O	−2391.8	−1420.9
n_y	−9146.0	−866.0

[a] UHF/6-31G*.
[b] Versus ground state S0 as a reference point.

TABLE 8.12 CASSCF/NBO Analysis of Electronic Configuration of **43T** Triplets[a,b]

Orbitals	Electronic Population[c]	
	Planar Triplet **43T**(n–π*)	Twisted Triplet **43T**(π–π*)
π_2^*	0.4198	0.2230
π_1^*	1.0141	0.9235
n	1.0073	1.9238
π_2	1.6568	0.9853
π_1	1.8929	1.8643

[a]Geometries were optimized with CASSCF(6,5)/6-31G* with NBO basis set.
[b]CASSCF energies: for **43T(n–π*)** –995.2822647 a.u., for **43T(π–π*)** –995.2796512 a.u.
[c]in electrons.

TABLE 8.13 Energies of Ground and Triplet States of *tert*-Butyl Enone **52**

Reaction Species[a]	*Ab Initio* Energy, a.u.[b]	Relative Energy, kcal/mol[b]
Ground state, relaxed geometry	–921.9353821	0
π-π* twisted triplet	–921.8662925	43.4
n-π* planar triplet	–921.8450080	56.7

[a] Geometry was optimized with the HF/6-31G* level.
[b] HF/6-31G*.

in Table 8.11). In contrast, the twisted triplet **43T(π-π*)** had mainly lost density from the α,β-carbon bond that categorizes the triplet as π–π*.

The CASSCF/NBO computations[25] used an NBO basis and an active space of (6,5) taken to include those eigenfunctions heavily participating in excitation. The results are outlined in Table 8.12. Triplet **43T(n-π*)** (planar) involves lone-pair electron excitation, confirming its structure as n-π*, whereas for triplet **43T(π-π*)** (twisted), a π-π* configuration is confirmed as dominant.

With knowledge that π-bond twisting is a factor in leading to the Type C rearrangement, we turned to study the 3-t-butylcyclohexenone **52** in Scheme 8.13. The introduction of the t-butyl substituent aimed at enhancing excited-state twisting by raising the steric repulsion energy of the planar geometry. That this occurred is seen from inspection of Table 8.13, where the twisted excited state is 13 kcal/mole lower in energy than the planar counterpart even without incorporation in the mini-crystal lattice.

Correspondingly, even in the solution irradiation of *tert*-butyl, enone **52** led to the Type C photoproducts only. However, the solid-state reaction was more selective in giving exclusively Type C photoproduct **53** (note Scheme 8.13). Additionally, the solid-state quantum yield was remarkably high

TABLE 8.14 Computational Results for the Enone **52** Type C Rearrangement Species[a]

	Ab Initio Energy, a.u	ΔE, kcal/mol[b]
Half-migrated intermediate optimized in gas phase	−921.8353739	0
Half-migrated intermediate optimized in crystal	−921.8289199	4.1
Photoproduct **54** optimized in crystal	−921.8465375	0
Photoproduct **53** optimized in crystal	−921.8224607	15.1

[a] Geometries were optimized with HF/6-31G* for gas phase and with ONIOM (HF/6-31G*:Dreiding) for crystal.
[b] Comparison between lines 1 and 2, and 3, and 4, respectively.

($\varphi = 0.017$) and not too different from the efficiencies often seen in solution rearrangements. Computations comparing the energies of the phenyl-bridged species in the gas phase and embedded in the mini-crystal lattice showed the crystal lattice species as only 4.1 kcal/mole higher in energy (see Table 8.14). This signifies relatively little steric interaction with the crystal-lattice cavity in the Type C reaction. The higher selectivity of the solid-state reaction, compared with the solution process, can be understood as a lattice effect of

15.1 kcal/mole (note entries 3 and 4 of Table 8.14), favoring formation of photoproduct **53**.

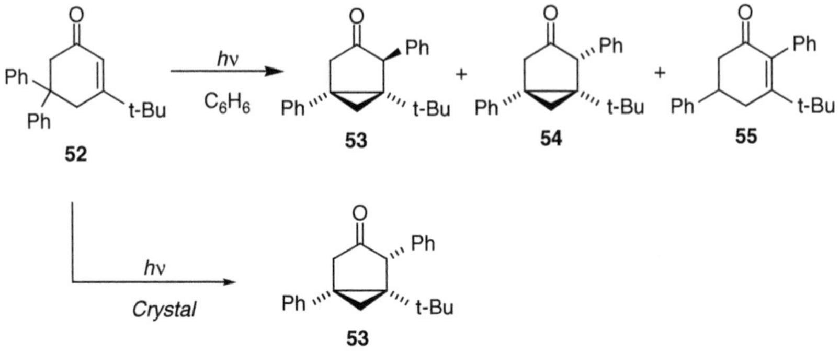

Scheme 8.13 Solution photochemistry of *tert*-butyl enone **52**.

8.11 CONCLUSION

This chapter has had several objectives. One is coverage and review of some of the author's solid-state photochemistry based on the mini-crystal lattice treatment. A second is the demonstration that the Cohen–Schmidt cavity can be put on a quantitative basis. However, probably the main objective is to provide evidence that a very broad spectrum of solid-state photochemistry can be understood and treated with theoretical methods.

ACKNOWLEDGMENT

Support of the author's research over the decades by the National Science Foundation is gratefully acknowledged. These days, with ubiquitous emphasis of support of research that will be of practical value in a year or two, such confidence in the value of basic research is particularly appreciated.

REFERENCES

1. Ciamican, G. and Silber, P., *Chem. Ber.*, 35, 4128, 1902.

2. Cohen, M.D. and Schmidt, G.M.J., *J. Chem. Soc.*, 1996–2000, 1964.

3. Chang, H.C., Popovitz-Biro, R., Lahav, M., and Leiserowitz, L., *J. Am. Chem. Soc.*, 109, 3883–3893, 1987.

4. Thomas, N.W., Ramdas, S., and Thomas, J.M.A., *Proc. R. Soc. London A*, 400, 219–227, 1985.

5. Warshel, A., Shakked, Z., Warshel, A., and Shakked, Z. *J. Am. Chem. Soc.*, 97, 5679–5684, 1975.

6. Gavezzotti, A., *J. Am. Chem. Soc.*, 105, 5220–5225, 1983.

7. McBride, J.M., *Acc. Chem. Res.*, 16, 304–312, 1983.

8. Ariel, S., Askari, S.H., Scheffer, J.R., Trotter, J., and Wireko, F., *J. Am. Chem. Soc.*, 109, 4623–4626, 1987.

9. This research began with the advent of Michael Zuraw on the scene as a graduate student; Zimmerman, H.E. and Zuraw, M.J., *J. Am. Chem. Soc.*, 111, 2358–2361, 1989; Zimmerman, H.E. and Zuraw, M.J., *J. Am. Chem. Soc.*, 111, 7974–7989, 1989.

10. Zimmerman, H.E. and Zhu, Z., *J. Am. Chem. Soc.*, 117, 5245–5262, 1995; Zimmerman, H.E. and Zhu, Z., *J. Am. Chem. Soc.*, 116, 9757–9758, 1994.

11a. Zimmerman, H.E., Binkley, R.W., Givens, R.S., and Sherwin, M.A., *J. Am. Chem. Soc.*, 89, 3932–3933, 1967.

11b. Zimmerman, H.E. and Grunewald, G.L., *J. Am. Chem. Soc.*, 88, 183–184, 1966.

11c. Zimmerman, H.E. and Armesto, D., *Chem. Rev.*, 96, 3065–3112, 1996.

11d. Zimmerman, H.E. and Cirkva, V., *Organ. Lett.*, 2, 2365–2367, 2000.

11e. Zimmerman, H.E. and Cirkva, V., *J. Org. Chem.*, 66, 1839–1851, 2001.

12. Zimmerman, H.E. and Grunewald, J.O., *J. Am. Chem. Soc.*, 89, 3354–3356, 1967; Zimmerman, H.E., Nasielski, J., Keese, R., and Swenton, J.S., *J. Am. Chem. Soc.*, 88, 4895–4903, 1966.

13. Zimmerman, H.E. and Sebek, P., *J. Am. Chem. Soc.*, 119, 3677–3690, 1997.

14. Zimmerman, H.E., Alabugin, I.V., Chen, W., and Zhu, Z., *J. Am. Chem. Soc.*, 121, 11930–11931, 1999.

15. Zimmerman, H.E., Rieke, R.D., and Scheffer, J.R., *J. Am. Chem. Soc.*, 89, 2033–2047, 1967; Zimmerman, H.E. and Wilson, J.W., *J. Am. Chem. Soc.*, 86, 4036–4042, 1964; Zimmerman, H.E. and Hancock, K.G., *J. Am. Chem. Soc.*, 90, 3749–3760, 1968.

16. Zimmerman, H.E., Alabugin, I.V., and Smolenskaya, V.N., *Tetrahedron, Lett.*, 56, 6821–6831, 2000.

17. Zimmerman, H.E., Sebek, P., and Zhu, Z., *J. Am. Chem. Soc.*, 120, 8549–8550, 1998.

18. Beck, A.K., Bastani, B., Plattner, D.A., Petter, W., Seebach, D., Braunschweiger, H., Gysi, P., and Vecchia, L.L. *Chimia*, 45, 238–244, 1991.

18b. Seebach, D., Beck, A.K., Imwinkelried, R., Roggo, S., and Wonnacott, A., *Helv. Chim. Acta*, 70, 954–974, 1987.

18c. Toda, F. and Tanaka, K., *Tetrahedron Lett.*, 29, 551–554, 1988.

19. Maseras, F. and Morokuma, K., *J. Comput. Chem.*, 16, 1170–1179, 1995; Matsubara, T., Sieber, S., and Morokuma, K., *Int. J. Quantum Chem.*, 60, 1101–1109, 1996.

20. Frisch, M.J., Trucks, G.W., Schlegel, H.B., Scuseria, G.E., Robb, M.A., Cheeseman, J.R., Zakrzewski, V.G., Montgomery, J.A., Jr., Stratmann, R.E., Burant, J.C., Millam, J.M., Daniels, A.D., Kudin, K.N., Strain, M.C., Farkas, O., Tomasi, J., Barone, V., Cossi, M., Cammi, R., Mennucci, B., Pomelli, C., Adamo, C., Clifford, C., Ochterski, J., Petersson, G.A., Ayala, P.Y., Cui, Q., Morokuma, K., Malick, D.K., Rabuck, A.D., Raghavachari, K., Foresman, J.B., Cioslowski, J., Ortiz, J.V., Stefanov, B.B., Liu, G., Liashenko, A., Piskorz, P., Komaromi, I., Gomperts, R., Martin, R.L., Fox, D.J., Keith, T., Al-Laham, M.A., Peng, C.Y., Nanayakkara, A., Gonzalez, C., Challacombe, M., Gill, P.M.W., Johnson, B., Chen, W., Wong, M.W., Andres, J.L., Gonzalez, C., Head-Gordon, M., Replogle, E.S., and Pople, J.A., *Gaussian 98, Revision A.6. Gaussian, Inc., Pittsburgh, PA, 1998.*

21. Zimmerman, H.E. and Nesterov, E.E.. *Organ. Lett.*, 2, 1169–1171, 2000.

22. Zimmerman, H.E. and Nesterov, E.E., *Acc. Chem. Res*, 20, 77–85, 2002.

23. Zimmerman, H.E. and Nesterov, E.E., *J. Am. Chem. Soc.*, 124, 2818–2830, 2002.

24. Zimmerman, H.E. and Solomon, R.D., *J. Am. Chem. Soc.*, 108, 6276–6289, 1986.

25. Nemukhin, A.V. and Weinhold, F., *J. Chem. Phys.*, 97, 1095–1108, 1992.

INDEX

Milton Keynes UK
Ingram Content Group UK Ltd.
UKHW020007071024
449327UK00031B/2683